Feature Papers in Section Biosensors 2023

Feature Papers in Section Biosensors 2023

Editor

Evgeny Katz

MDPI

Basel • Beijing • Wuhan • Barcelona • Belgrade • Novi Sad • Cluj • Manchester

Editor
Evgeny Katz
Clarkson University
Potsdam, NY
USA

Editorial Office
MDPI
St. Alban-Anlage 66
4052 Basel, Switzerland

This is a reprint of articles from the Special Issue published online in the open access journal *Sensors* (ISSN 1424-8220) (available at: https://www.mdpi.com/journal/sensors/special_issues/6TJDC2VD56).

For citation purposes, cite each article independently as indicated on the article page online and as indicated below:

Lastname, A.A.; Lastname, B.B. Article Title. *Journal Name* **Year**, *Volume Number*, Page Range.

ISBN 978-3-7258-0169-5 (Hbk)
ISBN 978-3-7258-0170-1 (PDF)
doi.org/10.3390/books978-3-7258-0170-1

Cover image courtesy of Evgeny Katz

Contents

About the Editor . . . . **vii**

Shubham S. Patil, Vijaykiran N. Narwade, Kiran S. Sontakke, Tibor Hianik and Mahendra D. Shirsat
Layer-by-Layer Immobilization of DNA Aptamers on Ag-Incorporated Co-Succinate Metal–Organic Framework for Hg(II) Detection
Reprinted from: *Sensors* **2024**, *24*, 346, doi:10.3390/s24020346 . . . . **1**

Siyuan Liao, Qi Chen, Haocheng Ma, Jingwei Huang, Junyang Sui and Haifeng Zhang
A Liquid Crystal-Modulated Metastructure Sensor for Biosensing
Reprinted from: *Sensors* **2023**, *23*, 7122, doi:10.3390/s23167122 . . . . **16**

Satheesh Natarajan, Jayaraj Joseph, Balamurugan Vinayagamurthy and Pedro Estrela
A Lateral Flow Assay for the Detection of *Leptospira lipL32* Gene Using CRISPR Technology
Reprinted from: *Sensors* **2023**, *23*, 6544, doi:10.3390/s23146544 . . . . **30**

Ali M. Almuhlafi and Omar M. Ramahi
A Highly Sensitive 3D Resonator Sensor for Fluid Measurement
Reprinted from: *Sensors* **2023**, *23*, 6453, doi:10.3390/s23146453 . . . . **39**

Anh Igarashi, Maho Abe, Shigeki Kuroiwa, Keishi Ohashi and Hirohito Yamada
Enhancement of Refractive Index Sensitivity Using Small Footprint S-Shaped Double-Spiral Resonators for Biosensing
Reprinted from: *Sensors* **2023**, *23*, 6177, doi:10.3390/s23136177 . . . . **55**

Sergey V. Stasenko, Andrey V. Kovalchuk, Evgeny V. Eremin, Olga V. Drugova, Natalya V. Zarechnova, Maria M. Tsirkova, et al.
Using Machine Learning Algorithms to Determine the Post-COVID State of a Person by Their Rhythmogram
Reprinted from: *Sensors* **2023**, *23*, 5272, doi:10.3390/s23115272 . . . . **72**

Darya V. Vokhmyanina, Olesya E. Sharapova, Ksenia E. Buryanovataya and Arkady A. Karyakin
Novel Siloxane Derivatives as Membrane Precursors for Lactate Oxidase Immobilization
Reprinted from: *Sensors* **2023**, *23*, 4014, doi:10.3390/s23084014 . . . . **89**

Artem Badarin, Vladimir Antipov, Vadim Grubov, Nikita Grigorev, Andrey Savosenkov, Anna Udoratina, et al.
Psychophysiological Parameters Predict the Performance of Naive Subjects in Sport Shooting Training
Reprinted from: *Sensors* **2023**, *23*, 3160, doi:10.3390/s23063160 . . . . **97**

Fatemeh Ahmadi Tabar, Joseph W. Lowdon, Soroush Bakhshi Sichani, Mehran Khorshid, Thomas J. Cleij, Hanne Diliën, et al.
An Overview on Recent Advances in Biomimetic Sensors for the Detection of Perfluoroalkyl Substances
Reprinted from: *Sensors* **2024**, *24*, 130, doi:10.3390/s24010130 . . . . **109**

Parmis Karimpour, James M. May and Panicos A. Kyriacou
Photoplethysmography for the Assessment of Arterial Stiffness
Reprinted from: *Sensors* **2023**, *23*, 9882, doi:10.3390/s23249882 . . . . **131**

Changyeop Lee, Chulhong Kim and Byullee Park
Review of Three-Dimensional Handheld Photoacoustic and Ultrasound Imaging Systems and Their Applications
Reprinted from: *Sensors* **2023**, *23*, 8149, doi:10.3390/s23198149 . . . . **164**

Lulu Tian, Cong Chen, Jing Gong, Qi Han, Yujia Shi, Meiqi Li, et al.
The Convenience of Polydopamine in Designing SERS Biosensors with a Sustainable Prospect for Medical Application
Reprinted from: *Sensors* **2023**, *23*, 4641, doi:10.3390/s23104641 . . . . **187**

About the Editor

Evgeny Katz

Prof. Evgeny Katz received a Ph.D. in Chemistry from the Frumkin Institute of Electrochemistry from the Russian Academy of Sciences in 1983. He was a senior researcher at the Institute of Photosynthesis from 1983 to 1991. Between 1992 and 1993, he conducted research at the München Technische Universität as a Humboldt Fellow. Later, between 1993 and 2006, Prof. Katz worked as a Research Associate Professor at the Hebrew University of Jerusalem. Since 2006, he has been a Milton Kerker Chaired Professor in the Department of Chemistry and Biomolecular Science at Clarkson University, USA. He has (co)authored over 530 papers, with a Hirsch index of 94. His scientific interests pertain to the broad areas of bioelectronics, biosensors, biofuel cells, and biocomputing.

sensors

Article

Layer-by-Layer Immobilization of DNA Aptamers on Ag-Incorporated Co-Succinate Metal–Organic Framework for Hg(II) Detection

Shubham S. Patil [1,2], Vijaykiran N. Narwade [1], Kiran S. Sontakke [2], Tibor Hianik [2,*] and Mahendra D. Shirsat [1,2,*]

[1] RUSA-Centre for Advanced Sensor Technology, Department of Physics, Dr. Babasaheb Ambedkar Marathwada University, Aurangabad 431004, India; shubhamspatil10297@gmail.com (S.S.P.); vkiranphysics@gmail.com (V.N.N.)
[2] Department of Nuclear Physics and Biophysics, Faculty of Mathematics, Physics and Informatics, Comenius University, 842 48 Bratislava, Slovakia; kiransontakke07@gmail.com
* Correspondence: tibor.hianik@fmph.uniba.sk (T.H.); mdshirsat.phy@bamu.ac.in (M.D.S.)

Abstract: Layer-by-layer (LbL) immobilization of DNA aptamers in the realm of electrochemical detection of heavy metal ions (HMIs) offers an enhancement in specificity, sensitivity, and low detection limits by leveraging the cross-reactivity obtained from multiple interactions between immobilized aptamers and developed material surfaces. In this research, we present a LbL approach for the immobilization of thiol- and amino-modified DNA aptamers on a Ag-incorporated cobalt-succinate metal–organic framework (MOF) (Ag@Co-Succinate) to achieve a cross-reactive effect on the electrochemical behavior of the sensor. The solvothermal method was utilized to synthesize Ag@Co-Succinate, which was also characterized through various techniques to elucidate its structure, morphology, and presence of functional groups, confirming its suitability as a host matrix for immobilizing both aptamers. The Ag@Co-Succinate aptasensor exhibited extraordinary sensitivity and selectivity towards Hg(II) ions in electrochemical detection, attributed to the unique binding properties of the immobilized aptamers. The exceptional limit of detection of 0.3 nM ensures the sensor's suitability for trace-level Hg(II) detection in various environmental and analytical applications. Furthermore, the developed sensor demonstrated outstanding repeatability, highlighting its potential for long-term and reliable monitoring of Hg(II).

Keywords: DNA aptamers; metal–organic frameworks; layer-by-layer immobilization; electrochemical sensors; heavy metal ions

Citation: Patil, S.S.; Narwade, V.N.; Sontakke, K.S.; Hianik, T.; Shirsat, M.D. Layer-by-Layer Immobilization of DNA Aptamers on Ag-Incorporated Co-Succinate Metal–Organic Framework for Hg(II) Detection. *Sensors* **2024**, *24*, 346. https://doi.org/10.3390/s24020346

Academic Editor: Evgeny Katz

Received: 28 November 2023
Revised: 2 January 2024
Accepted: 3 January 2024
Published: 6 January 2024

1. Introduction

Heavy metal ions (HMIs) are pervasive environmental pollutants known for their detrimental health effects and ecological consequences [1,2]. Heavy metals can enter the environment through industrial discharges, agricultural runoff, mining, and even natural erosion, and they tend to accumulate in soils, water bodies, and the food chain [3–5]. This persistence can lead to the bioaccumulation of heavy metals in living organisms, posing severe health risks to humans and wildlife. The development of advanced and sensitive detection methods for HMIs is of paramount importance to mitigate these concerns. Electrochemical sensors, with their high sensitivity and selectivity, have emerged as a promising avenue for HMI detection [6].

MOFs as electrode modifiers enhance HMI detection due to high surface area, tunable pores, and functionalizable structures [7–12]. MOFs offer a versatile and promising platform for HMI detection, contributing to environmental protection, public health, and safety by providing highly sensitive, selective, and efficient detection methods. Researchers continue to explore and develop new MOF-based materials and sensor designs to address the challenges of HMIs contamination. Various approaches have been explored before to inculcate the conductivity of the MOFs [13,14]. The incorporation of metal nanoparticles into an MOF array is one of the most prominent approaches to tune the conductivity of the MOF. MOFs modified with specific metal nanoparticles have shown great promise in HMI detection. These modified MOFs leverage the properties of both the MOFs and the incorporated metal nanoparticles to enhance the selectivity, sensitivity, and versatility of HMIs sensors. When managed appropriately, cross-reactivity in electrochemical detection can offer several positive advantages: enhanced sensitivity, reduced false negatives, multiplexed detection, etc. Even though cross-reactivity has numerous benefits, it should be carefully managed, validated, and controlled to avoid issues such as false positives, loss of specificity, and inaccurate quantification [15]. The advantages of cross-reactivity can be harnessed effectively when the recognition elements and assays are thoughtfully designed for the intended application. Utilizing various MOFs, their composites with metal nanoparticles, and DNA aptamers as an immobilizer, various practices have been explored previously to reduce cross-reactivity instead of inculcating its benefits [16]. Several modifications in the aptamer immobilization methodologies prevent cross-reactive effects on HMI detection adopted by various research groups [17–19]. However, no single attempt has been made to govern the impact of cross-reactivity on the electrochemical detection of HMIs.

The layer-by-layer (LbL) immobilization of biorecognition elements on nanomaterial surfaces is a well-established strategy in the field of biosensors and analytical chemistry [20]. This method employs dual biorecognition layers on the electrode surface, boosting selectivity and sensitivity. Each layer enhances specificity, introducing cross-reactivity that amplifies the analytical signal for cross-reactive HMIs more than the targeted one, by improving the binding properties of each aptamer layer [21]. Cross-reactivity also improves sensitivity by amplifying analytical responses for cross-reactive HMIs. In this study, we proffer an innovative methodology encompassing the LbL immobilization of thiol- (Apt-SH) and amino-modified aptamers (Apt-NH) onto the intricate surface matrix of silver nanoparticle-incorporated cobalt succinate MOFs (Ag@Co-Succinate).

DNA or RNA aptamers are short single-stranded nucleotides with highly specific target binding capabilities. They have gained significant attention as versatile HMIs recognition in sensor development [22,23]. Through LbL immobilization of two distinct aptamers, we seek to leverage the combined selectivity and sensitivity of these receptors for the precise detection of the target analyte [24]. To govern extensive cross-reactivity for electrochemical detection of HMIs, we proposed novel LbL immobilization of Apt-SH and Apt-NH onto the Ag@Co-Succinate. Firstly, Co-Succinate MOF was synthesized by a hydrothermal method and then chemically modified with Ag nanoparticles (AgNPs) to enhance its conductivity. The first layer of immobilization involves Apt-SH, which carries thiol functional groups for covalent attachment to the MOF surface [25]. This initial layer provides selectivity for one class of target analytes and forms a robust foundation. Apt-NH is then immobilized in a second layer, introducing an additional dimension of selectivity by targeting a different class of analytes. This approach shows great potential in situations where multiple analytes are present, often with varying chemical structures, and requires the accurate and simultaneous detection of a range of targets. The LbL immobilization of Apt-SH and Apt-NH on Ag@Co-Succinate MOF provides a versatile platform for detecting tailored HMIs, which can broaden the scope of analyte detection and increase the specificity and sensitivity of the detection system. This research sets the stage for the development of advanced HMIs sensors with a wide range of applications, including environmental monitoring, clinical diagnostics, and food safety assessment.

2. Materials and Methods

2.1. Chemicals and Reagents

Cobalt nitrate hexahydrate ($Co(NO_3)_3 \cdot 6H_2O$) and silver nitrate ($Ag(NO_3)$) were purchased from Sigma Aldrich (Darmstadt, Germany). Succinic acid ($C_4H_6O_4$) and N,N-dimethylformamide ($HCON(CH_3)_2$) (DMF) were purchased from Alfa Aesar (Waltham,

MA, USA). Deionized water (DI) was utilized as a solvent. As a precondition for making buffer solutions, sodium acetate (CH_3-COO-Na), acetic acid (CH_3COOH), sodium dihydro phosphate (NaH_2PO_4), potassium ferricyanide ($K_3[Fe(CN)_6]$), and sodium phosphate dibasic (Na_2HPO_4) were purchased from Molychem, Mumbai, India. Hg(II), Cd(II), Pb(II), Cu(II), Zn(II), Fe(II), and other HMI solutions were prepared by dissolving heavy metal chloride salts in DI water. In brief, a 1 M stock solution was prepared for each heavy metal by adding the necessary weight (in milligrams) of HMI salts in the desired volume (100 mL). Following the preparation of the stock solution, the concentration reduction procedure was used to generate a concentration of HMIs, ranging from 0.7 nM to 10 nM. DNA aptamers (5′-TTT TTT ACC CAG GGT GGG TGG GTG GGT-3′) [26] modified at 5′ terminal by thiol (Apt-SH) or amino-linker (Apt-NH) were purchased from GeneCust (Boynes, France).

2.2. Synthesis of Co-Succinate and Ag@Co-Succinate

With some adjustments made in accordance with previous literature [27], the solvothermal method was utilized to synthesize Co-succinate [27]. In a combination of 50 mL N,N-dimethylformamide (DMF) and 2.5 mL DI solvents, approximately 1.5 mmol of the 1.71 g succinic acid ligand was dissolved. The blend was swirled for 10 min. After that, the mixture was agitated for a further half hour while 1 mmol of $Co(NO_3)_2$ was added. The reaction mixture was subsequently heated at 120 °C for 20 h. Co-MOF produced purple crystals as a byproduct, which were separated and repeatedly cleaned using a 5:0.25 DMF and DI mixture. The resulting goods were subsequently heated at 60 °C to dry them out.

Moreover, 0.5 g of $Ag(NO)_3$ salt in proportion to Co-Succinate was used to modify each metal. The mixture was placed in 20 mL of DMF and annealed for 5 h at 60 °C. The brown-colored precipitate was dried using a vacuum filter and dried overnight at room temperature. The process of synthesis of Co-Succinate and Ag@Co-Succinate is illustrated in Scheme 1.

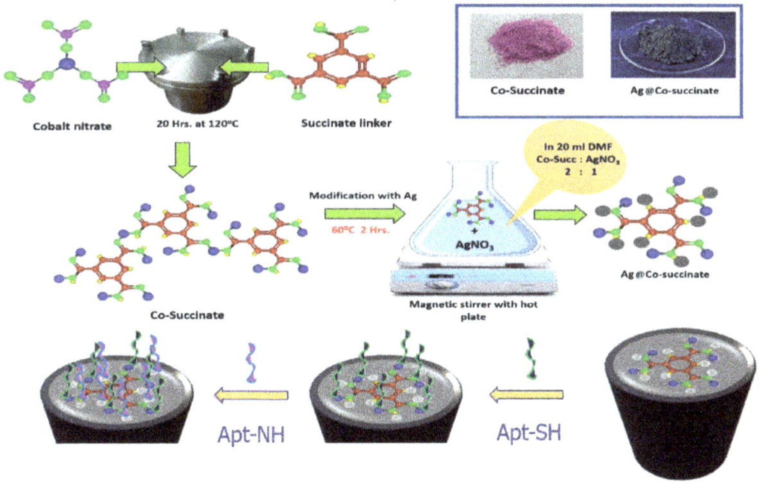

Scheme 1. Graphical representation of synthesis of Co-Succinate, Ag incorporation into Co-Succinate, and LbL immobilization of Apt-SH and Apt-NH of Ag@Co-Succinate.

2.3. Characterization Techniques

An X-ray diffractometer (XRD) 40.0 mA and 40.0 kV with a monochromatic CuK radiation source (λ = 1.54 Å) (Bruker D8 Advance, Bremen, Germany) was used for structural characterization of synthesized materials. Fourier-Transformed Infrared spectroscopy (FTIR) was taken with the Bruker Alpha model's ECO-ATR mode, which had an operating range of 600 cm^{-1} to 3000 cm^{-1} (Bruker, Bremen, Germany), and a Raman spectrophotometer (Xplora plus, Horiba Scientific, Paris, France) with a laser at 532 nm with an

1800 nm grating were used for spectroscopic characterization. Field emission scanning electron microscopy (FE-SEM) images of the materials were acquired using a TESCAN LYRA 3 (TESCAN, Brno, Czech Republic) operating at 30 kV. The electrochemical potentiostat and galvanostat CH-660C (CH-Insturments, Austin, TX, USA) was used for electrochemical characterization and electrochemical sensing.

2.4. *Immobilization of DNA Aptamers on Ag@Co-Succinate*

Firstly, 20 mg of Ag@Co-succinate was dispersed in 0.5 mL of DI. The suspended solution was prepared in a 1 mL Eppendorf tube. Then, 10 μL Nafion (Sigma-Aldrich, Saint Louis, MO, USA) was diluted in 1 mL of DI, and 5 μL of this diluted Nafion solution was added to an Eppendorf tube containing Ag@Co-Succinate. After 10 min. of ultrasonication, 5 μL of the above solution was drop-casted on a glassy carbon electrode (GCE) (CH Instruments, Austin, TX, USA). From 10 μM stock solution of the Apt-SH and Apt-NH, they were kept in small Eppendorf tubes to incubate in hot water around 95 °C until room temperature. Firstly, Ag@Co-Succinate modified GCE was incubated in Apt-SH for 5 h. And later on, electrodes were incubated with Apt-NH. Successive immobilization of Apt-SH and Apt-NH leads to the fabrication of Apt-SH-NH-Ag@Co-Succinate modified GCE electrode. LbL immobilization of Apt-SH and Apt-NH of Ag@Co-Succinate is illustrated in Scheme 1.

2.5. *Preparation of the Aptasensors and Their Electrochemical Characterization*

GCE, used for aptasensor preparation, was pre-treated by polishing it with 0.3 and 0.05-micron alumina slurry and sonicated in DI and ethanol for 2 min. and dried at room temperature. A total of 10 μL of each material (i.e., Co-Succinate, Ag@Co-Succinate) was cast over a 3 mm diameter GCE and left to dry at room temperature for 5 h. Also, the Apt-SH and Apt-NH were immobilized on Ag@Co-Succinate to determine their electrochemical response. A modified GCE was used as a working electrode, Ag/AgCl as a reference electrode, and platinum as a counter electrode. Differential pulse voltammetry (DPV) was used for the electrochemical detection of HMIs. In total, 10 mM phosphate-buffered saline (PBS), pH 7.4, was used as the electrolyte solution for setting up electrochemical cells. HMI solutions of varying concentrations ranging from 0.7 nM to 10 nM were introduced to an electrolyte solution. The selectivity measurements were performed using a constant concentration range of each HMI, i.e., 10 nM. Before the electrochemical sensing study, 0.1 M PBS containing 5 mM $Fe(CN)_6{}^{3-/4-}$ was used to record cyclic voltammograms (CVs). CV cycles were recorded at a scan rate of 0.1 V/s. Using an AC excitation signal and an open circuit potential of 300 mV in 10 mM PBS, electrochemical impedance spectroscopy (EIS) measurements were taken throughout a frequency range of 0.01 Hz to 100 kHz.

3. Results Discussion

3.1. *MOFs Characterizations*

3.1.1. Structural, Spectroscopic, and Morphological Characterizations

The XRD pattern in Figure 1a shows the Co-Succinate material's crystal lattice diffraction peaks at 2θ angles of 11.056° and 13.16°. These peaks are consistent with X-ray diffraction by the Co-Succinate crystal lattice. The pattern closely matches previously reported in the literature [27], indicating the synthesized Co-Succinate material's reliability and consistency. The observed sharp diffraction peaks at 2θ angles of 38.15°, 44.33°, 64.5°, and 77.43°, and their similarity to the JCPDS (Joint Committee on Powder Diffraction Standards) card No. 04-0783, confirm the presence of Ag nanoparticles in the Co-Succinate [28,29]. The excessive peaks suggest the extensive presence of silver in the Ag@Co-Succinate without affecting the Co-Succinate's phase. This means that the active sites for organic linkers remained unbonded, allowing specific HMIs to form a bond with these sites, as further explained by spectroscopic analysis.

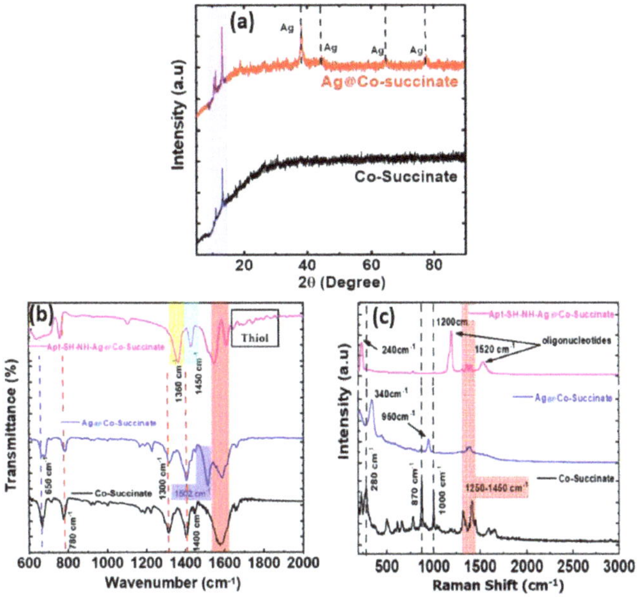

Figure 1. (**a**) X-ray diffraction patterns of Co-Succinate and Ag@Co-Succinate; (**b**) FTIR spectra of Co-Succinate, Ag@Co-Succinate, Apt-SH, and Apt-NH-Ag@Co-Succinate; (**c**) Raman spectra of Co-Succinate, Ag@Co-Succinate, Apt-SH, and Apt-NH-Ag@Co-Succinate.

The FTIR spectra of Co-Succinate and Ag@Co-Succinate in the range of 600 cm^{-1} to 2000 cm^{-1} are shown in Figure 1b. The peaks observed in Co-Succinate MOF at 650 cm^{-1} and 780 cm^{-1} are attributed to CH_3-metal groups [30], formed due to Co attachment to a methanoic group. The appearance of an additional peak for Ag/La-TMA at the spectral band of 680 cm^{-1} to 750 cm^{-1} indicates the formation of a CH_3-Ag bond. The presence of C=O stretching in both the Co-Succinate and guest Ag@Co-Succinate confirms the existence of the carboxylic group. An extra peak is observed for Ag/La-TMA in the same spectral range of 1500 cm^{-1}, signifying bond formation between Ag and the carboxylic group. Other peaks at 1300 cm^{-1} and 1580 cm^{-1} are attributed to C=O, C-O, and O-H bending vibrations of carboxylate ligands. These functional groups not involved in bond formation with guest metals may create an active site for HMI accumulation. Furthermore, after the incorporation of Apt-SH and Apt-NH onto Ag@Co-Succinate, the FTIR spectrum of Apt-SH-NH-Ag@Co-Succinate confirms the presence of thiols (SH) and amines (NH_2). A monosubstituted layer of thiol groups is observed in weak peaks ranging from 1600 cm^{-1} to 1900 cm^{-1} [31]. Similarly, a distinct peak is observed at approximately 1360 cm^{-1}, indicating the presence of amino groups. This peak indicates the successful immobilization of Apt-NH on the Ag@Co-Succinate material. The FTIR spectra clearly show the presence of amino and thiol functional groups, indicating the successful immobilization of Apt-SH and Apt-NH on the Ag@Co-Succinate material. This immobilization process plays a crucial role in customizing the material's recognition properties for specific target analytes, thus enhancing its overall performance and selectivity in various applications, including HMI detection.

The Raman spectra for Co-Succinate and Ag@Co-Succinate are shown in Figure 1c. The 100 to 340 cm^{-1} band is attributed to metal bonding to carboxylate groups. These bands were observed for both pristine systems and Ag@Co-Succinate. Ag@Co-Succinate displayed the benzonitrile group at 950 cm^{-1} due to the utilization of nitrate-based metal salt ($AgNO_3$) in the Co-Succinate modification process [32]. The one Raman active mode was observed for pristine Co-Succinate at the band of 1250 cm^{-1} to 1450 cm^{-1}, which was attributed to C-H deformations. Ag@Co-Succinate also exhibited Raman mode at this band. The band discerned at 1350 cm^{-1} for Ag@Co-Succinate belongs to benzene ring stretching vibrations due to the linker–metal bonding. Except for Ag@Co-Succinate, pristine Co-Succinate showcased stretching vibrations at 870 cm^{-1} (benzene ring stretching) [33]. This exclusion of the benzene ring stretching after modification of Co-Succinate with Ag metal due to blocking all active sites showed uniform attachment of Ag to all hydrocarbons. Moreover, Ag@Co-Succinate showed fewer Raman active bands than pristine Co-Succinate, confirming the Ag metal attachment to all functional groups. Additionally, the low-intensity band at 340 cm^{-1} was attributed to the symmetric stretching of Ag-coordinated oxygen. The Raman spectrum of Apt-SH revealed distinctive peaks, with a prominent peak at approximately 1200 cm^{-1}. The peak observed at 1200 cm^{-1} indicates the presence of thiol groups on the material surface, confirming the successful immobilization of Apt-SH [34]. The appearance of the 1520 cm^{-1} peak further reinforces the presence of amino groups in the immobilized Apt-NH, corresponding to the desired binding chemistry [34]. The distinct Raman spectra of Apt-SH and Apt-NH immobilized Ag@Co-Succinate materials validated the efficient incorporation of these aptamers onto the substrate. This confirmed the successful functionalization of the material surface with the desired molecular recognition elements. These findings contribute to the broader understanding of the material's biofunctionalization and its potential applications in selective HMI sensing.

In Figure 2a, a field emission scanning electron microscopy (FE-SEM) image of Co-Succinate displays a unique and intriguing structural feature. The Co-Succinate sample has a cauliflower-like shape, with numerous interconnected spherical structures resembling florets. These structures have a complex and highly textured surface, as seen in Figure 2b. The distinctive morphology of Co-Succinate is critical because it affects the properties of the material and its potential applications. The FE-SEM image of Ag@Co-Succinate (Figure 2c,d) reveals the presence of agglomerated silver nanoparticles on the surface of Co-Succinate crystals. This observation is crucial because it indicates the successful formation of a composite and suggests the potential for unique material properties resulting from this combination. The FE-SEM images illustrate the structural attributes of Co-Succinate and Ag@Co-Succinate and provide visual evidence of the surface modifications and interplay between the materials. These observations are pivotal in shedding light on the morphological and structural changes that may influence the materials' performance in HMI detection.

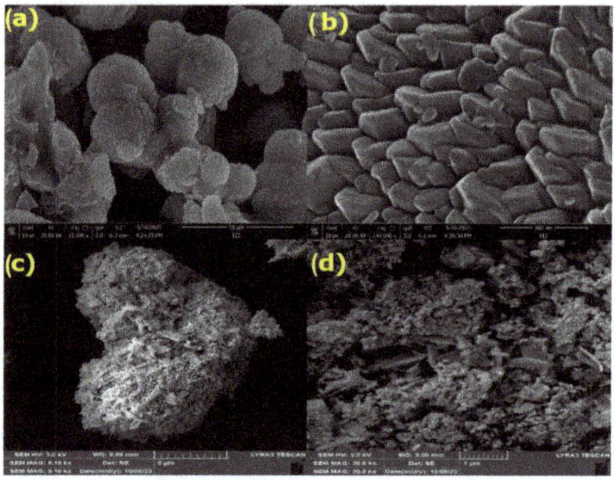

Figure 2. FE-SEM image of (**a**) Co-Succinate, (**b**) magnified image of Co-Succinate, (**c**) Ag@Co-Succinate, and (**d**) magnified image of Ag@Co-Succinate.

The phase contrast FE-SEM image (Figure 3a,b) revealed the distribution of Ag within the Ag@Co-Succinate material. Backscattered electrons are highly sensitive to the atomic number of elements, making them an excellent tool for discerning elemental composition and density variations. The presence of Ag in the image is distinctly highlighted, allowing us to visualize the spatial arrangement of Ag nanoparticles within the Co-Succinate matrix. Figure 4 comprehensively analyzed the Ag@Co-Succinate material's elemental distribution using FE-SEM mapping analysis. This technique allowed for the visual assessment of the spatial arrangement of Co and Ag elements in the composite material, providing insights into its structure and composition. The FE-SEM mapping image in Figure 4a clearly illustrates the distribution of Co within the Ag@Co-Succinate material. The image shows the spatial arrangement of Co atoms, highlighting the uniformity and consistent presence of cobalt throughout the material. Figure 4b showcases the Ag mapping results, offering insights into the spatial distribution of Ag within Ag@Co-Succinate. The image visually highlights the dispersion and distribution of Ag atoms, confirming the presence and arrangement of silver in the material. The combined mapping in Figure 4c reveals Co and Ag coexist within the Ag@Co-Succinate material. Importantly, it demonstrates the equal distribution of Ag throughout the material, emphasizing the successful incorporation of silver into the Co-Succinate matrix. Furthermore, the mapping results indicate the formation of strong and consistent bonds between silver and Co-Succinate. This observation is particularly significant, as it suggests the potential for enhanced material properties resulting from the well-established interaction between Ag and Co-Succinate.

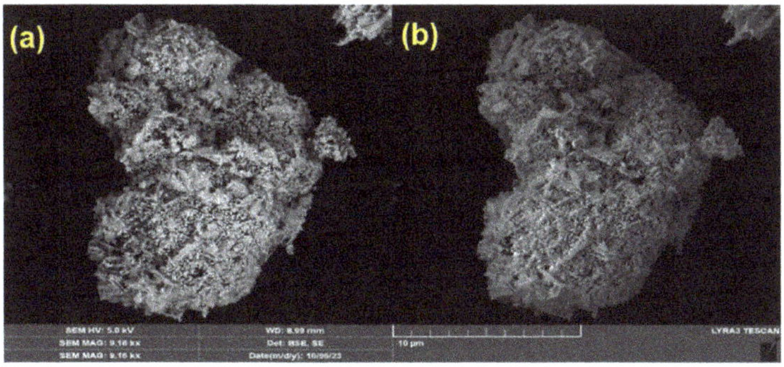

Figure 3. Phase contrast FE-SEM image of Ag@Co-Succinate. (**a**) Backscattered electron (BSE) scan and (**b**) normal scan.

3.1.2. Electrochemical Characterizations of the Aptasensors

Figure 5a shows a series of cyclic voltammograms obtained from different modified electrodes: Bare GCE, Co-Succinate-modified GCE, Ag@Co-Succinate-modified GCE, and Apt-SH-NH-Ag@Co-Succinate-modified GCE. Each modification step of the GCE resulted in unique redox peaks, with their intensities indicating the extent of electrochemical activity and redox reactions. A gradual improvement in the redox peak was observed as we progressed from bare GCE to Co-Succinate and then to Ag@Co-Succinate modifications. This enhancement indicates improved electrochemical performance and the ability of Ag@Co-Succinate to facilitate redox reactions. However, the most significant observation was seen in the cyclic voltammogram of the Apt-SH-NH-Ag@Co-Succinate-modified GCE. The redox peak was significantly enhanced, surpassing the enhancements seen in the previous modifications. This remarkable increase in the redox peak signified exceptional electrochemical activity, demonstrating the high electrochemical capabilities of the Apt-SH and Apt-NH for redox reactions of Ag@Co-Succinate. This finding is of paramount importance for the research since it suggests that the LbL immobilization of Apt-SH and Apt-NH on Ag@Co-Succinate electrodes is exceptionally well-suited for electrochemical sensing of HMIs.

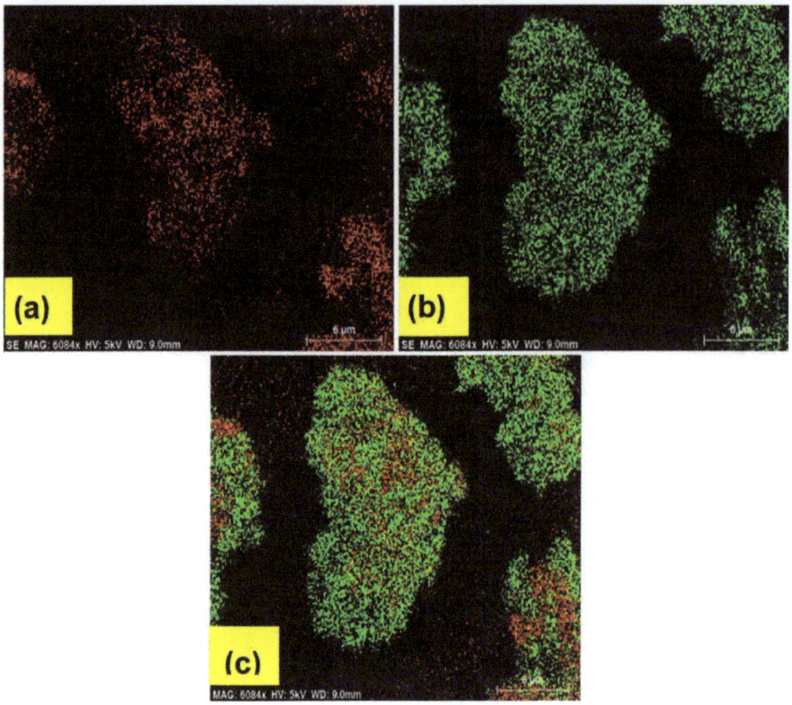

Figure 4. FE-SEM mapping images of Ag@Co-Succinate—(**a**) element Co, (**b**) element Ag, and (**c**) both Co and Ag elements.

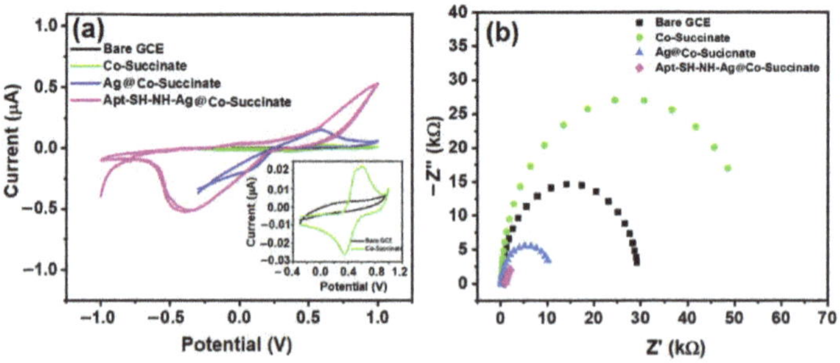

Figure 5. (**a**) Cyclic voltammogram. (**b**) Electrochemical impedance spectra of bare GCE, Co-Succinate, Ag@Co-Succinate, and Apt-SH-NH-Ag@Co-Succinate. The potential is vs. Ag/AgCl reference electrode.

Figure 5b shows the EIS analysis conducted on various modified electrodes, including bare GCE, Co-Succinate-modified GCE, Ag@Co-Succinate-modified GCE, and Apt-SH-NH-Ag@Co-Succinate-modified GCE. The EIS analysis reveals a significant trend in charge transfer resistance (R_{CT}) evolution as the GCE undergoes successive modifications. Notably, the R_{CT} was consistently decreased after each modification, except in the case of the Co-Succinate modification. The initial EIS analysis of bare GCE served as a baseline, with a specific R_{CT} value, i.e., 32 kΩ. Surprisingly, the modification of GCE with Co-Succinate showed significant enhancement in R_{CT} (59 kΩ), indicating the high charge transfer characteristics of Co-Succinate due to the blocking of the active surface area of GCE. The introduction of Ag into Co-Succinate results in a decrement in the R_{CT}, suggesting altered charge transfer dynamics due to presence of Ag. The most significant enhancement in R_{CT} was observed after the immobilization of Apt-SH-NH-Ag@Co-Succinate. This outcome indicated the introduction of the Apt-SH and Apt-NH, in combination with Ag@Co-Succinate, significantly affects the charge transfer resistance. The presence of both Apt-SH and Apt-NH appears to play a pivotal role in altering the charge transfer properties, potentially enhancing the sensitivity and selectivity of the modified electrode for HMI detection.

3.2. Electrochemical Sensing Responses of Aptasensors

Figure 6 shows the DPV analysis of different modified electrodes: Bare GCE, Co-Succinate-modified GCE, Ag@Co-Succinate-modified GCE, and Apt-SH-NH-Ag@Co-Succinate-modified GCE. The analysis provides crucial insights into the electrodes' capabilities for detecting HMIs, such as Hg(II), Pb(II), Cu(II), Fe(II), Zn(II), Ni(II), Cd(II), and Cr(II). The bare GCE electrode did not show any significant analytical signal for detecting HMIs, which indicates that it has limited applicability for this purpose. Co-succinate modification yielded a weak analytical response for Pb(II) and Hg(II). Although some response was observed, it was notably feeble. The Ag@Co-Succinate modification showed a response for both Pb(II) and Hg(II), but selectivity and sensitivity issues became evident. This suggests challenges in accurately discriminating between the two HMIs and achieving optimal sensitivity. The most remarkable outcome was observed in the Apt-SH-NH-Ag@Co-Succinate modification, which displayed a highly sensitive current response for Hg(II). This distinctive behavior underscores the selectivity of Apt-SH-NH-Ag@Co-Succinate for Hg(II). Thiol aptamer (Apt-SH) and amino aptamer (Apt-NH) are crucial in enhancing sensitivity and selectivity. Moreover, these results suggest that the interference of other HMIs in the electrochemical detection of Hg(II) does not adversely affect the selectivity and sensitivity of the sensor. This finding implies that the sensor is feasible for industries where the specific target is Hg, as it maintains its ability to selectively and sensitively detect Hg(II) in the presence of other HMIs.

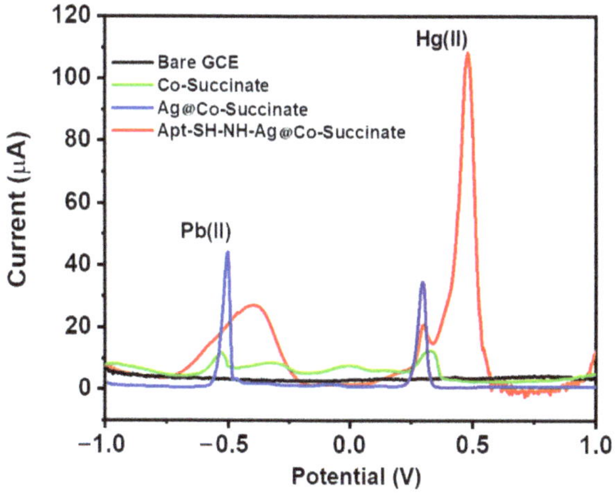

Figure 6. DPV responses for bare GCE, Co-Succinate, Ag@Co-Succinate, and Apt-SH-NH-Ag@Co-Succinate in PBS, pH 7.4, containing 10 nM of Hg(II), Pb(II), Cu(II), Fe(II), Zn(II), Ni(II), Cd(II), and Cr(II). The potential is vs. Ag/AgCl reference electrode.

3.3. Cross-Reactivity Obtained by LbL Immobilization of Aptamers

The LbL immobilization approach is identified as a potential cause of cross-reactivity, where the Apt-SH-NH-Ag@Co-Succinate modification shows a response for Hg(II) instead of Pb(II). This observation suggests the complex interplay of aptamers and materials in the detection process, leading to selective responses. The LbL immobilization technique has indeed proven effective in many cases, enhancing the sensor's selectivity for the intended HMIs. However, this approach can also be a double-edged sword, as this study highlights. The introduction of multiple layers, each with its specific recognition element, can lead to unexpected cross-reactivity. This research presents a case where layer-by-layer immobilization was designed for Pb(II) detection. However, to our surprise, and in contrast with the paper by Shamsipur et al. [23], the sensor exhibited a highly sensitive response to Hg(II) instead. This unexpected outcome prompted us to delve deeper into the factors contributing to this cross-reactivity. This investigation uncovers several factors that contribute to the observed cross-reactivity. The interaction between different aptamers in the LbL immobilization process can lead to unexpected recognition patterns. The properties of the substrate and materials used in the immobilization process can influence the binding affinities of the recognition elements. The electrochemical mechanisms involved in HMI detection can be complex, with multiple variables affecting the sensor's response.

Herein, both the aptamers (Apt-SH and Apt-NH) were developed to detect Pb(II) instead of Hg(II). The ligand exchange reaction occurs between the thiol functional group and amino functional groups. Due to this, the bonds between the metallic entities present in the Ag@Co-Succinate surface and an electron-donating end group ligand molecule, such as thiol and amine, undergo dynamic binding and unbinding processes [35,36]. Moreover, the ligand molecules on the surface can be exchanged by other ligands, possibly providing new properties or functionality to the particles. This may lead to aggregation of the particles. Such aggregation leads to disturbing the specific binding capacity of the aptamers to bind targeted HMIs.

3.4. DPV Response for Hg(II), Calibration Curve and Repeatability

Figure 7a shows the DPV response of Apt-SH-NH-Ag@Co-Succinate towards Hg(II) ranging from 0.7 nM to 10 nM. The presence of Hg(II) improves the charge transfer capabilities of Ag@Co-Succinate, as evidenced by a steady and dramatic increase in the current responses, as shown in Figure 7a when Hg(II) concentration is raised. Calibration plots (Figure 7b) show the linearity of the obtained analytical current responses. The minimum deviation in the error bar for each of the concentration levels of Hg(II) offers the accuracy and stability of the developed Apt-SH-NH-Ag@Co-Succinate sensor. Sensitivity was found to be 7.45 μA/nM with $R^2 = 0.989$. The limit of detection (LOD) has been calculated according to the formula $3\sigma/s$, where σ is the standard deviation of current responses obtained at lower HMIs concentration, and s is the slope of the calibration curve. The determined LOD = 0.3 nM was far below the maximum concentration level (MCL) suggested by the US-EPA [2]. Four independent experimental trials yielded consistent findings at 10 nM Hg(II) concentrations. Figure 7c shows the repeatable differential current overlapping signals. This confirmed the current output after a repeatable sort of experiment at a low molar concentration of Hg(II), as shown in Figure 7d.

The comparison between the electrochemical sensor based on aptamer-modified Ag@Co-Succinate (present work) and earlier reported aptamer-modified sensors with other materials for the detection of HMIs is illustrated in Table 1. It can be seen that the aptamer-modified Ag@Co-Succinate (present work) is better in terms of the range of detection and LOD.

3.5. Binding mechanism of DNA Aptamers and Hg (II)

The binding mechanism of DNA aptamers (Apt-SH and Apt-NH) with Hg(II) involves a series of coordinated interactions that contribute to the formation of a stable aptamer–metal complex. As depicted in Scheme 2, the interactions between aptamers and Hg(II) can be broadly categorized into coordination bonds, electrostatic interactions, and conformational changes in the aptamers upon binding. The thiol group in Apt-SH plays a crucial role in binding with Hg(II) ions. Mercury ions have a high affinity for thiol groups, forming strong coordination bonds. The interaction typically involves the displacement of water molecules from the Hg(II) coordination sphere by the sulfur atom of the thiol group.

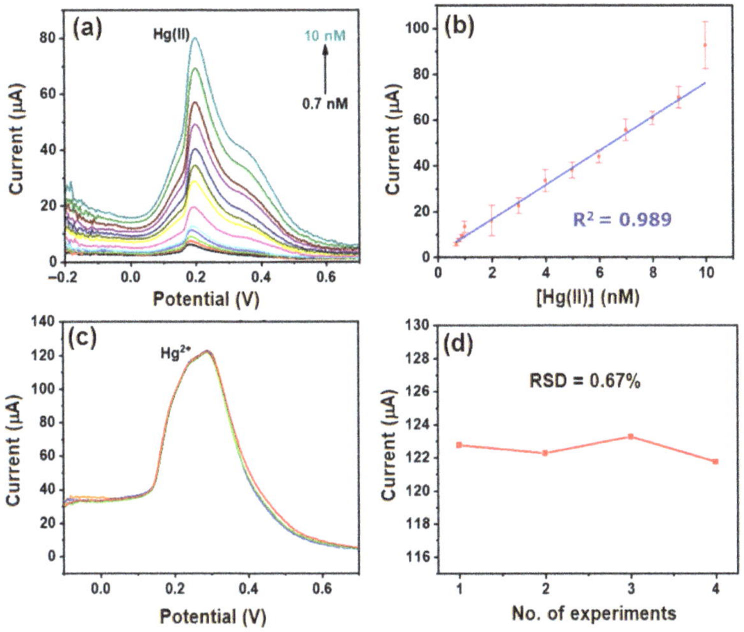

Figure 7. (**a**) DPV response of Apt-SH-NH-Ag@Co-Succinate in PBS containing Hg(II) concentration ranging from 0.7 nM to 10 nM, (**b**) calibration plot including error bars for each concentration range, (**c**) repeatability experiments for Apt-SH-NH-Ag@Co-Succinate in PBS containing 10 nM Hg(II), and (**d**) current response against number of experiments. The potential is vs. Ag/AgCl reference electrode.

Table 1. Comparison between the electrochemical sensor based on aptamer-modified Ag@Co-Succinate (present work) and earlier reported aptamer-modified sensors with other materials for detection of HMIs.

Materials for Aptamer Immobilization	HMI Detected	Range of Detection	LOD	Ref.
Ag-Au alloy NPs	AS(III)	0.01–10 μg/L	3 ng/L	[37]
Ag-Au alloy NPs	Pb(II)	0.01–10 μg/L	0.8 μM	[38]
T33/TOTO-3	Pb(II)	3–50 nM	1 nM	[39]
	Hg(II)	25–500 nM	10 nM	
Au nanoparticles	Hg(II)	1 nM–100 μM	0.6 nM	[40]
Au nanoparticles	Hg(II)	0.02–1 μM	16 nM	[41]
Graphene/Fe_3O_4-AuNP	Pb(II)	1–300 ng/mL	0.63 ng/mL	[42]
SiO_2@AuNPs	Hg(II)	10 nM–1 mM	10 nM	[43]
ZIF-8 MOF	Pb(II)	0.1–10 μg/L	0.096 μg/L	[44]
Ag@Co-Succinate	Hg(II)	0.7–10 nM	0.3 nM	This work

In Apt-NH, the amino groups contribute to the binding mechanism through coordination with Hg(II). The lone pair of electrons on the amino groups form coordination bonds with the metal ion, stabilizing the aptamer–metal complex. Both Apt-SH and Apt-NH may engage in electrostatic interactions with Hg(II) ions. The negatively charged phosphate backbone of DNA and the positively charged metal ions create favorable electrostatic attractions. These interactions can enhance the overall stability of the aptamer–metal complex. The synergistic effect of these interactions leads to the formation of a stable aptamer–metal complex, enabling selective and sensitive detection of Hg(II) ions.

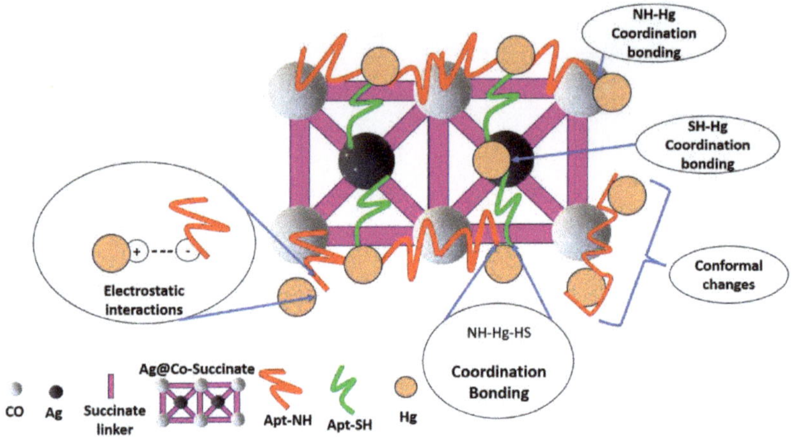

Scheme 2. Possible binding mechanism of DNA aptamers with Hg(II).

4. Conclusions

This research aimed to improve HMI detection methods by exploring innovative strategies such as LbL immobilization. The use of LbL immobilization showed the potential to enhance the sensitivity and selectivity of HMI detection. The process involved sequential attachment of aptamers (Apt-SH and Apt-NH) and Ag@Co-Succinate on the electrode surface, which allowed for tailored responses and introduced a dynamic dimension to the electrochemical sensing landscape. During the investigation, unexpected cross-reactivity was observed—instead of the anticipated response to Pb(II), the system exhibited sensitivity to Hg(II). This unexpected outcome challenges traditional assumptions and emphasizes the need for a deeper understanding of the factors influencing aptamer–material interactions. Before incorporating the effects of LbL immobilization, Ag@Co-Succinate MOF was used as a new electrode modifier for detecting HMIs as an electrochemical sensor. This study highlighted the crucial role of aptamers in determining selectivity. The inclusion of thiol aptamer (Apt-SH) and amino aptamer (Apt-NH) significantly influenced the system's response, with Apt-SH-NH-Ag@Co-Succinate displaying heightened sensitivity (7.45 μA/nM) and selectivity towards Hg(II) with an unprecedented detection limit of 0.3 nM. Moreover, the proposed sensor exhibited a repeatable analytical response with a minimum RSD of 0.67%. This emphasizes the complex interplay between aptamer sequences and the target ions.

The practical applications of Hg(II) detection presented in this work extend to diverse real-world environments, including environmental monitoring in industrial settings, water quality assessment, and compliance monitoring in regulatory contexts. Our sensor's selectivity and sensitivity make it a valuable tool for rapid, on-site detection of mercury contamination, aiding in early intervention and effective environmental management. Additionally, its portability and ease of use position it as a potential asset for fieldwork and emergency response scenarios, contributing to the broader goal of safeguarding human health and environmental well-being.

Author Contributions: Conceptualization, S.S.P., M.D.S. and T.H.; formal analysis, S.S.P., V.N.N., K.S.S., T.H. and M.D.S.; investigation, S.S.P.; writing—original draft preparation, S.S.P. and M.D.S.; writing—review and editing, S.S.P., V.N.N., T.H. and M.D.S.; supervision, M.D.S.; project administration, M.D.S.; funding acquisition, M.D.S. and T.H. All authors have read and agreed to the published version of the manuscript.

Funding: This research was funded by Inter-University Accelerator Center (IUAC), New Delhi, India (UFR no. 69330), University Grants Commission—Department of Atomic Energy (UGC—DAE) CSR, Indore (Project No. CRS/2021-22/ 01/456 dated 30 March 2022), Department of Science and Technology, Govt of India (DST—SERB), New Delhi (Project No. EEQ/2017/000645), University Grants Commission (UGC-SAP Programme) (F.530/16/DRS-I/2016 (SAP-II) Dt.16-04-2016), Department of Science and Technology, Govt of India (DST-FIST) (Project No. SR/FST/PSI-210/2016(c) dtd. 16 December 2016), and Rashtriya Uchachatar Shiksha Abhiyan (RUSA), Government of Maharashtra RUSA/Order/R&I/2016-2017/281. T.H. acknowledges the financial support of the European Union's Horizon 2020 research and innovation program through the Marie Skłodowska-Curie grant agreement no. 101007299 and the Scientific Grant Agency VEGA, project no. 1/0445/23.

Institutional Review Board Statement: Not applicable.

Informed Consent Statement: Not applicable.

Data Availability Statement: Data are contained within the article.

Acknowledgments: Mahendra D. Shirsat and Shubham S. Patil gratefully acknowledge the Slovak Academic Information Agency (SAIA) and Department of Nuclear Physics and Biophysics, Faculty of Mathematics, Physics and Informatics, Comenius University, Bratislava, Slovak Republic, for the sanction of scholarship under the framework of National Scholarship Program (NSP) of Slovak Republic. We are also grateful to Leonid Satrapinskyy (Faculty of Mathematics, Physics and Informatics, Comenius University in Bratislava) for help with the preparation of FE-SEM images.

Conflicts of Interest: The authors declare no conflicts of interest.

References

1. Patil, S.S.; Deore, K.B.; Narwade, V.N.; Peng, W.P.; Hianik, T.; Shirsat, M.D. Ultrasensitive and Selective Electrochemical Sensor Based on Yttrium Benzenetricarboxylate Porous Coordination Polymer (Y-BTC) for Detection of Pb^{2+} from Bio-Analytes. *ECS J. Solid State Sci. Technol.* **2023**, *12*, 057002. [CrossRef]
2. Sayyad, P.W.; Sontakke, K.S.; Farooqui, A.A.; Shirsat, S.M.; Tsai, M.-L.; Shirsat, M.D. A novel three-dimensional electrochemical Cd (II) biosensor based on l-glutathione capped poly (3, 4-ethylenedioxythiophene): Polystyrene sulfonate/carboxylated multiwall CNT network. *J. Sci. Adv. Mater. Devices* **2022**, *7*, 100504. [CrossRef]
3. Atchudan, R.; Perumal, S.; Edison, T.N.J.I.; Sundramoorthy, A.K.; Vinodh, R.; Sangaraju, S.; Kishore, S.C.; Lee, Y.R. Natural Nitrogen-Doped Carbon Dots Obtained from Hydrothermal Carbonization of Chebulic Myrobalan and Their Sensing Ability toward Heavy Metal Ions. *Sensors* **2023**, *23*, 787. [CrossRef] [PubMed]
4. Takahashi, F. The impact of cognitive aversion toward mercury on public attitude toward the construction of mercury wastes landfill site. *J. Mater. Cycles Waste Manag.* **2023**, *25*, 2642–2653. [CrossRef]
5. Chuai, X.; Yang, Q.; Zhang, T.; Xiao, R.; Cui, X.; Yang, J.; Zhang, T.; Chen, X.; Xiong, Z.; Zhao, Y.; et al. Migration and control of mercury in hazardous chemical waste incineration. *Fuel* **2023**, *334*, 126706. [CrossRef]
6. Bodkhe, G.A.; Hedau, B.S.; Deshmukh, M.A.; Patil, H.K.; Shirsat, S.M.; Phase, D.M.; Pandey, K.K.; Shirsat, M.D. Detection of Pb (II): Au Nanoparticle Incorporated CuBTC MOFs. *Front. Chem.* **2020**, *8*, 803. [CrossRef]
7. Narwade, V.N.; Rahane, G.K.; Bogle, K.A.; Tsai, M.-L.; Rondiya, S.R.; Shirsat, M.D. Bifunctional Supercapacitor and Photocatalytic Properties of Cuboid Ni-TMA MOF Synthesized Using a Facile Hydrothermal Approach. *J. Electron. Mater.* **2023**, *53*, 16–29. [CrossRef]
8. Wang, Y.; Wei, Y.; Li, S.; Hu, G. A Nitro Functionalized MOF with Multi-Enzyme Mimetic Activities for the Colorimetric Sensing of Glucose at Neutral pH. *Sensors* **2023**, *23*, 6277. [CrossRef]
9. Zheng, S.; Zhang, N.; Li, L.; Liu, T.; Zhang, Y.; Tang, J.; Guo, J.; Su, S. Synthesis of Graphene Oxide-Coupled CoNi Bimetallic MOF Nanocomposites for the Simultaneous Analysis of Catechol and Hydroquinone. *Sensors* **2023**, *23*, 6957. [CrossRef]
10. Tan, G.; Wang, S.; Yu, J.; Chen, J.; Liao, D.; Liu, M.; Nezamzadeh-Ejhieh, A.; Pan, Y.; Liu, J. Detection mechanism and the outlook of metal-organic frameworks for the detection of hazardous substances in milk. *Food Chem.* **2024**, *430*, 136934. [CrossRef]
11. Liu, X.; Yang, H.; Diao, Y.; He, Q.; Lu, C.; Singh, A.; Kumar, A.; Liu, J.; Lan, Q. Recent advances in the electrochemical applications of Ni-based metal organic frameworks (Ni-MOFs) and their derivatives. *Chemosphere* **2022**, *307*, 135729. [CrossRef] [PubMed]
12. Pan, Y.; Rao, C.; Tan, X.; Ling, Y.; Singh, A.; Kumar, A.; Li, B.; Liu, J. Cobalt-seamed C-methylpyrogallol [4] arene nanocapsules-derived magnetic carbon cubes as advanced adsorbent toward drug contaminant removal. *Chem. Eng. J.* **2022**, *433*, 133857. [CrossRef]
13. Deore, K.B.; Narwade, V.N.; Patil, S.S.; Rondiya, S.R.; Bogle, K.A.; Tsai, M.-L.; Hianik, T.; Shirsat, M.D. Fabrication of 3D bi-functional binder-free electrode by hydrothermal growth of MIL-101 (Fe) framework on nickel foam: A supersensitive electrochemical sensor and highly stable supercapacitor. *J. Alloys Compd.* **2023**, *958*, 170412. [CrossRef]

14. Dzikaras, M.; Barauskas, D.; Pelenis, D.; Vanagas, G.; Mikolajūnas, M.; Shi, J.; Baltrusaitis, J.; Viržonis, D. Design of Zeolitic Imidazolate Framework-8-Functionalized Capacitive Micromachined Ultrasound Transducer Gravimetric Sensors for Gas and Hydrocarbon Vapor Detection. *Sensors* **2023**, *23*, 8827. [CrossRef] [PubMed]
15. Castro, K.R.; Setti, G.O.; de Oliveira, T.R.; Rodrigues-Jesus, M.J.; Botosso, V.F.; de Araujo, A.P.P.; Durigon, E.L.; Ferreira, L.C.; Faria, R.C. Electrochemical magneto-immunoassay for detection of zika virus antibody in human serum. *Talanta* **2023**, *256*, 124277. [CrossRef]
16. Tavassoli, M.; Khezerlou, A.; Khalilzadeh, B.; Ehsani, A.; Kazemian, H. Aptamer-modified metal organic frameworks for measurement of food contaminants: A review. *Mikrochim. Acta* **2023**, *190*, 371. [CrossRef] [PubMed]
17. Zhu, C.; Liu, X.; Li, Y.; Yu, D.; Gao, Q.; Chen, L. Dual-ratiometric electrochemical aptasensor based on carbon nanohorns/anthraquinone-2-carboxylic acid/Au nanoparticles for simultaneous detection of malathion and omethoate. *Talanta* **2023**, *253*, 123966. [CrossRef]
18. Griem, P.; von Vultée, C.; Panthel, K.; Best, S.L.; Sadler, P.J.; Shaw III, C.F. T cell cross-reactivity to heavy metals: Identical cryptic peptides may be presented from protein exposed to different metals. *Eur. J. Immunol.* **1998**, *28*, 1941–1947. [CrossRef]
19. Feng, L.; Li, H.; Niu, L.-Y.; Guan, Y.-S.; Duan, C.-F.; Guan, Y.-F.; Tung, C.-H.; Yang, Q.-Z. A fluorometric paper-based sensor array for the discrimination of heavy-metal ions. *Talanta* **2013**, *108*, 103–108. [CrossRef]
20. Ariga, K.; Nakanishi, T.; Michinobu, T. Immobilization of biomaterials to nano-assembled films (self-assembled monolayers, Langmuir-Blodgett films, and layer-by-layer assemblies) and their related functions. *Int. J. Nanosci. Nanotechnol.* **2006**, *6*, 2278–2301. [CrossRef]
21. Goda, T.; Higashi, D.; Matsumoto, A.; Hoshi, T.; Sawaguchi, T.; Miyahara, Y. Dual aptamer-immobilized surfaces for improved affinity through multiple target binding in potentiometric thrombin biosensing. *Biosens. Bioelectron.* **2015**, *73*, 174–180. [CrossRef] [PubMed]
22. Hianik, T.; Ostatná, V.; Sonlajtnerova, M.; Grman, I. Influence of ionic strength, pH and aptamer configuration for binding affinity to thrombin. *Bioelectrochemistry* **2007**, *70*, 127–133. [CrossRef] [PubMed]
23. Hianik, T.; Wang, J. Electrochemical aptasensors–recent achievements and perspectives. *Electroanalysis* **2009**, *21*, 1223–1235. [CrossRef]
24. Gandotra, R.; Kuo, F.-C.; Lee, M.S.; Lee, G.-B. A Paper-Based Dual Aptamer Assay on an Integrated Microfluidic System for Detection of HNP 1 as a Biomarker for Periprosthetic Joint Infections. In Proceedings of the IEEE 36th International Conference on Micro Electro Mechanical Systems, Munich, Germany, 15–19 January 2023; pp. 1001–1004. [CrossRef]
25. Liu, S.; Bilal, M.; Rizwan, K.; Gul, I.; Rasheed, T.; Iqbal, H.M. Smart chemistry of enzyme immobilization using various support matrices—A review. *Int. J. Biol. Macromol.* **2021**, *190*, 396–408. [CrossRef] [PubMed]
26. Shamsipur, M.; Farzin, L.; Tabrizi, M.A.; Sheibani, S. Functionalized Fe_3O_4/graphene oxide nanocomposites with hairpin aptamers for the separation and preconcentration of trace Pb^{2+} from biological samples prior to determination by ICP MS. *Mater. Sci. Eng. C* **2017**, *77*, 459–469. [CrossRef] [PubMed]
27. Nurani, D.; Butar, B.; Krisnandi, Y. Synthesis and characterization of metal organic framework using succinic acid ligand with cobalt and iron metals as methylene blue dye adsorbent. In Proceedings of the IOP Conference Series: Materials Science and Engineering, Bali, Indonesia, 6–7 November 2019; p. 012055. [CrossRef]
28. Mehta, B.; Chhajlani, M.; Shrivastava, B. Green synthesis of silver nanoparticles and their characterization by XRD. In Proceedings of the Journal of Physics: Conference Series, Ujjain, India, 7–8 November 2016; p. 012050. [CrossRef]
29. Shameli, K.; Ahmad, M.B.; Zamanian, A.; Sangpour, P.; Shabanzadeh, P.; Abdollahi, Y.; Zargar, M. Green biosynthesis of silver nanoparticles using Curcuma longa tuber powder. *Int. J. Nanomed.* **2012**, *7*, 5603–5610. [CrossRef] [PubMed]
30. Ingle, N.; Sayyad, P.; Bodkhe, G.; Mahadik, M.; AL-Gahouari, T.; Shirsat, S.; Shirsat, M.D. ChemFET Sensor: Nanorods of nickel-substituted Metal–Organic framework for detection of SO_2. *Appl. Phys. A* **2020**, *126*, 723. [CrossRef]
31. Nagarjuna, R.; Saifullah, M.S.; Ganesan, R. Oxygen insensitive thiol–ene photo-click chemistry for direct imprint lithography of oxides. *RSC Adv.* **2018**, *8*, 11403–11411. [CrossRef]
32. Itoh, N.; Shirono, K.; Fujimoto, T. Baseline Assessment for the Consistency of Raman Shifts Acquired with 26 Different Raman Systems and Necessity of a Standardized Calibration Protocol. *Anal. Sci.* **2019**, *35*, 571–576. [CrossRef]
33. Takenaka, T. Infrared and Raman Spectra of TCNQ and TCNQ-d_4 Cry-stals (Commemoration Issue Dedicated to Professor Rempei Gotoh On the Occasion of his Retirement). *Bull. Inst. Chem. Res. Kyoto Univ.* **1969**, *47*, 387–400. Available online: https://cir.nii.ac.jp/crid/1050001202175269248 (accessed on 20 November 2023).
34. Park, T.; Lee, S.; Seong, G.H.; Choo, J.; Lee, E.K.; Kim, Y.S.; Ji, W.H.; Hwang, S.Y.; Gweon, D.-G.; Lee, S. Highly sensitive signal detection of duplex dye-labelled DNA oligonucleotides in a PDMS microfluidic chip: Confocal surface-enhanced Raman spectroscopic study. *Lab. Chip* **2005**, *5*, 437–442. [CrossRef] [PubMed]
35. Perumal, S. Mono-And Multivalent Interactions between Thiol and Amine Ligands with Noble Metal Nanoparticles. Ph.D. Thesis, Freie Universität Berlin, Berlin, Germany, 2012. [CrossRef]
36. Caragheorgheopol, A.; Chechik, V. Mechanistic aspects of ligand exchange in Au nanoparticles. *Phys. Chem. Chem. Phys.* **2008**, *10*, 5029–5041. [CrossRef] [PubMed]
37. Yadav, R.; Kushwah, V.; Gaur, M.; Bhadauria, S.; Berlina, A.N.; Zherdev, A.V.; Dzantiev, B. Electrochemical aptamer biosensor for As^{3+} based on apta deep trapped Ag-Au alloy nanoparticles-impregnated glassy carbon electrode. *Int. J. Environ. Anal. Chem.* **2020**, *100*, 623–634. [CrossRef]

38. Yadav, R.; Berlina, A.N.; Zherdev, A.V.; Gaur, M.; Dzantiev, B. Rapid and selective electrochemical detection of Pb^{2+} ions using aptamer-conjugated alloy nanoparticles. *SN Appl. Sci.* **2020**, *2*, 2077. [CrossRef]
39. Lin, Y.-W.; Liu, C.-W.; Chang, H.-T. Fluorescence detection of mercury (II) and lead (II) ions using aptamer/reporter conjugates. *Talanta* **2011**, *84*, 324–329. [CrossRef] [PubMed]
40. Li, L.; Li, B.; Qi, Y.; Jin, Y. Label-free aptamer-based colorimetric detection of mercury ions in aqueous media using unmodified gold nanoparticles as colorimetric probe. *Anal. Bioanal. Chem.* **2009**, *393*, 2051–2057. [CrossRef]
41. Tan, D.; He, Y.; Xing, X.; Zhao, Y.; Tang, H.; Pang, D. Aptamer functionalized gold nanoparticles based fluorescent probe for the detection of mercury (II) ion in aqueous solution. *Talanta* **2013**, *113*, 26–30. [CrossRef]
42. Tao, Z.; Zhou, Y.; Duan, N.; Wang, Z. A colorimetric aptamer sensor based on the enhanced peroxidase activity of functionalized graphene/Fe_3O_4-AuNPs for detection of lead (II) ions. *Catalysts* **2020**, *10*, 600. [CrossRef]
43. Lu, Y.; Zhong, J.; Yao, G.; Huang, Q. A label-free SERS approach to quantitative and selective detection of mercury (II) based on DNA aptamer-modified SiO_2@ Au core/shell nanoparticles. *Sens. Actuators B Chem.* **2018**, *258*, 365–372. [CrossRef]
44. Ding, J.; Zhang, D.; Liu, Y.; Zhan, X.; Lu, Y.; Zhou, P.; Zhang, D. An electrochemical aptasensor for Pb^{2+} detection based on metal–organic-framework-derived hybrid carbon. *Biosensors* **2020**, *11*, 1. [CrossRef]

sensors

MDPI

Article

A Liquid Crystal-Modulated Metastructure Sensor for Biosensing

Siyuan Liao, Qi Chen, Haocheng Ma, Jingwei Huang, Junyang Sui and Haifeng Zhang *

College of Electronic and Optical Engineering & College of Flexible Electronics (Future Technology), Nanjing University of Posts and Telecommunications, Nanjing 210023, China
* Correspondence: hanlor@163.com or hanlor@njupt.edu.cn

Abstract: In this paper, a liquid crystal-modulated metastructure sensor (MS) is proposed that can detect the refractive index (RI) of a liquid and change the detection range under different applied voltages. The regulation of the detection range is based on the different bias states of the liquid crystal at different voltages. By changing the sample in the cavity that is to be detected, the overall electromagnetic characteristics of the device in the resonant state are modified, thus changing the position of the absorption peaks so that different RI correspond to different absorption peaks, and finally realizing the sensing detection. The refractive index unit is denoted as RIU. The range of the refractive index detection is 1.414–2.828 and 2.121–3.464, and the corresponding absorption peak variation range is 0.8485–1.028 THz and 0.7295–0.8328 THz, with a sensitivity of 123.8 GHz/RIU and 75.6 GHz/RIU, respectively. In addition, an approach to optimizing resonant absorption peaks is explored, which can suppress unwanted absorption generated during the design process by analyzing the energy distribution and directing the current flow on the substrate. Four variables that have a more obvious impact on performance are listed, and the selection and change trend of the numerical values are focused on, fully considering the errors that may be caused by manufacturing and actual use. At the same time, the incident angle and polarization angle are also included in the considered range, and the device shows good stability at these angles. Finally, the influence of the number of resonant rings on the sensing performance is also discussed, and its conclusion has guiding value for optimizing the sensing demand. This new liquid crystal-modulated MS has the advantages of a small size and high sensitivity and is expected to be used for bio-detection, sensing, and so on. All results in this work were obtained with the aid of simulations based on the finite element method.

Keywords: metamaterial; liquid sensing; narrowband absorber; tunable device

Citation: Liao, S.; Chen, Q.; Ma, H.; Huang, J.; Sui, J.; Zhang, H. A Liquid Crystal-Modulated Metastructure Sensor for Biosensing. *Sensors* **2023**, *23*, 7122. https://doi.org/10.3390/s23167122

Academic Editor: Evgeny Katz

Received: 14 July 2023
Revised: 8 August 2023
Accepted: 10 August 2023
Published: 11 August 2023

1. Introduction

Metastructures, as artificially fabricated periodic topological subwavelength structures, have many excellent properties that are not available in artificial materials due to their specific three-dimensional special structures. They have gradually become a hot research topic across the globe. Metastructures can realize many extraordinary physical properties, such as a negative refractive index, an inverse Doppler effect, etc. Using the extraordinary physical properties of metastructures, sensors, filters, and couplers [1–3] can be designed, as well as novel functional devices that conventional materials cannot achieve, such as perfect absorbers [4–6], waveguides [7–9], photonic crystals [10], and metalenses [11,12]. One of the important research directions for metastructures in the current research is to transform them from simple planar structures to three-dimensional spatial structures. Such metastructures focus on the coupling relationship between spatial structures and can often achieve more flexible multifunctional design, thus becoming a hot field of multifunctional and integrated devices.

In the design of superstructures, tunable materials are often introduced, and liquid crystals are one of the important materials for voltage tuning. A liquid crystal (LC) is a material existing in a state of matter between the liquid and crystalline states. As an emerging tunable dielectric, LC has a tunable dielectric constant that can be continuously adjusted by applying a bias to orient the LC molecules. Due to its unique characteristics, including a low insertion loss, a wide operating band, low cost, and stable electromagnetic properties, especially at higher frequencies, it has advantages that other materials do not have. The material properties of LCs have been widely used in microwave and millimeter wave bands and are gaining attention in the electromagnetic field. Electromagnetic devices, including metastructures, have made outstanding progress with the help of LCs [13–17].

A common approach in the design of metamaterial-based sensors is to sense with the help of narrow-band absorption produced by an absorber. A perfect absorber is an important application of metastructures. Since it was first proposed in 2008, it has attracted considerable interest owing to its attractive characteristics, which encompass nearly perfect absorption, a flexible design, and a thin thickness. The application of the absorber is divided into two directions: broadband absorption [18–20] and narrowband absorption [21–23]. Among them, since the absorption peak of narrow-band absorption is often generated by resonance, it is easy to relate the frequency of the absorption peak to the external environment. This natural advantage can be exploited in applications such as sensing and detection [24–26], as well as broadband absorption, because it has irreplaceable advantages over traditional materials in terms of thickness, weight, etc. Therefore, it has a wide range of applications in various applications, including electromagnetic shielding, electromagnetic pollution prevention, multiplexing technology, and so on.

With the rapid development of the sensor industry, sensor devices have become indispensable components in various fields, including information technology, biotechnology, agriculture, etc. Due to their advantages of real-time detection, no pollution of the detected samples, etc., electromagnetic sensors in the terahertz band are considered to be a type of detector device with broad prospects. With the help of transmission spectrum analysis, absorption spectrum analysis, the optical rotation effect, the Hall effect, and other principles, metastructure sensors can be used to detect various physical quantities, such as the dielectric constant, thickness, angle, etc., greatly enriching the realization methods and detection means of sensors. In 2023, Guo et al. fabricated a sensor that could measure the concentration of glucose solution using an open square resonant ring [27]. Chen et al. used the principle of electromagnetic resonance coupling to produce a biosensor capable of detecting low concentrations of cytomas [28]. In 2022, Ismail et al. achieved sensing detection against coronaviruses using narrow-band absorption [29].

Considering the rapid growth of the bio-industry, there is a growing requirement for sensing and detection. There are still many interesting materials in the field of sensing that are not commonly utilized. Therefore, it is necessary to continue to diversify the use of sensing materials and device design. In this paper, a metastructure sensor (MS) modulated by a LC is proposed, which can realize refractive index (RI) detection in the THz band, enriching the design of LC sensors. The performance of the MS is given in Table 1. The orientation of the long axis of the LC can be controlled by the applied voltage. The orientation of the initial state is in the horizontal direction, and the voltage is applied in the vertical direction. The long axis of the LC points in the vertical direction when the full-bias state is applied. In the initial state, the detection range of the refractive index is 1.414–2.828, the absorption peak shifts from 1.028 THz to 0.8485 THz, and the sensitivity is 123.8 GHz/RIU. In the full-bias state, the detection range is 2.121–3.464, and the absorption peak shifts from 0.8328 THz to 0.7295 THz, with a sensitivity of 75.6 GHz/RIU. In addition, in the design process, it often faces the problem of multiple resonant peaks, that is, there will be redundant small absorption peaks around the absorption peak, so that the frequency and RI of the absorption peak cannot meet the injective relationship. This problem is discussed in this paper with the help of an energy density diagram and current flow direction, and a scheme to suppress the appearance of redundant absorption peaks is proposed. In addition, the effects of the errors in each parameter on performance are explored in detail, taking into account the possible errors in the manufacturing and actual use of the MS. The number of resonant rings, angles of incidence, and polarization are also equally indispensable parts used to fully consider the possibility of optimizing the device. The proposed MS has the advantages of a small size, tunability, a large measurement range, and high accuracy, which can be used for biosensing and other applications.

Table 1. The performance of the MS.

		Range	Sensitivity
Initial state	RI	1.414–2.828	123.8 GHz/RIU
	Frequency	0.8485–1.028 THz	
Full-bias state	RI	2.121–3.464	75.6 GHz/RIU
	Frequency	0.7295–0.8328 THz	

2. Theory and Model

In this paper, a high-frequency simulator structure (ANSYS Electronics Desktop) was selected for the simulation, and the finite element method was used for the solution. The *x*-*y* plane was set to periodic boundary conditions, and the *z*-direction was set to open space.

Figure 1 shows the structure of the MS, and the parameters of each dimension are given in Table 2. According to the structure diagram, the designed MS consists of a six-layer structure, in which the top layer is filled with the liquid sample to be measured, followed by the glass layer for confining the liquid crystal, the liquid crystal, the copper resonance ring in the liquid crystal, the liquid crystal, and the metal base plate, respectively. Among them, the resonant ring is in contact with the glass layer, thus creating a possibility for fabrication.

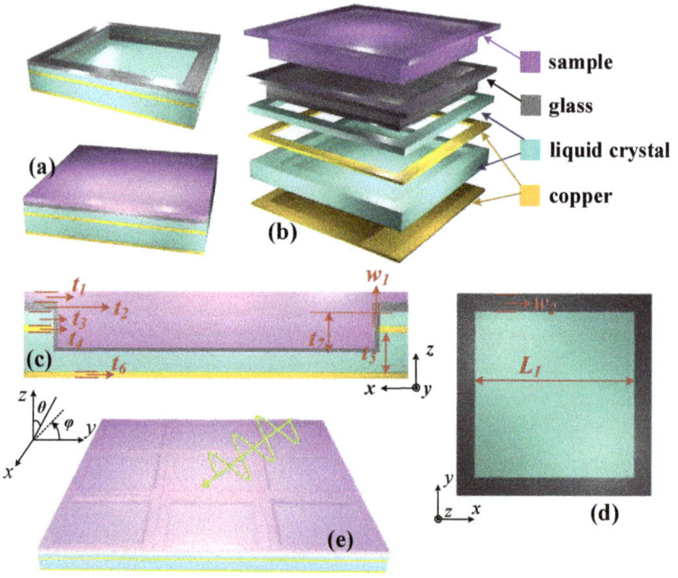

Figure 1. Schematic diagram of the MS structure: (**a**) unit of the sensor before and after filling with the liquid sample to be measured; (**b**) schematic diagram of the separated multi-layer structure; (**c**,**d**) size diagram of the unit, where t_7 is the depth of the filled cavity and w_1 is the thickness of the glass wall of the cavity; (**e**) schematic diagram of the array with incident electromagnetic waves (EMWs), with the incident angle θ and polarization angle φ indicated.

Table 2. Values of the parameters.

L_1 (μm)	w_1 (μm)	w_2 (μm)	t_1 (μm)
200	3	15	5
t_2 (μm)	t_3 (μm)	t_4 (μm)	t_5 (μm)
5	7	3	20
t_6 (μm)		t_7 (μm)	
3		20	

Among them, the conductivity of copper is $\sigma = 5.8 \times 10^7$ S/m [30]; the permittivity of glass is $\varepsilon = 4$ [31]; the long-axis RI of the liquid crystal is $n_e = 1.799$; and the short-axis RI of the liquid crystal is $n_o = 1.513$ [32]. In accordance with the axes defined in Figure 1, the initial orientation of the LC is defined in this paper to be towards the y-axis while pointing to the z-axis in the full-bias state. With such a definition, the RI matrix of the LC can be expressed by Equation (1), where $\boldsymbol{n_i}$ is the initial state and $\boldsymbol{n_F}$ is the full-bias state. For the RI of the intermediate state, it can be analyzed from an equivalent point of view, and for the incident EMW, the equivalent refractive index is given in Equation (2), where β represents the angle of rotation. In this paper, the default incident linear polarization wave is polarized in the y-direction.

$$\boldsymbol{n_i} = \begin{bmatrix} n_0 & 0 & 0 \\ 0 & n_e & 0 \\ 0 & 0 & n_0 \end{bmatrix} \quad \boldsymbol{n_F} = \begin{bmatrix} n_0 & 0 & 0 \\ 0 & n_0 & 0 \\ 0 & 0 & n_e \end{bmatrix} \tag{1}$$

$$n_{eff} = \sqrt{\frac{1}{\frac{\cos^2\beta}{n_e^2} + \frac{\sin^2\beta}{n_0^2}}} \tag{2}$$

As can be seen from Figure 2, the initial state of the liquid crystal is in the horizontal direction. After applying an external voltage, as the electric field strength gradually increases, the liquid crystal will gradually rotate and finally completely turn to the vertical direction. The specific full-bias voltage depends on the type of liquid crystal and the specific structure of the device. Based on this characteristic, for vertically incident EMW, the overall equivalent RI of the device will change under different voltages. In this paper, only the performance under zero voltage and full bias is discussed.

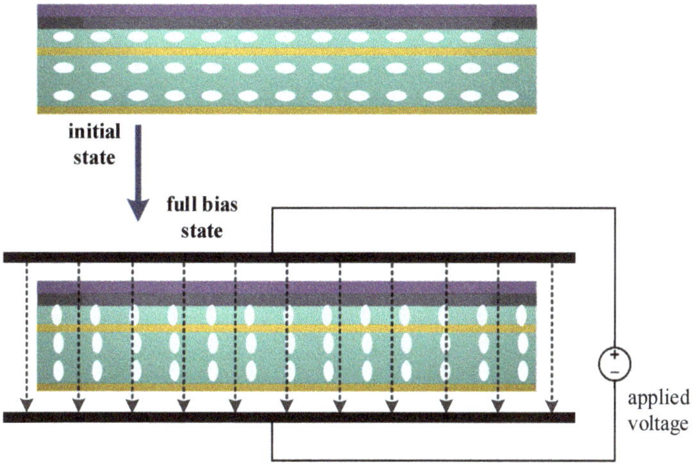

Figure 2. Schematic diagram of liquid crystal rotation with applied voltage.

For an incident EMW, it can be orthogonally decomposed into components in the x- and y-directions. Thus, for an incident EMW in any direction, the reflected and transmitted EMWs can be written in the following form:

$$\boldsymbol{E_{in}} = \begin{bmatrix} E_x \\ E_y \end{bmatrix} \tag{3}$$

$$\boldsymbol{E_{ref}} = \begin{bmatrix} r_{xx} & r_{xy} \\ r_{yx} & r_{yy} \end{bmatrix} \boldsymbol{E_{in}} \quad \boldsymbol{E_{tra}} = \begin{bmatrix} t_{xx} & t_{xy} \\ t_{yx} & t_{yy} \end{bmatrix} \boldsymbol{E_{in}} \tag{4}$$

$$r_{xx} = \frac{E_x^{ref}}{E_x^{in}} \quad r_{yx} = \frac{E_y^{ref}}{E_x^{in}} \tag{5}$$

$$r_{yy} = \frac{E_y^{ref}}{E_y^{in}} \quad r_{xy} = \frac{E_x^{ref}}{E_y^{in}} \tag{6}$$

$$t_{xx} = \frac{E_x^{tra}}{E_x^{in}} \quad t_{yx} = \frac{E_y^{tra}}{E_x^{in}} \tag{7}$$

In the above equation, $\boldsymbol{E_{in}}$ denotes the incident EMW, $\boldsymbol{E_{ref}}$ represents the reflected EMW, and $\boldsymbol{E_{tra}}$ is the transmitted EMW. And r_{xy} indicates the reflection coefficient of the EMW incident with y-polarization and reflected with x-polarization, namely, the cross-polarization reflection coefficient, and the other parameters are defined in the same way. Similarly, r_{xx} represents the co-polarization reflection coefficient, t_{xy} is the cross-polarization transmission coefficient, and t_{xx} corresponds to the co-polarization transmission coefficient. Based on the above, when incidence is along the y-direction, the absorptivity A of EMW can be defined as

$$A = 1 - |r_{yy}|^2 - |r_{xy}|^2 - |t_{yy}|^2 - |t_{xy}|^2 \tag{8}$$

Moreover, for the sensor, the Q-factor and figure of merit (FOM) are measures of sensitivity and detection accuracy. Defining the center frequency as f_0, the half-wave width as FWHM, and the sensitivity as S, the Q and FOM are calculated as follows [33]:

$$\mathrm{Q} = \frac{f_0}{\mathrm{FWHM}} \tag{9}$$

$$\mathrm{FOM} = \frac{S}{\mathrm{FWHM}} \tag{10}$$

Based on the above equation, the position of absorption peaks can be calculated as shown in Figure 3, which gives the relationship between the frequency where the absorption peak is located and the RI. The frequency f–RI equation is obtained by curve fitting. In the two states, R^2 equals 0.9922 and 0.9952, so the relationship between f and RI can be regarded as linear, which is convenient for the practical application of sensing.

$$d = \sqrt{\frac{2}{\omega u \sigma}} \tag{11}$$

On the other hand, skin depth is also extremely important in the sensing process. The general calculation formula for skin depth is demonstrated in Equation (11), where ω is the angular frequency, u is the magnetic permeability, and σ represents the conductivity. Considering that the response frequency of the device is around 1 THz, the skin depth should be around 0.066 μm when the frequency equals 1 THz. The copper material used in the device has a thickness of more than 3 μm, so it can be considered that EMWs will not penetrate the device.

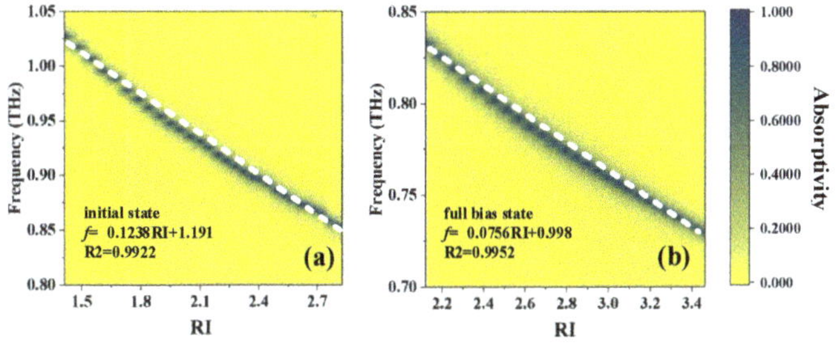

Figure 3. Schematic diagrams of the relationship between f and RI: (**a**) initial state and (**b**) full-bias state.

Based on the formula calculation, it can be seen that the Q-factor of the initial state varies between 130.54 and 178.73, and the main reason for the change is the shift in f_0, while the width change of a single resonance peak is not very obvious. Therefore, in this case, the FOM is more suitable to show the performance within the working frequency band. The FOM of the initial state is 20 RIU^{-1}. Similarly, the FOM of the full-bias state is around 13 RIU^{-1}, and the Q-factor will be in the range of 122.39 to 162.2. The change in value is caused by the variation in the absorption frequency in combination with small differences in the width of the absorption peaks.

Based on the above performance, a possible application scenario is given. Since different components in the blood will affect the overall RI of the blood, the concentration of creatinine in the blood can be measured by the change in refractive index. Table 3 shows the relationship between concentration and RI [33]. Apparently, RI values corresponding to different concentrations are within the detection range.

Table 3. RI of blood at different creatinine concentrations.

Creatinine Blood Sample Concentration (μmol L^{-1})	RI
80.9	2.661
81.43	2.655
82.3	2.639
83.3	2.610
84.07	2.589
85.28	2.565

3. Mechanism Analysis and Optimization

Figure 4 presents the energy distribution at the absorption peak (f = 0.8485 THz) and at f = 0.86 THz, where no absorption occurs when RI = 2.828. It can be seen that, compared with the frequency where no absorption occurs, a strong resonance appears clearly on the metal frame of the second layer between the two units at the absorption frequency. At the same time, the current on the copper ring (the third layer from the bottom up, as shown in Figure 1b) is also significantly enhanced, and the current on the bottom plate shows the current distribution of dipole resonance. Therefore, the absorption peak is caused by resonance. Therefore, the change in absorption frequency caused by different RI and LC orientations essentially changes the electromagnetic characteristics around the resonant part, thus changing the absorption frequency.

Although the absorption frequencies at different RI values present a good linear relationship, there are still excess absorption peaks at a few specific RI values. Although these excess absorption peaks do not reach 0.9 and their corresponding RI values are out of the measurement range, this brings risks for practical use due to the close proximity of the peaks and the overlap of the frequency bands. As demonstrated in Figure 5a, the absorption peaks that appear at the two values of RI in the initial state, 2.5 and 3.873, are quite close. The absorption peaks given in the figure are staggered, but since the blue absorption peak will move with the change in RI, a part of the blue peak must be located at 0.875 THz, corresponding to the RI value between about 0.25 and 2.65. Similarly, the position of the orange excess absorption peak will also change with the value of RI. According to the simulation results, the absorption peak that appears in the range of 3.6–4 for the RI will overlap with the detection band, which will cause errors in detection.

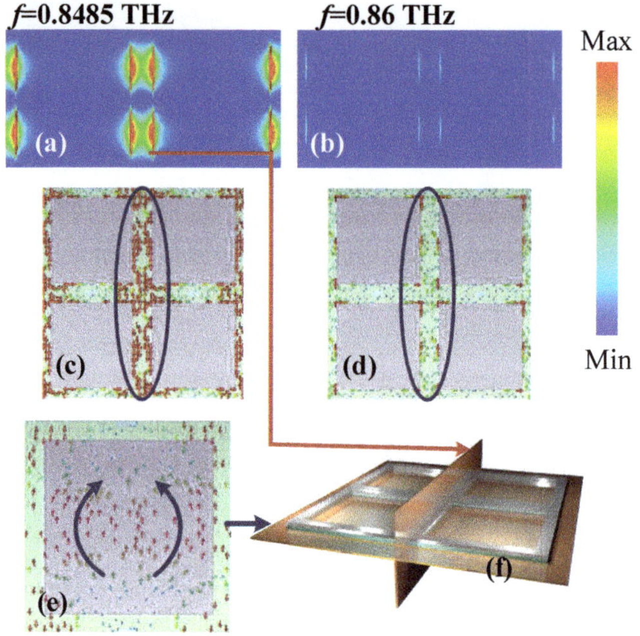

Figure 4. Energy density diagrams at (**a**) 0.8485 THz and (**b**) 0.86 THz when RI = 2.828. The surface current of the resonance ring at (**c**) 0.8485 THz and (**d**) 0.86 THz. (**e**) The surface current on substrate at 0.8485 THz. (**f**) The cross section diagram.

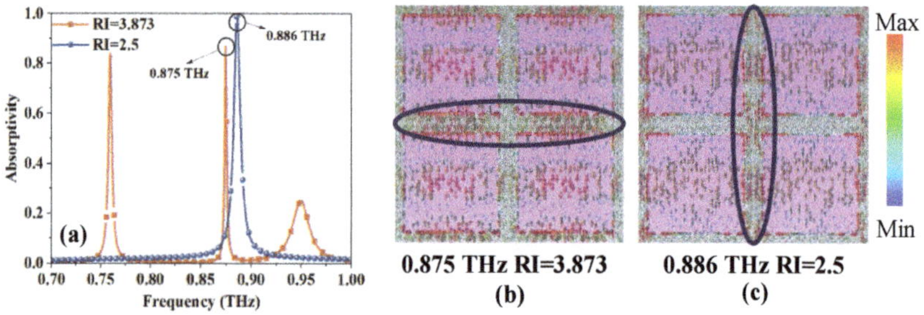

Figure 5. (**a**) Absorption peaks and excess absorption peaks at different RI values. Current diagram in different situations, with the strongest part of the current circled in the diagram: (**b**) 0.875 THz and RI = 3.873; and (**c**) 0.886 THz and RI = 2.5.

Figure 5b,c show the current maps in the two cases. It can be seen from the figure that the absorption peak at 0.875 THz is due to the vertical resonance, while the blue absorption peak used for detection is due to the horizontal inter-unit resonance. Therefore, the absorption can be weakened by cutting narrow slits on the bottom plate to block the current.

The specific method is to etch slits on the bottom copper plate, as shown in Figure 6a,b. The two sets of curves with different w_4 are shown in Figure 6c. As the value of w_4 increases, the excess absorption is clearly suppressed. Until w_4 = 40 μm is reached, the original absorption peak used for detection (RI = 2.5 group) did not show a significant decline. Only after continuing to increase to 60 μm does the absorptivity drop below 0.9; this is because the expansion of the slit will inevitably lead to an increase in transmittance and ultimately cause the absorptivity to decrease. Therefore, w_4 values in the range of 20–40 μm are acceptable. Based on this operation, the absorption caused by excess resonance can be effectively suppressed to ensure the accuracy of sensing.

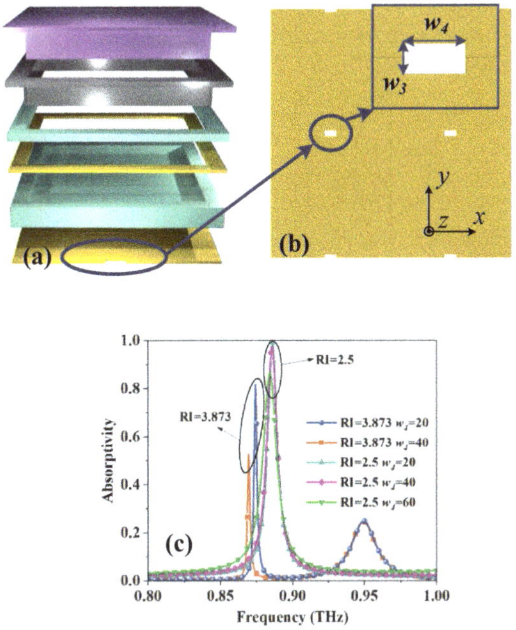

Figure 6. (**a**) A perspective view of the multi-layer structure. (**b**) Size diagram of the gap on the bottom plate, w_3 = 10 μm. (**c**) Influence on absorption peak under different RI and w_4.

4. Results and Discussion

The main reason for the shift of the absorption peak is that the measured object changes the electromagnetic characteristics of the MS as a whole, so the change of the structural parameters should have a similar effect on the absorption peaks at each frequency point. Considering this, at the critical point of the detection range, the absorptivity will slightly decrease, and the linear relationship of the fitting will become slightly worse. Therefore, all the following results are given in the initial state, when RI = 2.65. The fitting results here have better linearity, and the change in a single absorption peak can represent the change trend of all absorption peaks within a certain range.

The filling thickness of the sample is one of the most important indicators in practical applications. Figure 7a shows the effect of thickness t_1 on the absorption peak. It can be seen that in the region near 5 μm, the value of the absorptivity is relatively stable, and only the frequency shift is more obvious. This phenomenon can be intuitively understood as follows: when the thickness of the filler changes for a MS, its equivalent RI also changes. Therefore, such a property can also be applied to thickness detection and other aspects when the RI of the sample is known. When the thickness continues to increase, for EMWs, the influence of the MS gradually decreases, and the state becomes closer to EMWs directly incident to the sample interface; thus, the absorptivity decreases. According to the RI = 2.65 calculation, the incidence reflectivity is about 0.45, which means that the maximum absorptivity is 0.55.

It can be seen that as the thickness increases, the absorptivity decreases and becomes closer to this value.

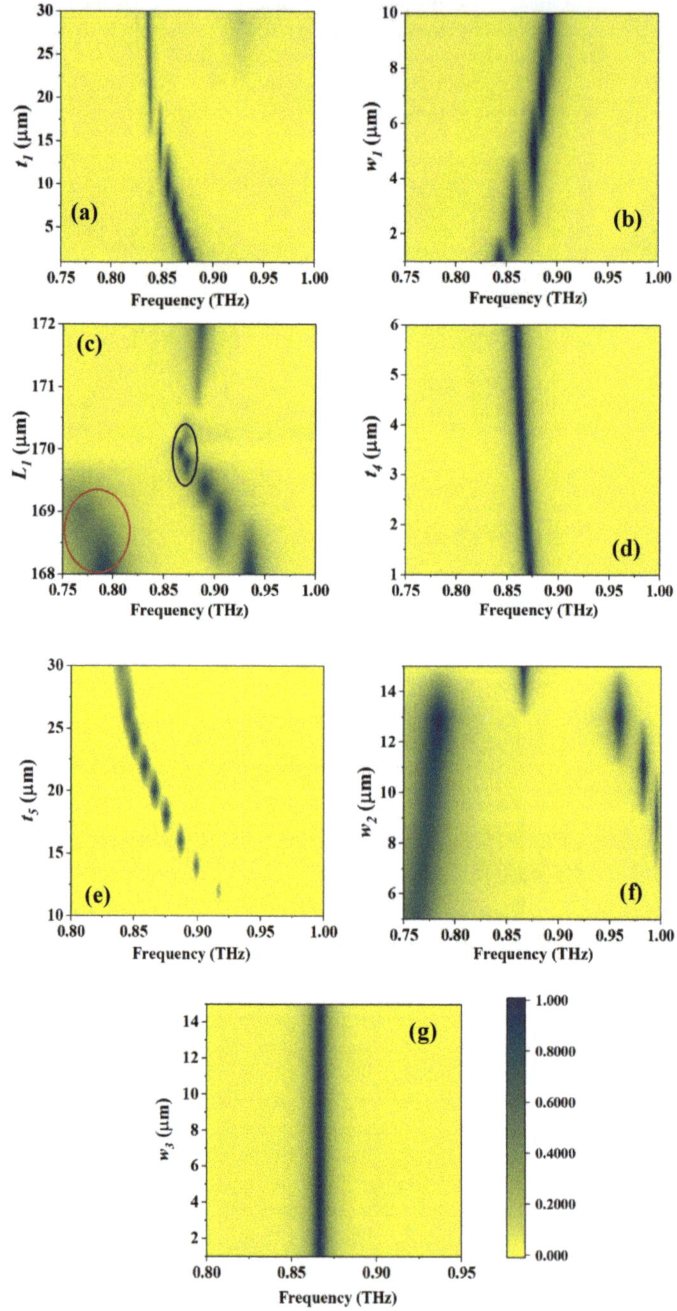

Figure 7. Effects of different parameters on the absorption peaks at RI = 2.65: (**a**) t_1, (**b**) w_1, (**c**) L_1, red circles are additional absorption peaks that appear at low frequencies, and blue circles are optimal selection values, (**d**) t_4, (**e**) t_5, (**f**) w_2, and (**g**) w_3.

The errors in the manufacturing process should also be fully considered. There are some sensitive parameters, which are the thickness w_1 of the glass wall, the side length L_1 of the filled groove, the thickness of the copper ring t_4, the distance from the substrate to the copper ring t_5, and the width of the copper ring w_2. The absorption peak is insensitive to other parameters, and at least no obvious difference is shown in a fairly large range of changes.

As illustrated in Figure 7b, the absorptivity corresponding to different w_1 is relatively stable, and the absorption peak gradually moves to a high frequency as the thickness increases. L_1 is a very sensitive variable, and only when it is near 170 μm can a stable absorption peak be formed, as marked in the center of Figure 7c. When the value is greater than 170 μm, the absorption quickly decreases, completely disappears at 170.6 μm, and then gradually increases, but it is difficult to return to above 0.9 in a large range. When L_1 is less than 170 μm, although it can maintain an absorptivity above 0.9 and is accompanied by some movement, it will excite an extra absorption peak in the low frequency range (circled in red), which overlaps with part of the detection range. Moreover, with the frequency shift, the Q-factor of the absorption peak also shows a more serious attenuation, which is not conducive to detection.

The influence of the thickness t4 of the copper ring on the absorption peak is shown in Figure 7d. From the offset of the absorption peak and the numerical value, it can be found that as the numerical value gradually increases, the absorption frequency point gradually shifts to a low frequency, and the absorption rate shows a trend of first increasing and then decreasing. Among them, t_4 = 3 μm is the maximum point, and then the absorption rate gradually decreases. At t_4 = 1 μm, although the absorption rate is not much different from that at 3 μm, its bandwidth is slightly increased, which will have a negative impact on the accuracy of sensing. Therefore, considering the situation comprehensively, choosing t_4 = 3 μm is the most suitable.

For t_5, the first thing to emphasize is that it corresponds to the distance between the copper ring and the bottom plate. The change in t_5 only indicates the up and down movement of the copper ring, and the sum of the LC thicknesses of its upper and lower layers is constant. Therefore, considering the manufacturing possibility and the overall thickness of the LC, the range of the change in t_5 is between 10 and 30 μm. Figure 7e shows the absorption peaks at different t_5 values. It can be seen that as t_5 increases, the absorption point first moves toward low frequency and stabilizes after increasing to 25 μm. On the other hand, t_5 = 20 μm is the optimum point of the absorption rate; the farther away from this value, the smaller the absorption rate, the larger the bandwidth, and the smaller the corresponding Q-factor.

Figure 7f shows the variation curve of the absorption peak with w_2. It should be noted that, limited by the size, the maximum value of w_2 can only be 15 μm, at which point the resonant rings between adjacent units are actually connected, which will weaken the absorption peak caused by the coupling between adjacent units. This can be confirmed from the figure. When the value is small, the absorption peak is divided into two parts: one part is the absorption concentrated at low frequency, which is characterized by a decrease as the value increases. The other part is the high-frequency absorption, which increases with the value and moves toward a low frequency. From the curve trend and shape, it can be determined that the final absorption peak used for detection at 15 μm is actually the initial high-frequency absorption, while the low-frequency absorption is completely suppressed as the ring width increases, making the gap between adjacent units' resonant rings gradually smaller until connected.

As mentioned above, the width of the gap on the bottom plate w_3 does not affect the absorption. This judgment is made considering that the value of w_4 is 20–40 μm, and that, without changing the shape of the gap, the width of w_3 should not exceed 20 μm. This is because if the width of w_3 is too large, the geometric structure essentially changes, and the gap exists only to cut off the current between the units rather than affect the current in the middle of the resonant unit.

The performance at different incidence angles θ and polarization angles φ is given in Figure 8. There is a minimal value near 40° as θ changes, and then it recovers. For φ, the absorption peak is stable, so the polarization is not sensitive. This feature allows for a wide range of signal source requirements for practical use.

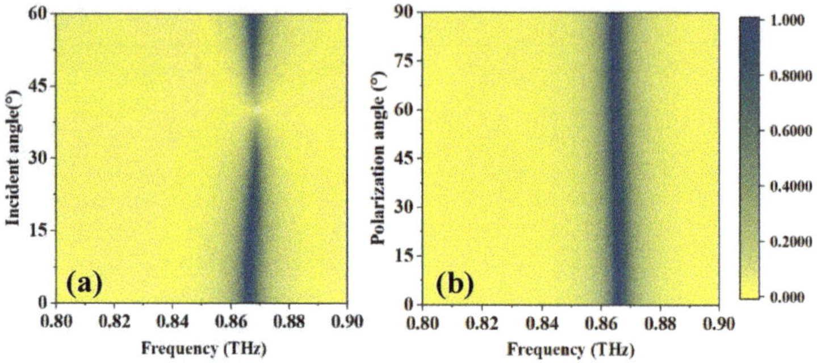

Figure 8. Absorptivity at different (**a**) θ and (**b**) φ.

As demonstrated in Figure 9a, a resonant ring of the same size is added at a position of 10 µm below the original resonant ring, which may excite new resonant peaks and provide absorption points with larger Q-factors. However, it may also produce redundant resonant peaks that cannot be suppressed. The absorption peak of the double-resonant ring structure in the initial state is shown in Figure 9b. Unfortunately, such a structure did not excite a steeper peak but destroyed the original resonance, causing the absorption rate to decrease overall, and it excited two absorption peaks with very close absorption peak values at RI = 2, further deteriorating the sensing effect, as displayed in Figure 9b. Therefore, for the existing structure, considering the simplicity of structure fabrication and effect, the single-resonant ring structure shows the best effect. Such a conclusion provides guidance for the optimization of sensing effects, that is, if further optimization of the Q-factor is desired, it can be optimized from the perspective of a planar structure and resonance principle, or similar. At least in terms of optimization complexity, vertical periodic structures are not preferred options.

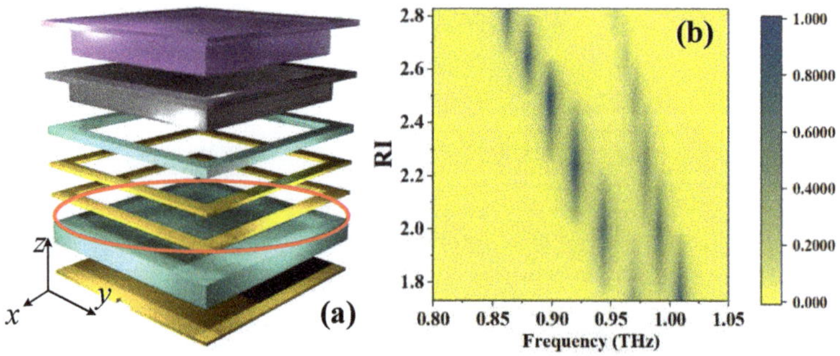

Figure 9. (**a**) Schematic diagram of double ring structure and (**b**) its performance.

Table 4 lists the comparison with other studies, and it can be found that although this work is not outstanding in terms of performance, the characteristics of liquid crystal tuning in the same field are not common.

Table 4. Comparison with other studies.

Ref.	Function	Tunable	Sensitivity	FOM	Liquid Crystal
[24]	Agricultural sensing	No	220.7 GHz/RIU	1.52	No
[26]	Thickness detection	No	6.61 GHz/mm	21.975	No
[29]	Biosensing	No	264 or 968 GHz/RIU	9.46	No
[33]	Biosensing	Yes	136.4–306.25 nm/RIU	1.5–10.3	No
This work	Biosensing	Yes	123.8 or 75.6 GHz/RIU	20 or 13	Yes

5. Conclusions

In summary, this paper proposes a liquid crystal-controlled metastructure sensor. Absorption can be caused by the resonance between adjacent resonant rings, and the absorption frequency point changes with the refractive index of the filling liquid. Sensing can be realized according to this phenomenon. With the help of liquid crystals, the detection range of the refractive index can be adjusted at different voltages. The device can be applied to the detection of biological solutions, such as creatinine concentrations in blood.

In the initial state, the detection range is 1.414–2.828, the frequency range of the absorption peak is 0.8485–1.028 THz, and the sensitivity is 123.8 GHz/RIU. When a voltage is applied to rotate the liquid crystal until it is fully biased, the detection range becomes 2.121–3.464, the frequency range is 0.7295–0.8328 THz, and the sensitivity is 75.6 GHz/RIU. The detailed performances are listed in Table 5. The refractive index of blood at different creatinine concentrations is usually between 2.5 and 2.7, which indicates that the existing detection range can meet the detection needs well and adapt to more application scenarios.

Table 5. Detailed performance of the device.

		Range	Sensitivity	FOM
Initial state	RI	1.414–2.828	123.8 GHz/RIU	20 RIU^{-1}
	Frequency	0.8485–1.028 THz		
Full bias state	RI	2.121–3.464	75.6 GHz/RIU	13 RIU^{-1}
	Frequency	0.7295–0.8328 THz		

In addition, this paper analyzes the cause of absorption through the current and energy density diagrams and proposes a method to suppress the additional absorption peak, which can avoid absorption at the same frequency point under different refractive indices.

Finally, the parameters of the device are discussed, including the effects of errors in the manufacturing process and the thickness errors of the sample liquid in actual use. The device is extremely stable at different angles; the incidence angle has a stable range of about 40°, and the polarization angle will not affect the performance. Furthermore, the number of resonant rings is also discussed, and it is clear that the optimal absorption can only be obtained at a single-resonant ring, and excess resonant rings will produce excess absorption peaks, which provides a guiding idea for the optimization of the structure.

This metastructure sensor has the advantages of real-time detection and a large detection range and can be used for biosensing, including, but not limited to, creatinine concentration measurements, providing new ideas for the design of tunable sensors.

Author Contributions: S.L. and Q.C.: Data curation, Formal analysis, Investigation, Writing—original draft, Visualization. H.M., J.H. and J.S.: Software, Validation. H.Z.: Conceptualization, Methodology,

Supervision, Writing—review and editing. All authors have read and agreed to the published version of the manuscript.

Funding: This research received no external funding.

Institutional Review Board Statement: Not applicable.

Informed Consent Statement: Not applicable.

Data Availability Statement: Samples of the compounds are available from the authors.

Acknowledgments: This work was supported by the National College Student Innovation Training Program (Grant No.202210293014Z).

Conflicts of Interest: The authors declare no conflict of interest.

References

1. Pligovka, A.; Lazavenka, A.; Turavets, U.; Hoha, A.; Salerno, M. Two-Level 3D Column-like Nanofilms with Hexagonally–Packed Tantalum Fabricated via Anodizing of Al/Nb and Al/Ta Layers—A Potential Nano-Optical Biosensor. *Materials* **2023**, *16*, 993. [CrossRef] [PubMed]
2. Ruan, J.-F.; Lan, F.; Tao, Z.; Meng, Z.-F.; Ji, S.-W. Tunable terahertz metamaterial filter based on applying distributed load. *Phys. Lett. A* **2021**, *421*, 127705. [CrossRef]
3. Pérez-Armenta, C.; Ortega-Moñux, A.; Luque-González, J.M.; Halir, R.; Reyes-Iglesias, P.J.; Schmid, J.; Cheben, P.; Molina-Fernández, Í.; Wangüemert-Pérez, J.G. Polarization-independent multimode interference coupler with anisotropy-engineered bricked metamaterial. *Photonics Res.* **2022**, *10*, A57–A65. [CrossRef]
4. Liu, Y.; Ma, W.-Z.; Wu, Y.-C.; Meng, D.; Cheng, Y.-Y.; Chen, Y.-S.; Liu, J.; Gu, Y. Multi-peak narrow-band metamaterial absorber for visible to near-infrared wavelengths. *Results Phys.* **2023**, *47*, 106374. [CrossRef]
5. Asgari, S.; Fabritius, T. Graphene-Based Multiband Chiral Metamaterial Absorbers Comprised of Square Split-Ring Resonator Arrays with Different Numbers of Gaps, and Their Equivalent Circuit Model. *IEEE Access* **2022**, *10*, 63658–63671. [CrossRef]
6. Zeng, X.; Rosenmann, D.; Czaplewski, D.A.; Gao, J.; Yang, X. Mid-infrared chiral metasurface absorbers with split-ellipse structures. *Opt. Commun.* **2022**, *525*, 128854. [CrossRef]
7. Cabo, R.F.; Vilas, J.; Cheben, P.; Velasco, A.V.; González-Andrade, D. Experimental characterization of an ultra-broadband dual-mode symmetric Y-junction based on metamaterial waveguides. *Opt. Laser Technol.* **2023**, *157*, 108742. [CrossRef]
8. Chen, H.; Zhang, Z.; Zhang, X.; Han, Y.; Zhou, Z.; Yang, J. Multifunctional Plasmon-Induced Transparency Devices Based on Hybrid Metamaterial-Waveguide Systems. *Nanomaterials* **2022**, *12*, 3273. [CrossRef]
9. Alibakhshikenari, M.; Ali, E.M.; Soruri, M.; Dalarsson, M.; Naser-Moghadasi, M.; Virdee, B.S.; Stefanovic, C.; Pietrenko-Dabrowska, A.; Koziel, S.; Szczepanski, S.; et al. A comprehensive survey on antennas on-chip based on metamaterial; metasurface; and substrate integrated waveguide principles for millimeter-waves and terahertz integrated circuits and systems. *IEEE Access* **2022**, *10*, 3668–3692. [CrossRef]
10. Pligovka, A.; Poznyak, A.; Norek, M. Optical Properties of Porous Alumina Assisted Niobia Nanostructured Films–Designing 2-D Photonic Crystals Based on Hexagonally Arranged Nanocolumns. *Micromachines* **2021**, *12*, 589. [CrossRef]
11. Sedaghat, A.; Mohajeri, F. Size reduction of a conical horn antenna loaded by multi-layer metamaterial lens. *IET Microw. Antennas Propag.* **2022**, *16*, 391–403. [CrossRef]
12. Zhao, F.; Li, Z.; Li, S.; Dai, X.; Zhou, Y.; Liao, X.; Cao, J.C.; Liang, G.; Shang, Z.; Zhang, Z.; et al. Terahertz metalens of hyper-dispersion. *Photon. Res.* **2022**, *10*, 886–895. [CrossRef]
13. Liu, S.; Xu, F.; Zhan, J.; Qiang, J.; Xie, Q.; Yang, L.; Deng, S.; Zhang, Y. Terahertz liquid crystal programmable metasurface based on resonance switching. *Opt. Lett.* **2022**, *47*, 1891–1894. [CrossRef]
14. Yu, H.; Wang, H.; Wang, Q.; Ge, S.; Hu, W. Liquid Crystal-Tuned Planar Optics in Terahertz Range. *Appl. Sci.* **2023**, *13*, 1428. [CrossRef]
15. Deng, G.; Mo, H.; Kou, Z.; Yang, J.; Yin, Z.; Li, Y.; Lu, H. A polyimide-free configuration for tunable terahertz liquid-crystal-based metasurface with fast response time. *Opt. Laser Technol.* **2023**, *161*, 109127. [CrossRef]
16. Chiang, W.; Silalahi, H.M.; Chiang, Y.; Hsu, M.; Zhang, Y.; Liu, J.; Yu, Y.; Lee, C.; Huang, C. Continuously tunable intensity modulators with large switching contrasts using liquid crystal elastomer films that are deposited with terahertz metamaterials. *Opt. Express* **2020**, *28*, 27676–27687. [CrossRef]
17. Wang, P.Y.; Sievert, B.; Svejda, J.T.; Benson, N.; Meng, F.Y.; Rennings, A.; Erni, D. A Liquid Crystal Tunable Metamaterial Unit Cell for Dynamic Metasurface Antennas. *IEEE Antennas Propag. Mag.* **2022**, *71*, 1135–1140. [CrossRef]
18. Xie, T.; Chen, D.; Xu, Y.; Wang, Y.; Li, M.; Zhang, Z.; Yang, J. High absorption and a tunable broadband absorption based on the fractal Technology of Infrared Metamaterial Broadband Absorber. *Diam. Relat. Mater.* **2022**, *123*, 108872. [CrossRef]
19. Yao, X.; Huang, Y.; Li, G.; He, Q.; Chen, H.; Weng, X.; Liang, D.; Xie, J.; Deng, L. Design of an ultra-broadband microwave metamaterial absorber based on multilayer structures. *Int. J. RF Microw. Comput. Aided Eng.* **2022**, *32*, e23222. [CrossRef]
20. Yao, B.; Zeng, Q.; Duan, J.; Wei, L.; Kang, J.; Zhang, B. A transparent water-based metamaterial broadband absorber with a tunable absorption band. *Phys. Scr.* **2023**, *98*, 015507. [CrossRef]

21. Liao, Y.L.; Zhao, Y. Ultra-narrowband dielectric metamaterial absorber with ultra-sparse nanowire grids for sensing applications. *Sci. Rep.* **2020**, *10*, 1480. [CrossRef] [PubMed]
22. Chao, C.C.; Kooh, M.R.R.; Lim, C.M.; Thotagamuge, R.; Mahadi, A.H.; Chau, Y.F.C. Visible-range multiple-channel metal-shell rod-shaped narrowband plasmonic metamaterial absorber for refractive index and temperature sensing. *Micromachines* **2023**, *14*, 340. [PubMed]
23. Fan, Y.; Yang, R.; Li, Z.; Zhao, Y.; Tian, J.; Zhang, W. Narrowband metamaterial absorbers based on interlaced T-shaped all-dielectric resonators for sensing application. *J. Opt. Soc. Am. B* **2022**, *39*, 2863–2869. [CrossRef]
24. Lang, T.; Zhang, J.; Qiu, Y.; Hong, Z.; Liu, J. Flexible terahertz Metamaterial sensor for sensitive detection of imidacloprid. *Opt. Commun.* **2023**, *537*, 129430. [CrossRef]
25. Wang, Y.; Wang, T.; Yan, R.; Yue, X.; Wang, L.; Wang, H.; Zhang, J.; Yuan, X.; Zeng, J. A Tunable Strong Electric Field Ultra-Narrow-Band Fano Resonance Hybrid Metamaterial Sensor Based on LSPR. *IEEE Sens. J.* **2023**, *23*, 14662–14669. [CrossRef]
26. Khalil, M.A.; Yong, W.H.; Islam, M.T.; Hoque, A.; Islam, M.S.; Leei, C.C.; Soliman, M.S. Double-negative metamaterial square enclosed QSSR for microwave sensing application in S-band with high sensitivity and Q-factor. *Sci. Rep.* **2023**, *13*, 7373. [CrossRef]
27. Guo, W.; Zhai, L.H.; El-Bahy, Z.M.; Lu, Z.; Li, L.; Elnaggar, A.Y.; Ibrahim, M.M.; Cao, H.; Lin, J.; Wang, B. Terahertz metamaterial biosensor based on open square ring. *Adv. Compos. Hybrid Mater.* **2023**, *6*, 92. [CrossRef]
28. Chen, K.L.; Ruan, C.; Zhan, F.; Song, X.; Fahad, A.K.; Zhang, T.; Shi, W. Ultra-sensitive terahertz metamaterials biosensor based on luxuriant gaps structure. *Iscience* **2023**, *26*, 105781. [CrossRef]
29. EL-Wasif, Z.; Ismail, T.; Hamdy, O. Design and optimization of highly sensitive multi-band terahertz metamaterial biosensor for coronaviruses detection. *Opt. Quant. Electron.* **2023**, *55*, 604. [CrossRef]
30. Luo, X.; Xiang, P.; Yu, H.; Huang, S.; Yu, T.; Zhu, Y.F. Terahertz Metamaterials Broadband Perfect Absorber Based on Molybdenum Disulfide. *IEEE Photon. Technol. Lett.* **2022**, *34*, 1100–1103. [CrossRef]
31. Zeng, L.; Zhang, H.F.; Liu, G.B.; Huang, T. A three-dimensional Linear-to-Circular polarization converter tailored by the gravity field. *Plasmonics* **2019**, *14*, 1347–1355. [CrossRef]
32. Chiang, W.F.; Lu, Y.Y.; Chen, Y.P.; Lin, X.Y.; Lim, T.S.; Liu, J.H.; Le, C.R.; Huang, C.Y. Passively tunable terahertz filters using liquid crystal cells coated with metamaterials. *Coatings* **2021**, *11*, 381. [CrossRef]
33. Arafa, H.A.; Mohameda, D.; Mohasebab, M.A.; El-Gawaadc, N.S.A.; Trabelsi, Y. Biophotonic sensor for the detection of creatinine concentration in blood serum based on 1D photonic crystal. *RSC Adv.* **2020**, *10*, 31765–31772.

sensors

Communication

A Lateral Flow Assay for the Detection of *Leptospira lipL32* Gene Using CRISPR Technology

Satheesh Natarajan [1,*], Jayaraj Joseph [2], Balamurugan Vinayagamurthy [3] and Pedro Estrela [4,5]

1 Healthcare Technology Innovation Centre, Indian Institute of Technology Madras, Chennai 600113, India
2 Department of Electrical Engineering, Indian Institute of Technology Madras, Chennai 600036, India; jayaraj@ee.iitm.ac.in
3 Indian Council of Agricultural Research-National Institute of Veterinary Epidemiology and Disease Informatics (ICAR-NIVEDI), Bangalore 560064, India; b.vinayagamurthy@icar.gov.in
4 Department of Electronic and Electrical Engineering, University of Bath, Bath BA2 7AY, UK; p.estrela@bath.ac.uk
5 Centre for Bioengineering & Biomedical Technologies, University of Bath, Bath BA2 7AY, UK
* Correspondence: satheesh@htic.iitm.ac.in

Abstract: The clinical manifestation of leptospirosis is often misdiagnosed as other febrile illnesses such as dengue. Therefore, there is an urgent need for a precise diagnostic tool at the field level to detect the pathogenic *Leptospira lipL32* gene at the molecular level for prompt therapeutic decisions. Quantitative polymerase chain reaction (qPCR) is widely used as the primary diagnostic tool, but its applicability is limited by high equipment cost and the lack of availability in every hospital, especially in rural areas where leptospirosis mainly occurs. Here, we report the development of a CRISPR dFnCas9-based quantitative lateral flow immunoassay to detect the *lipL32* gene. The developed assay showed superior performance regarding the lowest detectable limit of 1 fg/mL. The test is highly sensitive and selective, showing that leptospirosis diagnosis can be achieved with a low-cost lateral flow device.

Keywords: *Leptospira*; CRISPR; dFnCas9; lateral flow device

Citation: Natarajan, S.; Joseph, J.; Vinayagamurthy, B.; Estrela, P. A Lateral Flow Assay for the Detection of *Leptospira lipL32* Gene Using CRISPR Technology. *Sensors* **2023**, *23*, 6544. https://doi.org/10.3390/s23146544

Academic Editor: Evgeny Katz

Received: 23 June 2023
Revised: 13 July 2023
Accepted: 18 July 2023
Published: 20 July 2023

1. Introduction

Early and accurate detection of pathogens is crucial for promptly administering therapeutic decisions. The most common zoonotic disease in humans is leptospirosis, caused by the pathogenic bacteria *Leptospira*, with a million cases reported yearly, which includes 60,000 deaths [1,2]. The clinical manifestations of human leptospirosis range from subclinical infection to severe fatal disease [2,3]. The symptoms are not disease-specific, so laboratory confirmation is required to diagnose leptospirosis [4]. Though the microscopic agglutination test (MAT) is the gold standard method, serological complications, time consumption, and live strain maintenance requirements have made this method less than ideal [5]. Due to their complicated isolation methodologies, slow growth, low recovery, and high contamination, the culture methods also make them unsuitable for widespread clinical diagnosis [6]. Immunofluorescence assays targeting *OmpL54* can also be used [7], but they provide many false-negative results [8].

Although molecular techniques such as real-time polymerase chain reaction (qPCR) are the most validated diagnostic methods for genes such as *lipL32* [9] and *lfb1* [10], they fail the recognition of the species from "intermediate" clade or pathogenic *Leptospira* spp. group II clade [11,12]. This issue has been suggested as a potential factor contributing to the underreporting of *Leptospira* spp. in several geographic regions [11]. To overcome this problem, precise diagnostic tools are proposed and developed in this work to detect nucleic acids utilizing genome editing techniques such as CRISPR-Cas9 endonucleases [13]. Here, we applied and optimized inactive *Francisella novicida* Cas9 (dFnCas9)-based technology [14] to diagnose the pathogenic biomarker *lipL32* gene for *Leptospira* detection. These tools use the specific properties of the Cas enzymes, i.e., upon recognition of the foreign pathogenic DNA, the non-specific DNase/RNase activity of the Cas enzyme activity will be stimulated for its recognition of the foreign DNA. It is also a rapid, cost-effective method for nucleic acid detection in clinical samples with excellent sensitivity and specificity, making them an ideal tool for point-of-care (POC) diagnosis [15] and, in recent years, the method has been implemented with lateral flow assays [16–18].

The dFnCas9 has a very high intrinsic specificity to the target [14]; we speculated that the enzyme could also be used for detecting the *lipL32* gene on a paper strip with high accuracy. Herein, we demonstrate a dFnCas9-based quantitative lateral flow assay to detect the *lipL32* gene. Upon DNA loading, an automatic sandwiched assay was completed within 15 min, allowing for ultra-specific assay compared to other sensors enabling on-field analysis or bedside testing. The lateral flow paper strip was used on the quantitative strip reader iQuant developed at the Healthcare Technology Innovation Centre, Indian Institute of Technology, Madras, India [19–21]. The advantages of the reader and performance comparison with other systems can be found in the literature [19].

We explored the dFnCas9 with PCR amplification and lateral flow assay for cost-effective *Leptospira* detection with high sensitivity and avoided the need for sophisticated instrumentation. To enable such diagnosis, we used our quantitative lateral flow assay strip to capture RNP-bound biotinylated substrate molecules on a test line coated with the streptavidin of the paper strip using FITC-labelled chimeric gRNA (Figure 1). To make the strip readout reproducible in point-of-care settings, we chemically modified the chimeric gRNA using synthetic backbone modifications (phosphorothioate) to increase the stability and robustness of the readout. In the conjugation pad, we conjugated the FITC antibodies with the Alexafluor-674 to make the kit quantitative. Finally, using an optimized PCR protocol followed by dFnCas9, we developed an assay to detect *Leptospira lipL32* gene sequences from DNA samples within an hour.

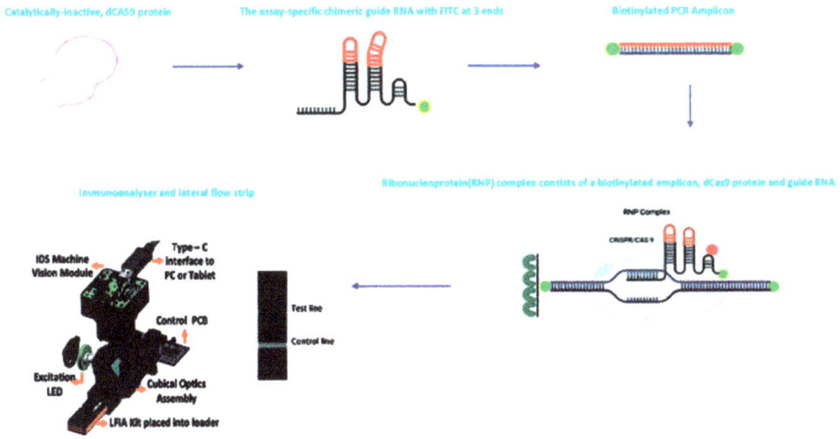

Figure 1. Experimental flow diagram starting from the dFnCas9 to the lateral flow assay.

2. Materials and Methods

Nitrocellulose membranes (HiFlow135) were procured from Merck Millipore (Bedford, MA, USA). Sample pad (CF-4), conjugation pad (Standard-17), and absorbent pad (CF-6) were obtained from Cytiva, UK. The anti-FITC monoclonal antibodies, goat anti-chicken IgY antibody (AF 010), and normal chicken IgY control (AB-101-C) were bought from R&D Systems (Minneapolis, MN, USA). Biotin-BSA, Streptavidin, and Alexa Fluor 647 NHS ester were procured from Thermo Fisher (Waltham, MA, USA); Sephadex G20 column from GE Healthcare (Uppsala, Sweden). PBS (137 mm NaCl, 2.7 mm KCl, 10 mm Na_2HPO_4, 1.8 mm

KH_2PO_4) and PB (75.4 mm $Na_2HPO_4{\cdot}7H_2O$, 24.6 mm $NaH_2PO_4{\cdot}H_2O$) buffers, NaOH, $NaHCO_3$, NaN_3, BSA, and Tween-20, were purchased from Sigma-Aldrich (St. Louis, MO, USA).

2.1. Study Design

The study intended to develop robust CRISPR diagnostics that can perform with high accuracy for the *lipL32* gene of the *Leptospira* at a significantly low cost and time. For the *lipL32* gene detection using *Leptospira* positive DNA extracted from a reference, *Leptospira interrogans* serovar Icterohaemorrhagiae was received from the National Institute of Veterinary Epidemiology and Diseases Informatics, Bangalore.

2.2. dFnCas9 Protein

The plasmids containing dFnCas9 (catalytically inactive, dead) [14] were kindly gifted by Dr. Debojyoti Chakraborty, IGIB, New Delhi. The sequences specific to the protein were transformed and expressed in *Escherichia coli* Rosetta 2 (DE3) (Novagen). Rosetta 2 (DE3) cells were cultured in Luria Bertani (LB) broth medium (with 50 mg/mL kanamycin) at 37 °C and induced using 0.5 mM Isopropyl β-D-thiogalactopyranoside (IPTG) when OD_{600} reached 0.6. After overnight culturing, *E. coli* cells were harvested and resuspended in a lysis buffer (20 mM HEPES, 500 mM NaCl, 5% glycerol) supplemented with 1× PIC (Roche) containing 100 mg/mL lysozyme. After cell lysis by sonication, the lysate was put through Ni-NTA beads (Roche). The eluted protein was purified by Superdex 200 16/60 column (GE Healthcare) in a buffer solution of 20 mM HEPES pH 7.5, 150 mM KCl, 10% glycerol, and 1 mM DTT. The purified proteins were quantified by a BCA protein assay kit (Thermo Fisher Scientific, Waltham, MA, USA) and stored at −80 °C until further use.

2.3. LipL32 Gene Detection PCR

The *lipL32* gene regions were PCR amplified using 5′ biotin-labelled primers. The chimeric gRNA (crRNA: TracrRNA) was prepared by mixing the crRNA (*lipL32* gene) and the synthetic 3′-FITC-labelled TracrRNA in an equimolar ratio with the help of the annealing buffer (100 mM NaCl, 50 mM Tris-HCl pH 8, and 1 mM $MgCl_2$), and heated at 95 °C for 2–5 min and then allowed to cool at room temperature for 15–20 min. The chimeric gRNA-dead FnCas9 RNP complex was prepared by mixing it equally (protein:sgRNA molar ratio, 1:1) in buffer (20 mM HEPES, 150 mM KCl, 1 mM DTT, 10% glycerol, 10 mM $MgCl_2$) and incubating for another 10 min at room temperature. The target biotinylated amplicons were then incubated in the RNP complexes for 10 min at 37 °C. The detailed protocol for the CRISPR/CAS and the PCR for *lig A*, *lig B*, and *lipL41* are also given in the supplementary file.

2.4. Antibody Conjugation

The monoclonal mouse anti-FITC antibody and the anti-chicken IgY were conjugated with the fluorescent organic dye Alexa Fluor 647, which contains a succinimidyl ester to react with the primary amines present in the antibodies. Briefly, the detection antibody, both monoclonal mouse anti-FITC antibody and the anti-chicken IgY @ 1 mg/mL in 10 mM phosphate solution (pH 7.4), was mixed with the Alexa Fluor 647 (20 molar excess) for 1 h at RT. The conjugates were purified by Sephadex G20 gel chromatography column by centrifuging at 4000 rpm for 5 min and collecting the labelled antibody in 1× PBS buffer containing 1% BSA, 0.1% Tween 20. The conjugate was mixed with glycerol to the final concentration of 10 μg/mL and stored at 4 °C until further use.

2.5. Lateral Flow Immunoassay (LFIA)

The LFIA strip contains a sample pad (10 mm length), a polyester fiber conjugate pad (13 mm length), a nitrocellulose membrane (25 mm length), and an absorbent pad (15 mm length). The sample pad was pre-treated with sample pad pre-treatment buffer containing (1× PBS, 5% BSA, 0.1% Tween-20), followed by air drying at 1 h at RT. The conjugate pad, after pre-treatment, was dispensed with 0.3 ng/mL antibody-dye conjugate in 10 mm PBS buffer with 0.1% Tween, 0.1% BSA and subsequently dried for 1 h at 40 °C. The NC membrane was dispensed with 0.5 mg/mL of streptavidin, and 0.2 mg/mL anti-chicken IgY was dispensed over the analytical strips at a 0.6 μL/cm rate using the Claremont antibody dispenser with the chemryx syringe pump in the Test and Control lines at the rate of 1 μL/cm. Following dispensing, the analytical NC membrane was airdried at 37 °C for 12 h. Finally, all the pads were laminated with a partial overlapping of 2 mm and cut with a width of 3.1 mm with the Werfen guillotine cutter. The assembled LFIA strips were stored at 4 °C until further use.

This experiment was designed to investigate the CRISPR-based dFnCas9 detection of the *Leptospira lipL32* gene using the lateral flow immunoassay strip, which involved dispensing capture biomolecules onto the test lines of the NC membrane, the addition of Alexa-labelled biomolecules on to the conjugate pads, cartridge assembly, and running of samples containing with the adequate sample biomolecules. The typical test procedure involved the addition of 90 μL of the appropriate lateral flow chase buffer to the sample pad of the LFIA cartridge, which was subsequently inserted into the iQuant analyzer [21]. The process was monitored for about 15 min. The LED-based instrument scanned the signal generated in the Nitro Cellulose membrane as a two-dimensional pixel map for the quantification of the Alexa Fluor 647 labelled component. The pixel map was processed by the NI LabView software [20]. All the quantitative data were assessed by GraphPad Prism 6.0 (GraphPad Software, La Jolla, CA, USA). The intensity data were used for the calculation of the pixel volume of the test, V_T, and control, V_C, lines [20]. The mean volume ratio, V_R, defined as the ratio V_T/V_C, was also calculated when required. The immunoassays were performed in triplicate for each of the samples.

2.6. *Leptospira Standards and Samples*

The calibration curves were constructed for the range of 0–10 pg/mL concentration using *Leptospira* standards prepared in 1× PBS, 1% BSA, and 0.1% Tween-20. After the analysis, the mean volume ratio, V_r, was plotted vs. the *lipL32* gene standard concentration to construct the calibration curves. The standard deviation was utilized to calculate the coefficient of variation (CoV) according to CoV = SD/mean × 100%. The calibration curve data were analyzed by linear regression, and the limits of detection (LOD) and quantitation (LOQ) were determined based on the residual standard deviation and slope of the calibration curves obtained. Standard samples with the same concentrations were also prepared in the buffer. These diluted samples were then analyzed for LFIA.

3. Results and Discussion

3.1. *LFIA and Fluorescence Assay Validation with lipL32 Gene Leptospira Genomic DNA*

Genomic DNA from the reference strain *Leptospira* interrogans serovar Icterohaemorrahae was diluted for the PCR with the designated primer employed for the *lipL32* gene (Table 1).

Figure 2 shows the purified protein dFnCas9 on the SDS page and the gel electrophoresis pictures of the *lipL32* gene. This amplicon was then used to establish the limit of detection (LOD) of the LFIAs. The fluorescence signal above the background was observed for each of the samples, indicating that the dFnCas9-based assay limit of detection (LOD) was similar to that of the RT-PCR. We further quantified the *lipL32* gene amplicons and performed a copy number LOD. Fluorescence above the background was observed with 0.001 pg/mL copies of target DNA (Figure 3). Under the 35-cycle PCR amplification parameters, this represents the copies less than 5 are the starting material. LFIA was optimized with the soak DNA and irrelevant sgRNA. Under these conditions, test lines were observable at a sensitivity that was below that of RT-PCR or Cas9 fluorescence. These results showed that the LFIA with Cas9 fluorescence can detect the nucleic acid from the *Leptospira lipL32* gene.

Table 1. Oligonucleotides were used for the development of the dFnCas9 lateral flow assay.

Primer	Dir	Sequence
lipL32 gene	F	B-GAA GTG AAA GGA TCT TTC GTT GCA
lipL32 gene	R	B-CGT CAG AAG CAG CTT TTT TCA AAG
lipL32 gene	F	B-GGT ATT CCA GGT GTG AGC CC
lipL32 gene	R	B-CGC GTC AGA AGC AGC TTT TT
crRNA IVT DNA oligo	F	5′ TAA TAC GAC TCA CTA TAC TCA AAT CCT GAA GAA TTG CGT TTC AGT TGC TGA ATT AT 3′
crRNA IVT DNA oligo	R	5′ ATA ATT CAG CAA CTG AAA CGC AAT TCT TCA GGA TTT GAG TAT AGT GAG TCG TAT TA 3′
dFnCas9-Syn-tracrRNA		5′ G*U*A AUU AAU GCU CUG UAA UCA UUU AAA AGU AUU UUG AAC GGA CCU CUG UUU GAC ACG UC*U* U – FITC 3′

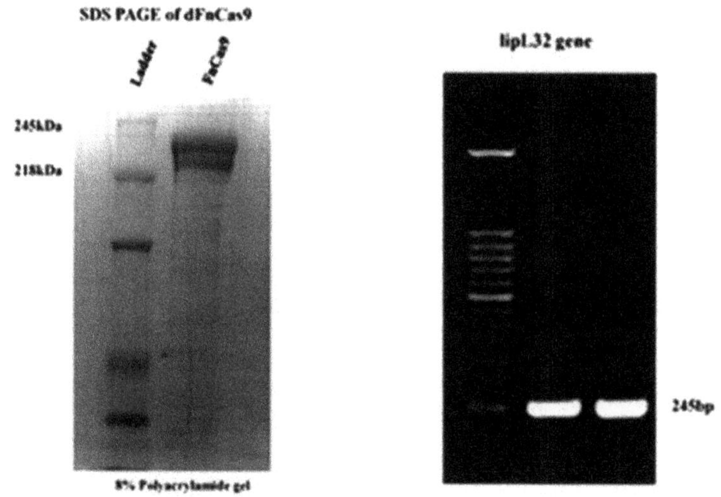

(**a**) (**b**)

Figure 2. (**a**) SDS PAGE of purified en31dFnCas9; (**b**) PCR results for the *Leptospira lipL32* gene.

3.2. Optimization

In the present work, several parameters impacting the performance of the *Leptospira* test strip (including the concentration of lateral flow assay chase buffer and NC membrane) were evaluated to obtain the maximal sensing efficacy. Details of the assay optimization are presented in Figures 4 and 5. The highest peak signal of V_T/V_C indicates the best conditions. The chase buffer with PBS + 0.5% BSA + 0.5% Tween 20 + 2% PVP-40 + 5 mm EDTA yielded the best results, as seen in Figure 4, and hence was used in the subsequent experiments. Different NC membranes were tested using this buffer with NC HF180 giving the best results (Figure 5). This membrane was then used for all subsequent experiments.

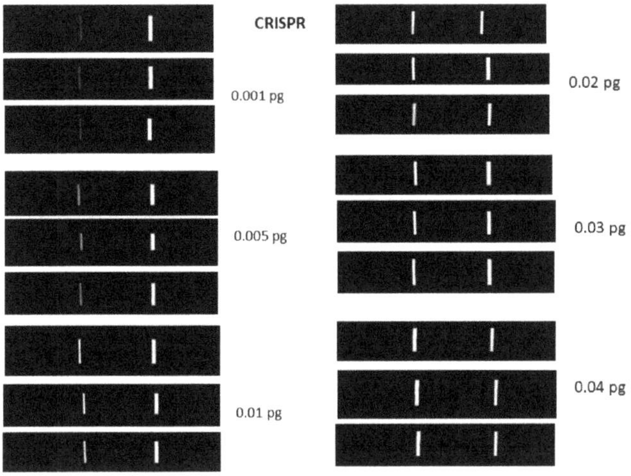

Figure 3. Pictures of the lateral flow devices for target DNA in the concentrations 0.001, 0.005, 0.01, 0.02, 0.03, and 0.04 pg/mL. Three independent devices are shown for each concentration.

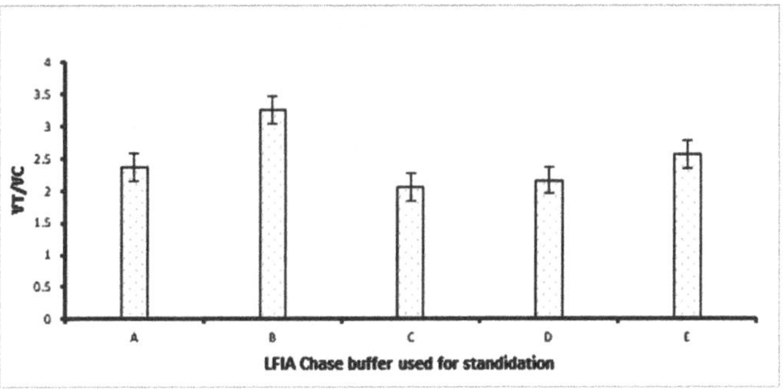

Figure 4. Performance of different mixtures of chase buffer. A: PBS + 0.75% BSA + 0.75% Tween 20 + 1% PVP40 + 1mm EDTA; B: PBS + 0.5% BSA + 0.5% Tween 20 + 2% PVP40 + 5mm EDTA; C: PBS + 1.25% BSA + 1% Tween 20 + 1.5% PVP40 + 3mm EDTA; D: PBS + 1.25% BSA + 1.25% Tween 20 + 1.5% PVP40 + 2mm EDTA; E: PBS + 1.75% BSA + 1.5% Tween 20 + 2.5% PVP40 + 2mm EDTA.

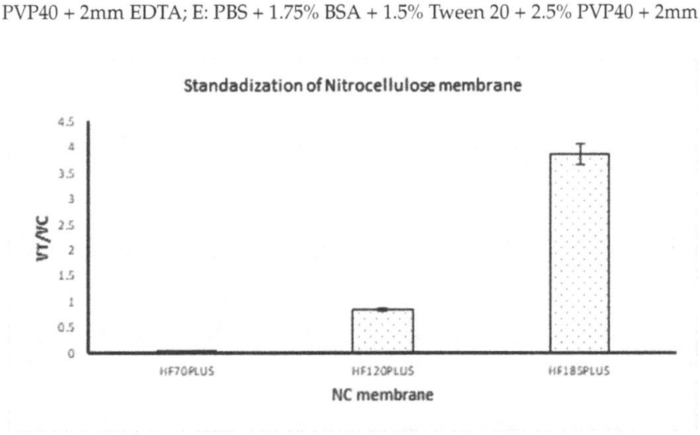

Figure 5. Effect of the different NC membranes on the signal: HF70 Plus, HF120 Plus, and HF180 Plus.

3.3. *Analytical Performance*

The performance of the CRIPSR-based dFnCas9 lateral flow assay for detecting the *lipL32* gene was evaluated by the quantitative immunoanalyzer iQuant developed by our institute (HTIC, IIT Madras). As seen in Figure 6, the sensor exhibited an ultrasensitive response toward the *lipL32* gene. As this is a sandwich type of assay, the response proportionally increased with the concentration of the *lipL32* gene, where a linear dynamic response (Figure 6) was observed in the 0−10 pg/mL range with a limit of detection (LOD) of 0.001 pg/mL (calculated from 3 SD_{blank}/slope). Furthermore, the analytical performance of the CRIPSR-based dFnCas9 LFIA test strip exhibited the lowest LOD and widest linear dynamic range among the previously reported methods for the *lipL32* gene available so far. Therefore, the proposed CRIPSR-based dFnCas9 LFIA test strip confirms a strong potential for real sample analysis. In addition, the reproducibility of the test strip (the % relative standard deviation (RSD), n = 3) of the present method was examined. It was observed that the reproducibility for detecting the *lipL32* gene was lower than 5%, which is highly accepted for LFIA-based point-of-care techniques. This value showed that the newly developed immunosensor has a remarkably very low standard deviation with excellent fabrication repeatability. To demonstrate the stability of the immunosensor, the storage lifetime was then investigated. For the stability test of the strip, the new CRIPSR-based LFIA sensor was stored in a desiccator at room temperature for the stipulated time until use. The strips were stable for up to 2 weeks, with the percentage change in the signal less than 5% (% RSD = 4.31). This result indicated that the developed CRIPSR-based LFIA sensor demonstrated very satisfactory stability.

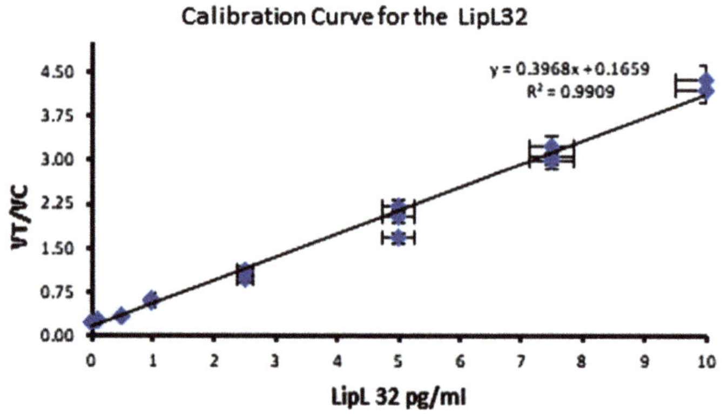

Figure 6. Calibration curve of the dFnCas9 LFIA in the concentration range 0–10 pg/mL. Linear fit shows an R^2 value of 0.9909.

3.4. *Specificity*

The cross-reactivity of the CRIPSR-based LFIA test strip was investigated for the non-specific reactions to other genes coding proteins/antigens. Three unrelated genes of *Leptospira* (*ligB*, *lipL41*, and *ligA*) were tested with the CRIPSR-based dFnCas9 LFIA test strip. As expected, the results were positive for the pathogenic *lipL32* gene only, while negligible responses were obtained for the other genes tested (Figure 7). Therefore, it can be concluded that the developed CRIPSR-based LFIA test strip sensor has excellent specificity to *lipL32*.

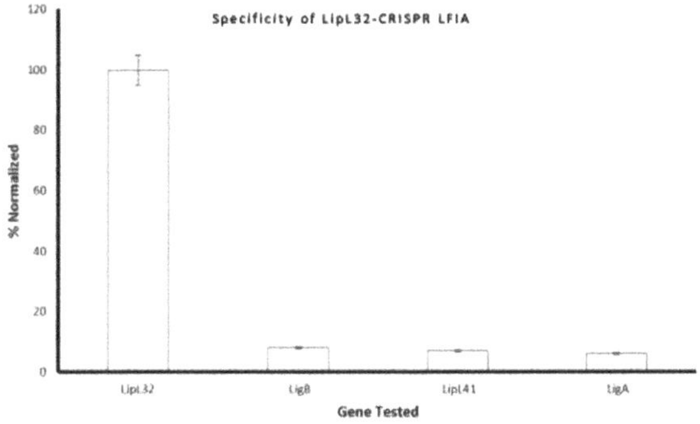

Figure 7. Specificity of the *Leptospira* lateral flow assay: *lipL32*, *ligB*, *lipL41*, *ligA*.

4. Conclusions

PCR is the gold standard for the specific and sensitive quantification of many pathogens. However, due to the requirement of sophisticated instrumentation and the relatively high cost of the technique when compared to that needed for rapid point-of-care tests, PCR has not been widely used as an early diagnostic tool for *Leptospira*. The CRISPR-based dFnCas9 LFIA test strip is a new nucleic acid detection platform able to diagnose many infectious diseases. This study is the first report for *Leptospira* detection using the dFnCas9 LFIA assay targeting the *lipL32* gene.

In this study, we developed a dFnCas9 lateral flow immunoassay for diagnosing leptospirosis targeting *Leptospira lipL32* gene detection. This lateral flow test strip was made portable and thus allowed for on-field testing. Interestingly, the developed test strip exhibited a LOD of 1 fg/mL with a linear dynamic range between 0 and 10 pg/mL, the lowest LOD among other reports available thus far. In summary, we believe that this developed lateral flow platform can be further extended for other biomarker detection, whereby a sensitive yet straightforward approach is of primary concern. The developed test strip can be used as a sensitive approach to diagnosing leptospirosis using DNA extracted from clinical samples from patients and samples from animals associated with the reproductive disorder.

Supplementary Materials: The following supporting information can be downloaded at: https://www.mdpi.com/article/10.3390/s23146544/s1.

Author Contributions: Conceptualization, P.E., J.J., B.V. and S.N.; methodology, S.N.; formal analysis, P.E., J.J., B.V. and S.N.; investigation, P.E., B.V. and S.N.; writing—original draft preparation, P.E., B.V. and S.N.; writing—review and editing, P.E. and S.N. All authors have read and agreed to the published version of the manuscript.

Funding: This research received no external funding.

Institutional Review Board Statement: *Leptospira* cultures procured from the Indian Council of Medical Research (ICMR)-Regional Medical Research Centre (RMRC), Port Blair, are being maintained in the Leptospirosis Research Laboratory, ICAR-NIVEDI, Bengaluru, and DNA extracted from reference *Leptospira* culture was used in the study (Institutional Biosafety Committee (IBSC)-F. No. 6-52/NIVEDI/Biosafety/2019/10/05 dated on 7 September 2021).

Informed Consent Statement: Not applicable.

Data Availability Statement: Data are available from corresponding authors, which can be obtained on formal request.

Acknowledgments: We thank Debojyoti Chakraborty (IGIB, New Delhi, India) for providing the dFnCas9. The authors thank ICAR and ICAR-National Institute of Veterinary Epidemiology and Diseases Informatics, Bengaluru, for their encouragement and support for collaborative research work.

Conflicts of Interest: The authors declare no conflict of interest.

References

1. Costa, F.; Hagan, J.E.; Calcagno, J.; Kane, M.; Torgerson, P.; Martinez-Silveira, M.S.; Stein, C.; Abela-Ridder, B.; Ko, A.I. Global morbidity and mortality of leptospirosis: A systematic review. *PLoS Negl. Trop. Dis.* **2015**, *9*, e0003898. [CrossRef] [PubMed]
2. Torgerson, P.R.; Hagan, J.E.; Costa, F.; Calcagno, J.; Kane, M.; Martinez-Silveira, M.S.; Goris, M.G.A.; Stein, C.; Ko, A.I.; Abela-Ridder, B. Global burden of leptospirosis: Estimated in terms of disability adjusted life years. *PLoS Negl. Trop. Dis.* **2015**, *9*, e0004122. [CrossRef] [PubMed]
3. Chin, V.K.; Lee, T.Y.; Lim, W.F.; Wan Shahriman, Y.W.Y.; Syafinaz, A.N.; Zamberi, S.; Maha, A. Leptospirosis in human: Biomarkers in host immune responses. *Microbiol. Res.* **2018**, *207*, 108–115.
4. Musso, D.; La Scola, B. Laboratory diagnosis of leptospirosis: A challenge. *J. Microbiol. Immunol. Infect.* **2013**, *46*, 245–252. [CrossRef]
5. Nilofar, R.; Fernando, N.; de Silva, N.L.; Karunanayake, L.; Wickremasinghe, H.; Dikmadugoda, N.; Premawansa, G.; Wickramashinghe, R.; de Silve, H.J.; Premawansa, S.; et al. Diagnosis of leptospirosis: Comparison between microscopic agglutination test, IgM-ELISA and IgM rapid immunochromatography test. *PLoS ONE* **2015**, *10*, e0129236.
6. Nisansala, G.G.T.; Muthusinghe, D.; Gunasekara, T.D.C.P.; Weerasekera, M.M.; Fernando, S.S.N.; Ranasinghe, K.N.P.; Marasinghe, M.G.C.P.; Fernando, P.S.; Koizumi, N.; Gamage, C.D. Isolation and characterization of Leptospira interrogans from two patients with leptospirosis in Western Province, Sri Lanka. *J. Med. Microbiol.* **2018**, *67*, 1249–1252. [CrossRef] [PubMed]
7. Pinne, M.; Haake, D. Immuno-fluorescence assay of leptospiral surface-exposed proteins. *J. Vis. Exp.* **2011**, *53*, 2805.
8. Goris, M.G.A.; Hartskeerl, R.A. Leptospirosis serodiagnosis by the microscopic agglutination test. *Curr. Protoc. Microbiol.* **2014**, *32*, 12E.5. [CrossRef] [PubMed]
9. Stoddard, R.A. Detection of pathogenic Leptospira spp. through real-time PCR (qPCR) targeting the LipL32 gene. *Methods Mol. Biol.* **2013**, *943*, 257–266. [PubMed]
10. Merien, F.; Portnoi, D.; Bourhy, P.; Charavay, F.; Berlioz-Arthaud, A.; Baranton, G. A rapid and quantitative method for the detection of Leptospira species in human leptospirosis. *FEMS Microbiol. Lett.* **2005**, *249*, 139–147. [CrossRef] [PubMed]
11. Tsuboi, M.; Koizumi, N.; Hayakawa, K.; Kanagawa, S.; Ohmagari, N.; Kato, Y. Imported *Leptospira licerasiae* infection in traveler returning to Japan from Brazil. *Emerg. Infect. Dis.* **2017**, *23*, 548–549. [CrossRef] [PubMed]
12. Thibeaux, R.; Girault, D.; Bierque, E.; Soupé-Gilbert, M.E.; Rettinger, A.; Douyère, A.; Meyer, M.; Iraola, G.; Picardeau, M.; Goarant, C. Biodiversity of environmental Leptospira: Improving identification and revisiting the diagnosis. *Front. Microbiol.* **2018**, *9*, 816. [CrossRef] [PubMed]
13. Kaminski, M.M.; Abudayyeh, O.O.; Gootenberg, J.S.; Zhang, F.; Collins, J.J. CRISPR-based diagnostics. *Nat. Biomed. Eng.* **2021**, *5*, 643–656. [CrossRef]
14. Kumar, M.; Gulati, S.; Ansari, A.H.; Phutela, R.; Acharya, S.; Azhar, M.; Murthy, J.; Kathpalia, P.; Kanakan, A.; Maurya, R.; et al. FnCas9-based CRISPR diagnostic for rapid and accurate detection of major SARS-CoV-2 variants on a paper strip. *eLife* **2021**, *10*, e67130. [CrossRef]
15. Puig-Serra, P.; Casado-Rosas, M.C.; Martinez-Lage, M.; Olalla-Sastre, B.; Alonso-Yanez, A.; Torres-Ruiz, R.; Rodriguez-Perales, S. CRISPR approaches for the diagnosis of human diseases. *Int. J. Mol. Sci.* **2022**, *23*, 1757. [CrossRef]
16. Osborn, M.J.; Bhardwa, A.; Bingea, S.P.; Knipping, F.; Feser, C.J.; Lees, C.J.; Collins, D.P.; Steer, C.J.; Blazar, B.R.; Tolar, J. CRISPR/Cas9-based lateral flow and fluorescence diagnostics. *Bioengineering* **2021**, *8*, 23. [CrossRef]
17. Ali, Z.; Sánchez, E.; Tehseen, M.; Mahas, A.; Marsic, T.; Aman, R.; Rao, G.S.; Alhamlan, F.S.; Alsanea, M.S.; Al-Qahtani, A.A.; et al. Bio-SCAN: A CRISPR/dCas9-based lateral flow assay for rapid, specific, and sensitive detection of SARS-CoV-2. *ACS Synth. Biol.* **2022**, *11*, 406–419. [CrossRef]
18. Wang, H.; Wu, Q.; Zhou, M.; Li, C.; Yan, C.; Huang, L.; Qin, P. Development of a CRISPR/Cas9-integrated lateral flow strip for rapid and accurate detection of Salmonella. *Food Control* **2022**, *142*, 109203. [CrossRef]
19. Joseph, J.; Vasan, J.K.; Shah, M.; Sivaprakasam, M.; Mahajan, L. iQuant™ Analyser: A rapid quantitative immunoassay reader. In Proceedings of the Annual International Conference of the IEEE Engineering in Medicine and Biology Society, Jeju, Republic of Korea, 11–15 July 2017; pp. 3732–3736.
20. Shah, M.I.; Rajagopalan, A.; Joseph, J.; Sivaprakasam, M. An improved system for quantitative immunoassay measurement in ImageQuant. In Proceedings of the 2018 IEEE Sensors, New Delhi, India, 28–31 October 2018; pp. 645–648.
21. Kumarasami, R.; Vasan, J.K.; Joseph, J.; Sithambaram, P.; Pandidurai, S.; Sivaprakasam, M. iQuant Auto: Automated rapid test platform for immunodiagnostics. In Proceedings of the Annual International Conference of the IEEE Engineering in Medicine and Biology Society, Online, 20–24 July 2020; pp. 6131–6134.

sensors

MDPI

Article

A Highly Sensitive 3D Resonator Sensor for Fluid Measurement

Ali M. Almuhlafi [1,*] and Omar M. Ramahi [2]

[1] Electrical Engineering Department, King Saud University, Riyadh 11421, Saudi Arabia
[2] Department of Electrical and Computer Engineering, University of Waterloo, Waterloo, ON N2L 3G1, Canada; oramahi@uwaterloo.ca
* Correspondence: aalbishi@ksu.edu.sa

Abstract: Planar sub-wavelength resonators have been used for sensing applications, but different types of resonators have different advantages and disadvantages. The split ring resonator (SRR) has a smaller sensing region and is suitable for microfluidic applications, but the sensitivity can be limited. Meanwhile, the complementary electric-LC resonator (CELCR) has a larger sensing region and higher sensitivity, but the topology cannot be easily designed to reduce the sensing region. In this work, we propose a new design that combines the advantages of both SRR and CELCR by incorporating metallic bars in a trapezoid-shaped resonator (TSR). The trapezoid shape allows for the sensing region to be reduced, while the metallic bars enhance the electric field in the sensing region, resulting in higher sensitivity. Numerical simulations were used to design and evaluate the sensor. For validation, the sensor was fabricated using PCB technology with aluminum bars and tested on dielectric fluids. The results showed that the proposed sensor provides appreciably enhanced sensitivity in comparison to earlier sensors.

Keywords: microwaves; split-ring resonators; complementary split-ring resonators; complementary electric-LC resonator; fluid-level detection; sensitivity enhancement

Citation: Almuhlafi, A.M.; Ramahi, O.M. A Highly Sensitive 3D Resonator Sensor for Fluid Measurement. *Sensors* **2023**, *23*, 6453. https://doi.org/10.3390/s23146453

Academic Editors: Evgeny Katz and Michael Kraft

Received: 25 May 2023
Revised: 26 June 2023
Accepted: 14 July 2023
Published: 17 July 2023

1. Introduction

Nowadays, sensitive and affordable measurement systems are important for a wide range of applications such as lab-on-a-chip, point-of-care lab testing, rapid testing for quality control, environmental monitoring, and public health and safety [1]. These applications require highly sensitive and inexpensive measurement systems. Planar sub-wavelength resonators have emerged as strong candidates for designing sensitive near-field sensors due to their ability to concentrate electromagnetic fields in a small volume surrounding the resonator [2]. By perturbing these fields through topological or material changes, the impedance of the sensors can change, leading to physical measurable changes that reflect changes in the magnetic or dielectric properties of the surroundings [3].

Split-ring resonators (SRRs) and complementary split-ring resonators, which are sub-wavelength resonators, have been utilized to develop various sensing modalities [4,5]. Examples include biochemical applications [6–9], anomaly detection in metals and dielectric materials [10–12], microfluidic applications [13–15], dielectric characterization [6,16–23], and fluid concentration detection [24]. However, planar sub-wavelength resonators have a limitation in their effective sensing regions, which affects the sensor's sensitivity. This is because the material under test (MUT) does not interact effectively with the highest field concentration [25], and so the shift in resonance frequencies and overall sensitivity of the sensor are minimized. To overcome this limitation, sharp tips can be introduced and symmetry can be broken in the resonators (e.g., split-ring resonators) to increase the interaction with MUT [26]. Despite the fact that most of the energy is stored in the substrate, in [27], a significant enhancement in the resonators' sensitivity can be achieved by creating a substrate channel that accommodates MUT and enables it to interact more effectively with the resonators. But, designing a channel in the sensitive volume can be challenging, and optimization may be needed to locate the optimal sensitive position of the channel inside the substrate [27]. Thus, it is essential to explore and investigate other techniques for sensitivity enhancement.

The feasibility of using the three-dimensional capacitor-based technique for sensitivity enhancement was reported in [28]. This technique can be combined with other resonators that are suitable for specific applications, especially those requiring smaller sensing regions. For instance, in [29], this technique was employed to detect fluids using split-ring resonators (SRRs). Complementary structures such as CSRRs have higher sensitivity [16]; however, they have larger effective sensing regions. Therefore, a new design that can effectively combine the advantages of smaller sensing regions and higher sensitivity is necessary to improve the overall performance of these resonators.

In this work, we present a new sensor design for achieving maximum sensitivity without requiring the material under test to interact with the entire sensing volume of resonators. The proposed design utilizes a trapezoid-shaped resonator (TSR) with metallic bars (which is effectively a three-dimensional capacitor) to enhance the electric field in the sensing region and increase the overall sensitivity. The trapezoid shape allows for the sensing region to be reduced while maintaining high sensitivity, while the metallic bars enhance the electric field in the sensing region. The TSR design combines the advantages of both split-ring resonators (SRRs) and complementary structures such as complementary electric-LC resonators (CELCRs). The geometry of the trapezoid shape has two bases that can be chosen to create smaller sensing regions, similar to SRRs, or even smaller yet with higher sensitivity. Compared to SRRs, TSRs can be designed in the ground plane of microstrip lines, making them advantageous for microfluidic applications where there will be no direct interaction with the strip lines if a micro-channel is placed in the sensing region. Thus, such designs can be adopted for microfluidic applications [3,30–32].

2. Trapezoid Shape and Three-Dimensional Capacitor for Sensitivity Enhancement

Complement structure-based sensors, such as CSRRs, have shown higher sensitivity [16,33], which can be associated with their larger physical sensing regions compared to SRRs. Indeed, SRRs have been adopted for applications that require smaller sensing regions such as microfluidic applications [3,30–32]. However, other complement structures, with more control in their topological structures, can be considered, such as CELCRs, which can be designed using a trapezoid shape to create smaller sensing regions. A trapezoid shape has two bases that can be chosen, so using two trapezoid shapes, one can design a complement structure-based resonator with a smaller sensing region. For understanding the benefit of trapezoid shapes, we will start by analyzing rectangular shapes.

Figure 1 shows a CELCR composed of two rectangular shapes, denoted as R_1 and R_2 [28]. The area that is critical for sensing is a × (L − 2a), denoted as S in Figure 1. To enable a high-sensitivity response, the sensing region must be fully covered with MUTs. However, for technologies that require smaller sensing regions or MUTs' volumes on a micro-scale, CELCRs in the current topological form will not be suitable for such applications. For example, in microfluidic-based technology, the localization and selectivity of electromagnetic energy to a small region are essential if utilized for heating and sensing individual droplets [31,32,34].

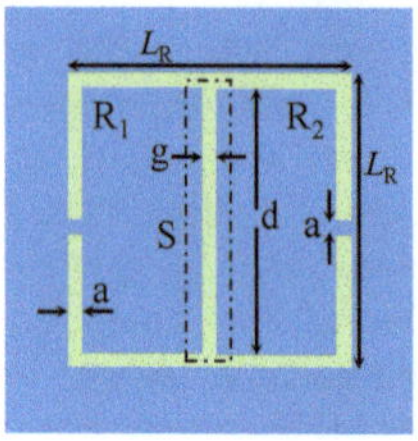

Figure 1. CELCR designed using two-rectangular shapes denoted as R_1 and R_2.

Compared with the traditional CELCR shape, the trapezoidal shape can address the challenge of achieving smaller and more sensitive sensing areas. Thus, the trapezoidal resonator design combines the advantages of SRR and CELCR in terms of having smaller sensing regions and being highly sensitive. A micro-scale-based region can be achieved using two-trapezoid shapes denoted as T_1 and T_2, as shown in Figure 2a. In fact, at the limit where one of the bases of a trapezoid shape is approaching zero, the shape will become triangular; hence, a smaller region can be further achieved. The trapezoid-shaped resonator illustrated in Figure 2a is designed in the ground plane of a microstrip line through an etching process. When operating at the resonance frequency, there is a significant increase in the amount of electric energy stored in the proximity of the TSR [2]. The electric field in the substrate and the surrounding air can be modeled using an effective capacitance C_{sub} (representing the capacitance due to the electric fields in the substrate) and C_{air} (representing the capacitance due to the electric fields in the air) (see Figure 2b). Therefore, the total effective capacitance of the TSR can be obtained by adding these two capacitances together. If the resonator effective capacitance is denoted as C_R, then $C_R = \alpha_1 C_{sub} + \alpha_2 C_{air}$, where α_1 and α_2 are real numbers accounting for the contribution factor of C_{sub} and C_{air} to the total C_R; hence, $\alpha_1 + \alpha_2 = 1$.

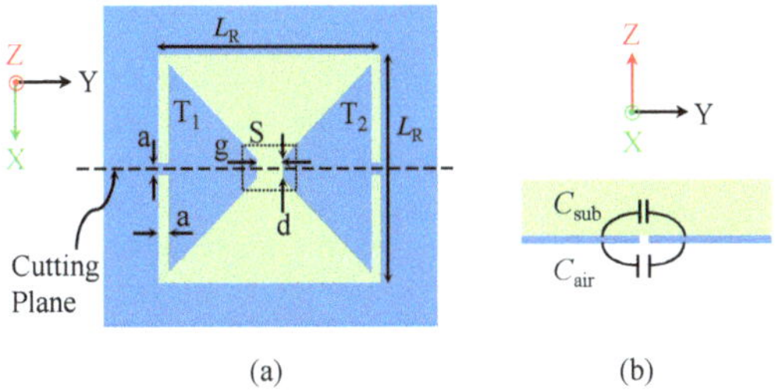

Figure 2. (**a**) TSR designed using trapezoid shapes denoted as T_1 and T_2. (**b**) Side view of TSR through the cutting plane shown in Figure 2a.

Since the CELC resonator is a sub-wavelength resonator, it can be modeled and analyzed using lumped circuit elements. However, the model is only accurate near the resonance frequency. Figure 3 shows the diagram of a microstrip line used to excite the TSR. Figure 4 shows a circuit model for the proposed TSR. L_{TL} and C_{TL} represent the equivalent inductance and the capacitance per unit length of the transmission line (TL); whereas C_R, R_R, and L_R represent the equivalent capacitance, resistance, and inductance of the resonator, respectively. Since the circuit model is similar to the circuit reported in [35], the resonance frequency is given as

$$f_z = \frac{1}{2\pi\sqrt{L_R(C_R + C_{TL})}} \tag{1}$$

In terms of C_{sub} and C_{air}, the resonance frequency can be expressed as

$$f_z = \frac{1}{2\pi\sqrt{L_R(\alpha_1 C_{sub} + \alpha_2 C_{air} + C_{TL})}} = \frac{1}{2\pi\sqrt{L_R((1-\alpha_2) C_{sub} + \alpha_2 C_{air} + C_{TL})}} \tag{2}$$

Thus, if α_2 is approaching 1, the contribution of C_{air} will be increased; hence, MUTs will be interrogated effectively with the electric field, leading to higher sensitivity.

Since the electric field will be focused in the material with the highest dielectric constant (the substrate), the total capacitance is largely determined by the effective capacitance C_{sub}, as noted in [25]. Thus, it is expected that α_1 will be higher. By placing MUT on the opposite side of the substrate in free space, the interaction with the resonators is minimized, resulting in a minimal shift in the resonance frequency and therefore minimizing the overall sensitivity of the sensor. To overcome this limitation, one can adopt the three-dimensional capacitor reported in [28,29]. Figure 5 shows the diagram of the sensor where a 3D capacitor is included by adding metallic bars that are extended into the free space and vertically from the sensor plane while being attached to the resonator. The metallic bars, denoted by their physical length as L_P, can be modeled as an additional capacitor (C_{PP}). Figure 6 illustrates the corresponding circuit model updated with the additional element of C_{PP}. C_{PP} is now a function of L_P, consequently, as C_{PP} will be added in parallel to C_{R0}, the resonance frequency, at the transmission zero, will be a function of L_P, given as

$$f_z(L_P, \epsilon_{MUT}, g, d) = \frac{1}{2\pi\sqrt{L_R(C_{R0} + C_{TL} + C_{PP}(L_P, \epsilon_{MUT}, g, d))}} \quad (3)$$

where C_{R0} is the original capacitor of the resonator without the metallic bars, and ϵ_{MUT} is the permittivity of MUT between the parallel bars. Note that $C_R = C_{R0} + C_{pp}$. Therefore, the metallic bars will increase the contribution of C_{air} to the total capacitance of the resonator, C_R. Hence, the interaction is increased with MUT, which is expected to increase the sensitivity. Furthermore, based on the circuit model shown in Figure 6, the proposed resonator can be classified as a parallel resonator. The quality factor can be expressed as [14]

$$Q = R_R\sqrt{\frac{C_{R0} + C_{TL} + C_{PP}}{L_R}} \quad (4)$$

From (4), it can be observed that C_{PP} will enhance the quality factor. However, since C_{PP} is a function of ϵ_{MUT}, which will be added in parallel to C_{R0} and C_{TL}, its equivalent losses will lower the overall losses, and consequently, lower the quality factor. In addition, by loading C_{PP} with MUT, the electric field will be more focused in MUT than the substrate, which will cause the lowering of the coupling (smaller C_{TL}) between TL and the resonator. Thus, C_{TL} depends on C_{PP}. Consequently, from (4), one can expect a trade-off between a higher quality factor and a higher coupling factor (represented in higher C_{TL}).

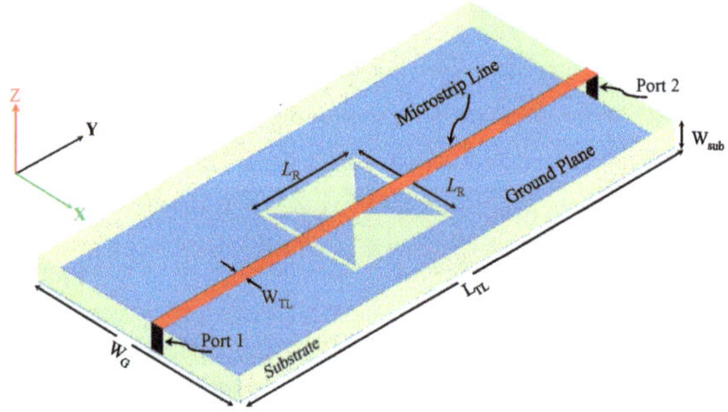

Figure 3. A perspective view of the TSR sensor geometry.

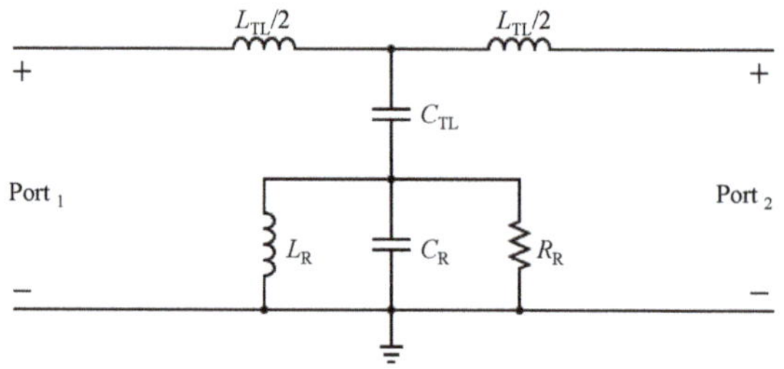

Figure 4. Two-port circuit model of the microstrip line used to excite the TSR, where L_{TL} and C_{TL} are the equivalent inductance and the capacitance per unit length of the transmission line (TL), and C_R, R_R, and L_R are the equivalent capacitance, resistance, and inductance of the resonator, respectively.

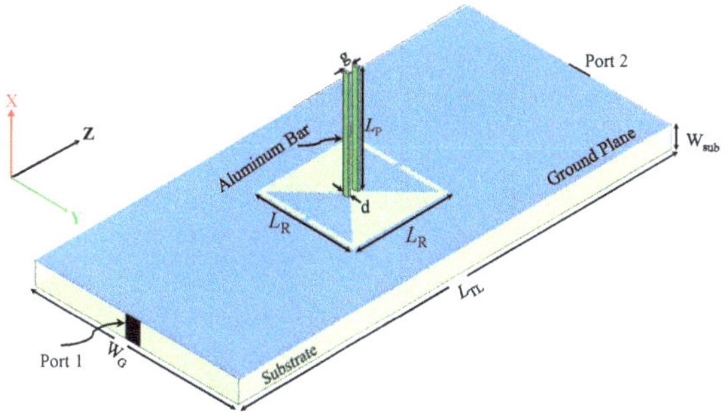

Figure 5. Perspective view of a TSR sensor loaded with aluminum bars.

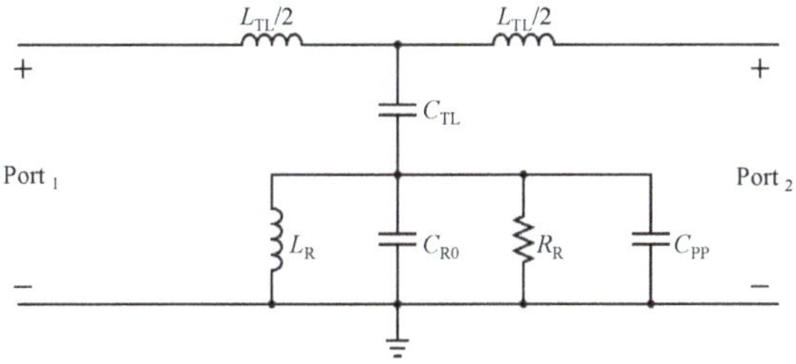

Figure 6. Circuit lumped-element model of a transmission line (TL) used to excite TSR, where L_{TL} and C_{TL} are the equivalent capacitance and the inductance of TL, L_R, C_R ($C_R = C_{R0} + C_{pp}$), and R_R are the equivalent inductance, capacitance, and resistance of the resonator, respectively. C_{PP} is the equivalent capacitance of the metallic bars.

3. Sensor Design, Numerical Analysis, and Discussion

Sub-wavelength resonators such as the trapezoid-shaped resonator can be excited using a quasi-TEM mode that can be generated using a microstrip line. The resonator is designed in the ground plane by etching out the topology depicted in Figure 2. For the purpose of comparisons with the resonator in [29], the dimensions are a = g = 0.5 mm, L = 7.5 mm, S = a × a (a^2 = 0.25 mm^2).

Since the sensor response will be evaluated using scattering parameters ($|S_{21}|$), which will be recorded using a vector network analyzer (VNA) with a 50 Ω input impedance, a 50 Ω microstrip line with a strip width of 1.63 mm was designed on a Rogers substrate (RO4350) with a thickness of W_{TL} = 0.76 mm, a loss tangent of 0.004, and a relative permittivity of 3.66. The dimensions that are shown in Figure 2 are L_{TL} = 100 mm and W_G = 50 mm. The bars' length (L_p) will be studied extensively by exploring different values to investigate its effects on the resonance frequency. The metallic bars are aluminum. Table 1 presents a summary of the design specification used in this work.

Table 1. Design Criteria.

Sensor Type	a (mm)	L_R (mm)	W_{TL} (mm)	W_G (mm)	L_{TL} (mm)	W_{sub} (mm)	L_P (mm)	g (mm)	d (mm)
CELCR (No Bars)	0.5	7.5	1.63	50	100	0.762	NA	0.5	0.5
TSR (No Bars)	0.5	7.5	1.63	50	100	0.762	NA	0.5	0.5
TSR (with Bars)	~	~	~	~	~	~	Vari.	0.5	0.5

Figure 7a shows the $|S_{21}|$ of the sensor without the bars, over the frequency range of 4.5 to 7.5 GHz. The frequency at which $|S_{21}|$ is minimum is the resonance frequency, which has a value of 6.28 GHz (the resonance frequency was chosen for its suitability to our measurement setup and to facilitate comparison with the sensor reported in reference [29]).

The calculated quality factor of the resonator without the bars is found to be approximately 43. Since the circuit model shown in Figure 4 is used to predict the resonance frequency, it is important to validate such a circuit. The procedure is as follows: The response of the sensor ($|S_{21}|$ and $|S_{11}|$) in the form of a touchstone file was extracted from the full-wave numerical simulation performed using ANSYS HFSS [36]. Then, the response was imported to the circuit simulator Keysight-ADS [37]. By using the optimization toolbox provided in ADS, the circuit elements shown in Figure 4 were extracted. Table 2 shows the extracted elements.

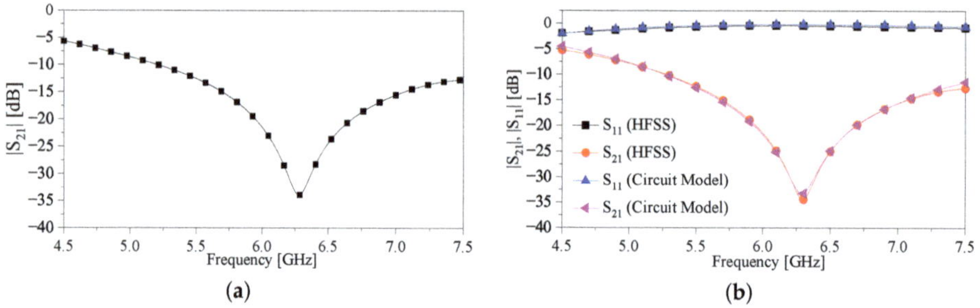

Figure 7. (**a**) The response ($|S_{21}|$) of the TSR sensor without the metallic bars extracted using numerical simulation. (**b**) A comparison of the response obtained from the numerical simulation and the circuit model.

Table 2. Extracted circuit elements

Sensor Type	L_{TL} (nH)	C_{TL} (pF)	L_R (nH)	C_R (pF)	R_R (kΩ)
The trapezoid-shaped resonator (No Bars)	1.176	1.288	0.252	1.25	0.376
The trapezoid-shaped resonator (with Bars)	1.332	1.782	0.401	3.341	0.934

With the metallic bars, the resonance frequency becomes a function of the length of bars (L_P), as well as the gap between the bars (denoted as g in Figure 5). To investigate the effects of the bars' length on the resonance frequency and the quality factor, L_P was varied from 5 to 65 mm in increments of 5 mm. Since there will be an overlap between the response curves at different values of L_P, some values were selected for the plot as shown in Figure 8. However, one can plot the response ($|S_{21}|$) as a 2D image as shown in Figure 9. The image color ranges from deep yellow to deep blue, where deep yellow represents a full transmission, while deep blue represents a transmission zero. On the y-axes, the distance (in frequency) between the deep blue colors represents the changes in the resonance frequencies (the frequencies at transmission zero) due to the changes in the bars' length (L_P). An example of Δf_z is shown in Figure 9.

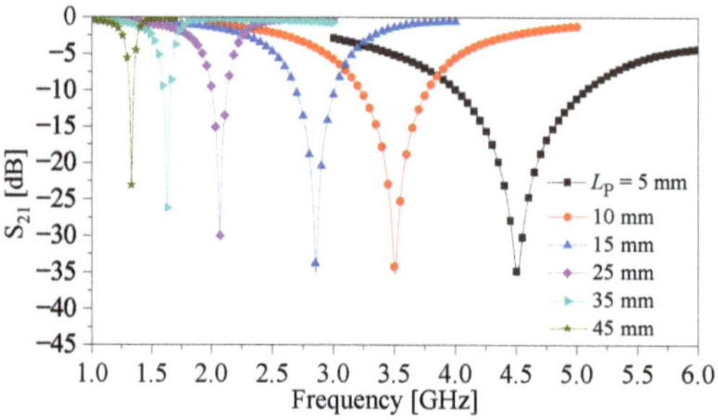

Figure 8. Response ($|S_{21}|$) of the sensors at different values of L_P = 5, 10, 15, 25, 35, and 45 mm.

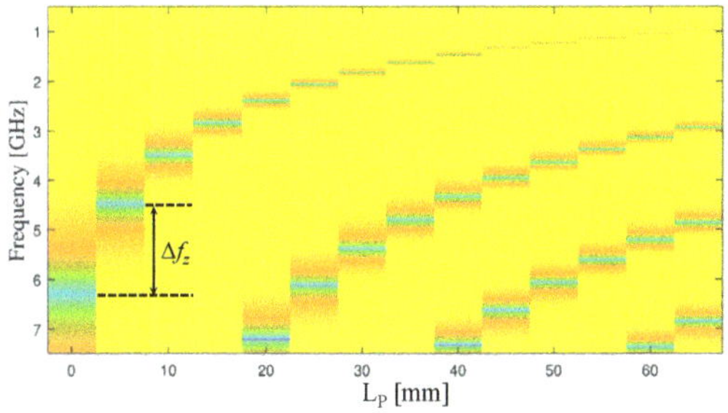

Figure 9. 2D image of the response ($|S_{21}|$) at different values of L_P = 0, 5, 10, 15, 25, 35, and 45 mm.

Plotting $|S_{21}|$ as an image will help to observe the width of the minimum transmission (in frequency) at which the quality factor of the resonator can be observed. As was expected from (4), and quantitatively proven using the numerical simulation, it is evident that the bars contribute to the enhancement of the quality factor. For the ranges of L_P from 5 to 65 mm, the quality factor is given in Figure 10.

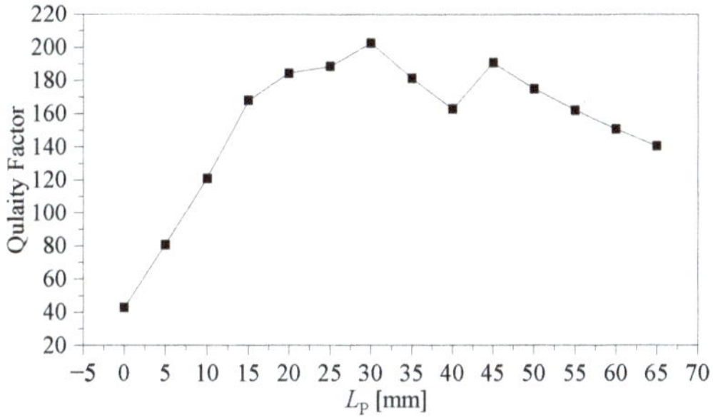

Figure 10. Quality factor as a function of the bars' length, L_P.

The circuit model shown in Figure 6 was validated using the same previous procedure. Note that the chosen L_P was 10 mm (other values can be alternatively chosen). The extracted circuit elements are presented in Table 2. The response of the circuit model versus the numerical simulation is shown in Figure 11. From Table 2, one can observe that the values of L_R and C_R are increased by 59% and 167%, respectively. The increment in both values caused the resonance frequencies to shift down to lower frequencies. In the case of L_P = 10, the resonance frequency was shifted from 6.28 GHz (no bars) to 3.51 GHz, which is approximately 44%.

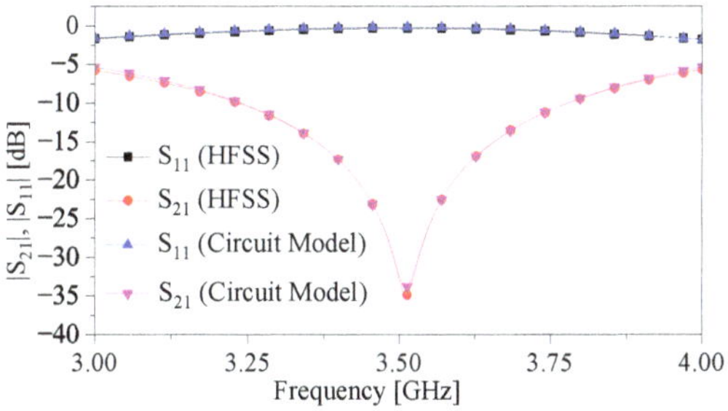

Figure 11. Response of the circuit model and the numerical simulation with the bar's length of L_P = 10.

It is expected that the total electric field will be enhanced between the bars. To quantify this enhancement, the field was calculated using HFSS. The electric field was calculated along the virtual line shown in Figure 12a for L_P = 30 (which corresponds to the highest quality factor). The total calculated electric field of the TSR with and without the metallic bars is shown in Figure 13. It can be easily observed that the metallic bars concentrate the electric field in the sensing region. Therefore, as predicted by (3), the material between the metallic bars will strongly affect the resonance frequency.

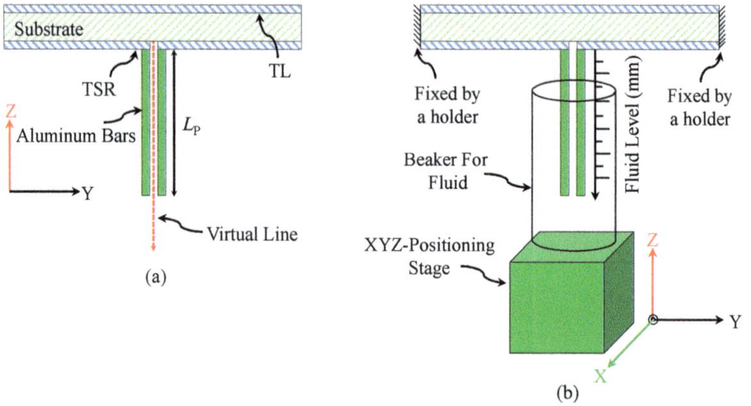

Figure 12. Side view of the sensor presenting (**a**) the virtual line used to plot the total electric field between the bars (**b**) the schematic of the experimental setup used for the measurement.

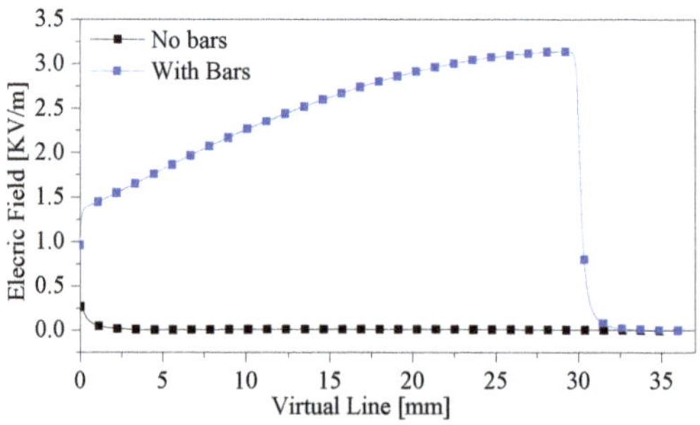

Figure 13. Total electric field between the bars along the virtual line shown in Figure 12a.

The detection mechanism of the sensing system is through observing the changes in the resonance frequency due to the presence of MUT in the sensing region. To compare the sensitivity of the proposed sensor with other planar sensors such as the one recently reported in [29], the following equations were used,

$$S = \frac{\Delta f_z}{f_{z0}(\epsilon_r - 1)} \times 100 \quad (5)$$

$$\text{Norm } \Delta f = \frac{\Delta f_z}{f_{z0}} \times 100 \quad (6)$$

where Δf_z represents the relative change in the resonance frequency compared to the reference case (i.e., air), f_{z0} is the resonance frequency of the reference case, $\epsilon_r - 1$ is the relative difference in the dielectric constant of the materials being studied and Norm Δf is the normalized resonance frequency shift.

To simulate a real-world application, the arms were immersed completely inside a cylindrical dielectric material with its permittivity being varied from 1 to 10 in increments of 0.5. Figure 14a,b show the resonance frequency shift (Δf_z) and the sensitivity of the sensor with and without the metallic bars versus the relative permittivity. Table 3 gives a performance comparison for the sensor with and without the bars.

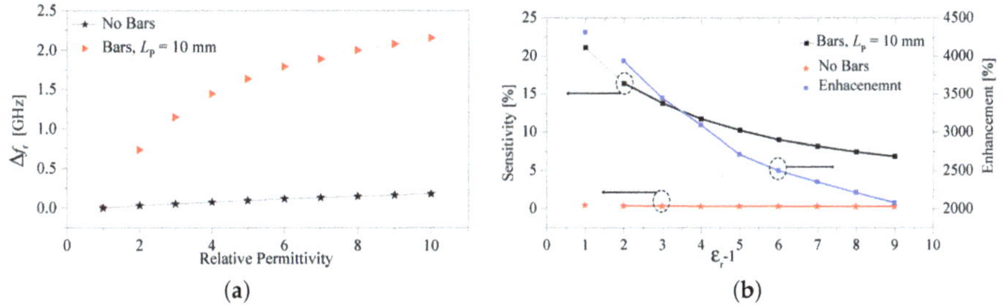

Figure 14. (**a**) Resonance frequency shift (Δf_z). (**b**) The sensitivity, and the enhancement in the sensitivity of the proposed sensor with and without the metallic bars.

Table 3. Sensitivity comparison between the sensor with and without metallic bars.

$\epsilon_r - 1$	Trapezoid-Shaped Resonator (No Bars)	Trapezoid-Shaped Resonator (with Bars)
1	0.477%	21.094%
2	0.406%	16.405%
3	0.387%	13.749%
4	0.366%	11.701%
5	0.363%	10.228%
6	0.345%	8.979%
7	0.332%	8.152%
8	0.320%	7.418%
9	0.313%	6.835%

The sensor was examined to detect the dielectric constant when the bars are immersed at different levels from 1 to 10 mm with an increment of 1 mm. Simultaneously, the dielectric constant was varied from 1 to 100 with an increment of 1 for the range 1 to 11, and an increment of 10 for the range 11 to 100. Figure 15 shows the resonance frequency shift (Δf_z) versus the relative permittivity at different immersion levels.

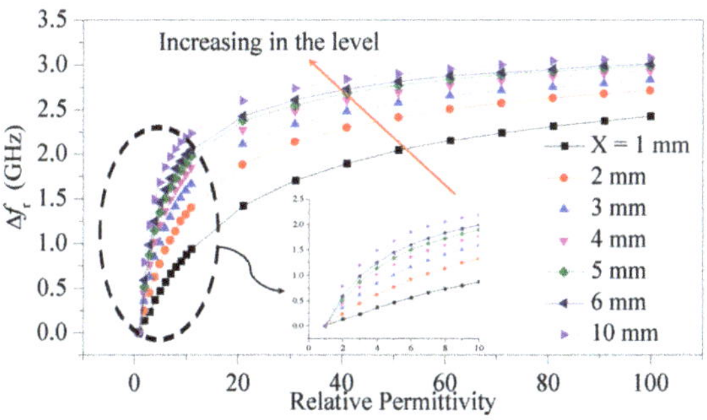

Figure 15. Resonance frequency shift as a function of the relative permittivity for different immersion levels and at different dielectric constants.

The proposed sensor was compared with the sensor reported in [29], where both resonators have identical bars' lengths. Equations (5) and (6) will be used for the evaluation and comparison. Figure 16a,b show the resonance frequency shift and sensitivity of SRR and TSR with metallic bars of length 4 mm. Note that the dimensions of the SRR and the bars' length were taken from [29]. By comparing the sensitivity of SRR and TSR, the enhancement in the sensitivity was quantified and presented in Figure 16b. It is evident that the TSR combines the advantages of SRR and CELCR of having smaller sensing regions and being highly sensitive. Table 4 provides a comprehensive comparison between different microwave sub-wavelength microwave sensors. Furthermore, from Figure 16a considering the case of TSR, there are three regions where Δf_r as a function of ϵ_r behaves differently with the increment of ϵ_r, where region$_3$ indicates the approaching to a saturation region. To quantify the minimum detectable range, we utilized a fitting function technique to mathematically model Δf_r as a function of ϵ_r, which can be expressed as,

$$\Delta f_r(\epsilon_r) = 3.2154 - 1.7812e^{(\frac{-\epsilon_r}{1.19663})} - 2.8787e^{(\frac{-\epsilon_r}{6.05237})} \tag{7}$$

The change in Δf_r due to a small change in ϵ_r can be expressed as,

$$\frac{d\Delta f_r(\epsilon_r)}{d\epsilon_r} = 1.48851e^{(\frac{-2\epsilon_r}{1.19663})} + 0.4756e^{(\frac{-2\epsilon_r}{6.05237})} \tag{8}$$

Thus, one can use Equation (8) such that

$$min\{\frac{d\Delta f_r(\epsilon_r)}{d\epsilon_r}\} = min\{1.48851e^{(\frac{-2\epsilon_r}{1.19663})} + 0.4756e^{(\frac{-2\epsilon_r}{6.05237})}\} \tag{9}$$

and solve it numerically to find ϵ_r at which Δf_r is minimum. Since region$_3$ is approaching saturation, it is expected that the minimum detectable values will depend on the resolution of the apparatus, e.g., VNA, which is used to measure the response of the system. At the limit where the value of Equation (9) is zero, the system cannot distinguish between two close values of ϵ_r.

Table 4. A comprehensive comparison between microwave sub-wavelength microwave sensors.

Ref.	f_r [GHz]	Tunable	F. Level	MAX. S [%]	L. Dielectric	H. Dielectric	Relative Size	Geometry
[38]	20	NO	No	0.347	—	Yes	0.086 λ_0	Planar
[39]	1.35	NO	No	—	Yes	Yes	0.25 λ_0	Planar
[40]	3.1	NO	No	0.187	Yes	Yes	0.65 λ_0	Planar
[41]	1.91	NO	No	—	—	Yes	0.06 λ_0	Planar
[13]	2.1	NO	No	—	—	Yes	0.049 λ_0	Planar
[29]	2.828	NO	Yes	13.719	Yes	No	0.0707 λ_0	Planar with metallic bars
[T.W]	**4.812**	**NO**	**Yes**	**21.094**	**Yes**	**Yes**	**0.12 λ_0**	**Planar with metallic bars**

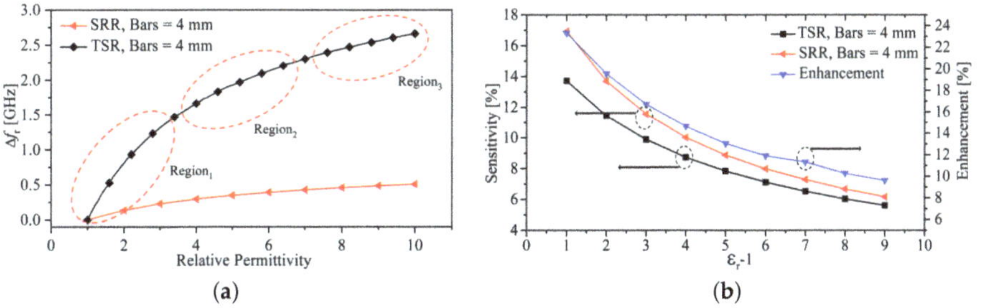

Figure 16. (**a**) Resonance frequency shift (Δf_Z). (**b**) Sensitivity, and the enhancement in the sensitivity of the proposed sensor with and without metallic bars.

4. Fluid-Level Measurements

This part focuses on the fabrication and testing of TSR sensors for fluid detection and fluid-level measurement. The sensors were fabricated using printed circuit board (PCB) technology. The design specifications of the sensors are presented in Table 1. To enhance the sensitivity, metallic bars with a length of (L_P) = 30 mm were utilized as this parameter was found to exhibit the highest quality factor. While other metallic materials with higher conductivity, such as gold and silver, could be employed, aluminum material was chosen for these bars due to its low cost and ease of fabrication. Since it is difficult to solder aluminum to copper, a conductive glue from the manufacturer MG Chemicals (8331-14G) was used. The glue is a 2-part epoxy with a 1:1 mix ratio, a working time of 10 min, and a resistivity of 7.0×10^{-3} Ω·cm. It is recommended by the manufacturer to allow 24 h to cure at room temperature or to use an oven by following one of these options (time/temperature): 15 min at 65 Celsius or 7 min at 125 Celsius. The glue was used to attach the bars to the surface plane of the sensor. Note that from (3), it is evident that a smaller value of the distance between the bars (g) will lead to a higher capacitance of C_{PP}, which will shift the resonance frequency to a lower frequency. With the use of the conductive glue, one can tune the resonance frequency to an intended one by optimizing the distance between the bars, g. To assess the enhanced sensitivity, the two resonators, with and without the aluminum bars, were tested. Figure 17 shows the fabricated TSR sensor.

The procedure for the fluid detection involved immersing the sensor bars inside two different fluids, chloroform and dichloromethane, with permittivities of 4.81 and 8.93, respectively (taken from [42]). We also tested other materials with higher dielectric constants, such as distilled water, to evaluate the sensor's performance under different conditions. The immersion levels of the bars were controlled using an XYZ-positioning stage with a resolution of 0.01 mm in the Z-direction, as shown in Figure 17d. The data were collected using a 50 Ω vector network analyzer. The network analyzer was calibrated first using a short-open-load-thru calibration kit to ensure accurate measurements. The collected data were then analyzed to determine the sensor's sensitivity and accuracy in measuring fluid levels. The response of the sensor with the bars placed in the air is shown in Figure 18. The results show a strong agreement between the simulation and the measurements.

To quantify the enhanced sensitivity of the sensor, first, the TSR sensor was tested without the bars (shown in Figure 17c). The plastic container (encompassing the sensing region) was filled with the fluids under test. The results of this test are given in Figure 19 showing a sensitivity of 0.92% for chloroform and 0.66 % for dichloromethane. The experiment was repeated with the TSR sensor with the bars using the same fluids and glass beaker, as shown in Figure 12b. The bars were completely immersed in the fluids. The result of this test is presented in Figure 20, showing a sensitivity of 14.1% for chloroform and 8.3% for dichloromethane. Therefore, the enhancement in sensitivity of the TSR sensor is 1435% for chloroform and 1168% for dichloromethane.

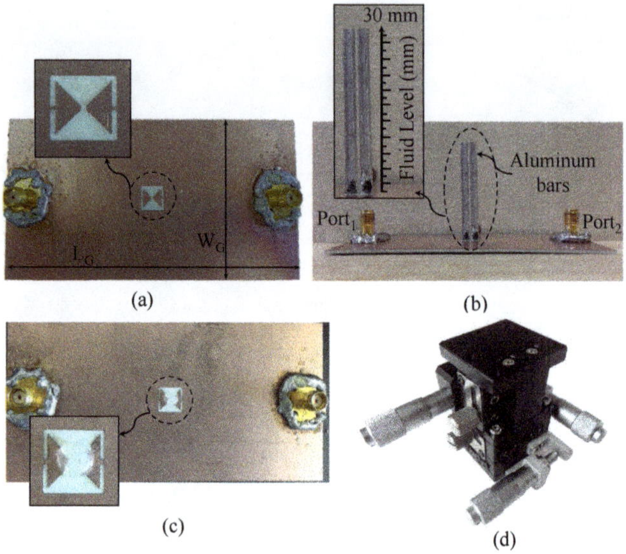

Figure 17. The fabricated TSR sensors. (**a**) Without the bars. (**b**) With the bars. (**c**) The sensor without the bars in a plastic fluid container. (**d**) Manual-XYZ-positioning stage.

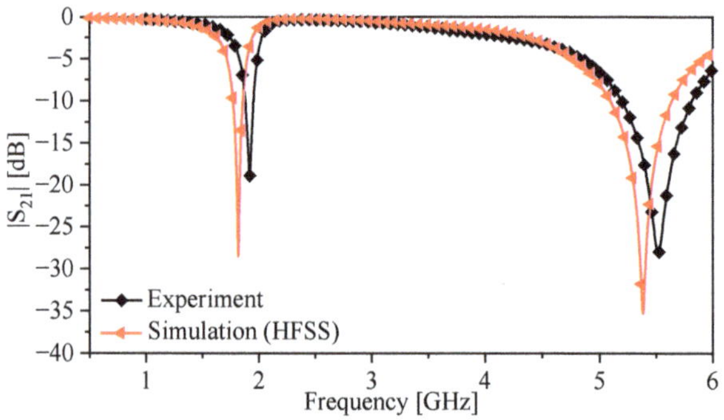

Figure 18. Measured $|S_{21}|$ over the frequency range 0.3 to 6 GHz for the TSR sensor with bars.

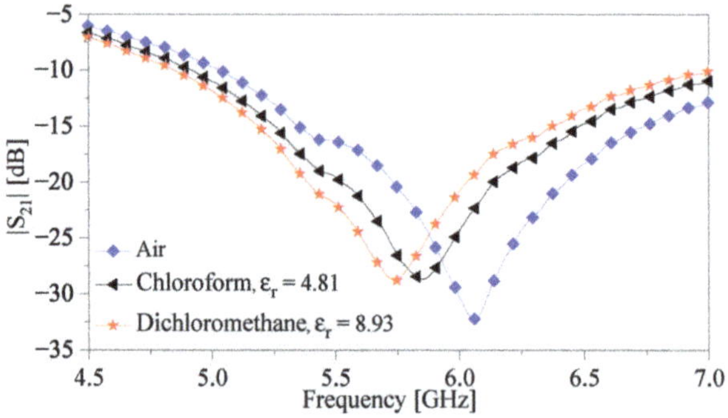

Figure 19. Measured $|S_{21}|$ for the sensor without the bars, in the presence of air, chloroform, and dichloromethane.

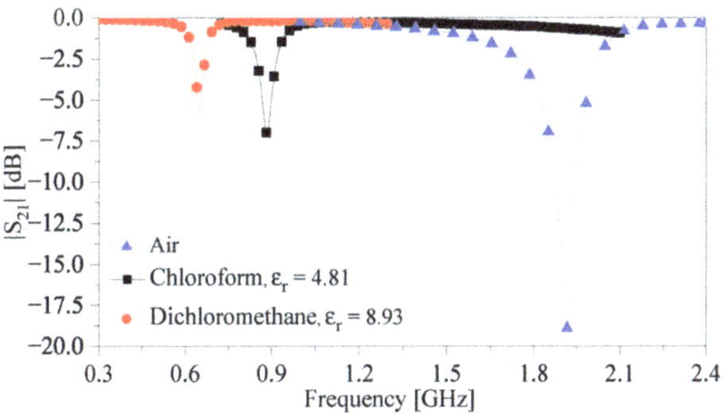

Figure 20. Measured $|S_{21}|$ for the sensor with the bars, in the presence of air, chloroform, and dichloromethane.

For the fluid-level measurement, the bars were gradually immersed inside a glass beaker containing chloroform, with an immersion increment of 0.76 mm (note that the first step was 0.5 mm). The resonance frequency at each step versus the step values was used to create a calibration curve, as shown in Figure 21. Calibration curves were also generated for dichloromethane and distilled water using the same procedure, as illustrated in Figure 22.

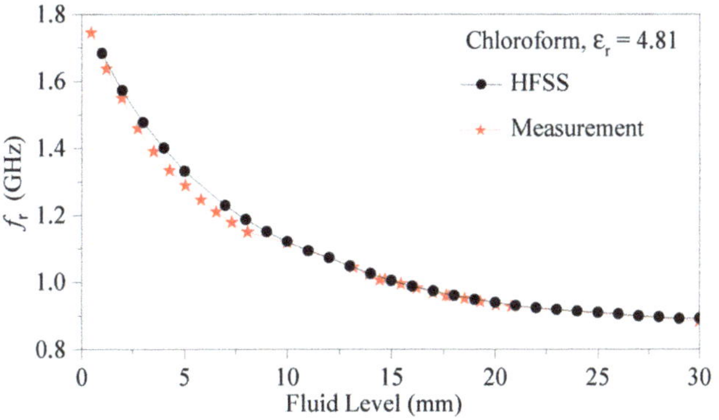

Figure 21. Calibration curve generated by the experiment and simulation for measurement of chloroform.

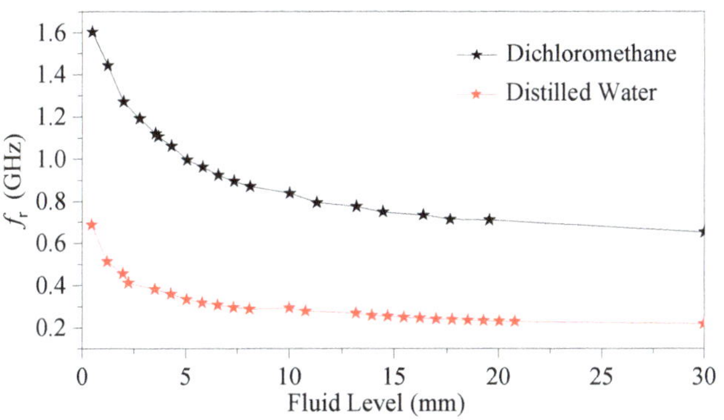

Figure 22. Calibration curves for the dichloromethane and distilled water generated by the experiment.

5. Conclusions

In this work, we proposed a new design for planar sub-wavelength resonators that combines the advantages of split ring resonators (SRR) and complementary electric-LC resonators (CELCR) by incorporating metallic bars in a trapezoid-shaped resonator. The trapezoid shape allows for a smaller sensing region, while the metallic bars enhance the electric field in the sensing region, resulting in higher sensitivity. Numerical simulations were used to design and evaluate the sensor. The sensor was fabricated and validated using PCB technology with aluminum bars. The testing of the sensor's sensitivity demonstrated that it can be used for microfluidic and level detection applications, with the enhanced sensitivity achieved through the use of metallic bars. The calibration curves generated for fluid-level measurement can serve as a basis for future applications in fluid-level detection and related fields. One of the main advantages of this new design is that it offers higher sensitivity than previous designs while also having a smaller sensing region, making it well-suited for microfluidic applications. Additionally, the use of aluminum bars makes the sensor more affordable and easier to fabricate.

Author Contributions: The idea behind this work was developed A.M.A. and O.M.R. A.M.A. designed the sensors, prepared the samples, conducted the experiment, collected the data, and analyzed the data. The manuscript was written by A.M.A. and O.M.R. All authors have read and agreed to the published version of the manuscript.

Funding: This research was funded by Researchers Supporting Project number (RSP2023R482), King Saud University, Riyadh, Saudi Arabia.

Institutional Review Board Statement: Not applicable.

Informed Consent Statement: Not applicable.

Data Availability Statement: Data generated during the study are contained within the article.

Conflicts of Interest: The authors declare no conflicts of interest.

Abbreviations

The following abbreviations are used in this manuscript:

MUT	Material under test
TL	Transmission line
SRR	Split-ring resonator
CSRR	Complementary split-ring resonator
TSR	Trapezoid-shaped resonator

References

1. Gascoyne, P.R.C.; Vykoukal, J.V. Dielectrophoresis-based sample handling in general-purpose programmable diagnostic instruments. *Proc. IEEE* **2004**, *92*, 22–42. [CrossRef]
2. Pendry, J.; Holden, A.; Robbins, D.; Stewart, W. Magnetism from conductors and enhanced nonlinear phenomena. *IEEE Trans. Microw. Theory Tech.* **1999**, *47*, 2075–2084. [CrossRef]
3. Boybay, M.S. Behavior of metamaterial-based microwave components for sensing and heating of nanoliter-scale volumes. *Turk. J. Electr. Eng. Comput. Sci.* **2016**, *24*, 3503–3512. [CrossRef]
4. Salim, A.; Lim, S. Review of Recent Metamaterial Microfluidic Sensors. *Sensors* **2018**, *18*, 232. [CrossRef]
5. Chen, T.; Li, S.; Sun, H. Metamaterials application in sensing. *Sensors* **2012**, *12*, 2742–2765. [CrossRef]
6. Lee, H.J.; Yook, J.G. Biosensing using split-ring resonators at microwave regime. *Appl. Phys. Lett.* **2008**, *92*, 254103. [CrossRef]
7. Lee, H.J.; Lee, J.H.; Moon, H.S.; Jang, I.S.; Choi, J.S.; Yook, J.G.; Jung, H.I. A planar split-ring resonator-based microwave biosensor for label-free detection of biomolecules. *Sens. Actuators B Chem.* **2012**, *169*, 26–31. [CrossRef]
8. Lee, H.J.; Lee, H.S.; Yoo, K.H.; Yook, J.G. DNA sensing using split-ring resonator alone at microwave regime. *J. Appl. Phys.* **2010**, *108*, 014908. [CrossRef]
9. Torun, H.; Top, F.C.; Dundar, G.; Yalcinkaya, A. An antenna-coupled split-ring resonator for biosensing. *J. Appl. Phys.* **2014**, *116*, 124701. [CrossRef]
10. Albishi, A.; Boybay, M.S.; Ramahi, O.M. Complementary Split-Ring Resonator for Crack Detection in Metallic Surfaces. *IEEE Microw. Wirel. Compon. Lett.* **2012**, *22*, 330–332. [CrossRef]

11. Albishi, A.; Ramahi, O.M. Detection of Surface and Subsurface Cracks in Metallic and Non-Metallic Materials Using a Complementary Split-Ring Resonator. *Sensors* **2014**, *14*, 19354–19370. [CrossRef]
12. Albishi, A.M.; Ramahi, O.M. Microwaves-Based High Sensitivity Sensors for Crack Detection in Metallic Materials. *IEEE Trans. Microw. Theory Tech.* **2017**, *65*, 1864–1872. [CrossRef]
13. Withayachumnankul, W.; Jaruwongrungsee, K.; Tuantranont, A.; Fumeaux, C.; Abbott, D. Metamaterial-based microfluidic sensor for dielectric characterization. *Sens. Actuators A Phys.* **2013**, *189*, 233–237. [CrossRef]
14. Ebrahimi, A.; Withayachumnankul, W.; Al-Sarawi, S.; Abbott, D. High-sensitivity metamaterial-inspired sensor for microfluidic dielectric characterization. *IEEE Sens. J.* **2014**, *14*, 1345–1351. [CrossRef]
15. Caglayan, H.; Cakmakyapan, S.; Addae, S.A.; Pinard, M.A.; Caliskan, D.; Aslan, K.; Ozbay, E. Ultrafast and sensitive bioassay using split ring resonator structures and microwave heating. *Appl. Phys. Lett.* **2010**, *97*, 093701. [CrossRef]
16. Boybay, M.; Ramahi, O.M. Material Characterization Using Complementary Split-Ring Resonators. *IEEE Trans. Instrum. Meas.* **2012**, *61*, 3039–3046. [CrossRef]
17. Boybay, M.S.; Ramahi, O.M. Non-destructive thickness Measurement using quasi-static resonators. *IEEE Microw. Wirel. Compon. Lett.* **2013**, *23*, 217–219. [CrossRef]
18. Yang, C.L.; Lee, C.S.; Chen, K.W.; Chen, K.Z. Noncontact measurement of complex permittivity and thickness by using planar resonators. *IEEE Trans. Microw. Theory Tech.* **2016**, *64*, 247–257. [CrossRef]
19. Saadat-Safa, M.; Nayyeri, V.; Ghadimi, A.; Soleimani, M.; Ramahi, O.M. A pixelated Microwave near-field Sensor for precise characterization of Dielectric Materials. *Sci. Rep.* **2019**, *9*, 13310. [CrossRef]
20. Velez, P.; Munoz-Enano, J.; Grenier, K.; Mata-Contreras, J.; Dubuc, D.; Martin, F. Split Ring Resonator-Based Microwave Fluidic Sensors for Electrolyte Concentration Measurements. *IEEE Sens. J.* **2019**, *19*, 2562–2569. [CrossRef]
21. Lee, C.; Bai, B.; Song, Q.; Wang, Z.; Li, G. Open Complementary Split-Ring Resonator Sensor for Dropping-Based Liquid Dielectric Characterization. *IEEE Sens. J.* **2019**, *19*, 11880–11890. [CrossRef]
22. Chuma, E.L.; Iano, Y.; Fontgalland, G.; Roger, L.L.B.; Loschi, H. PCB-integrated non-destructive microwave sensor for liquid dielectric spectroscopy based on planar metamaterial resonator. *Sens. Actuators A Phys.* **2020**, *312*, 112112. [CrossRef]
23. Lobato-Morales, H.; Choi, J.H.; Lee, H.; Medina-Monroy, J.L. Compact Dielectric-Permittivity Sensors of Liquid Samples Based on Substrate-Integrated-Waveguide With Negative-Order-Resonance. *IEEE Sens. J.* **2019**, *19*, 8694–8699. [CrossRef]
24. Velez, P.; Grenier, K.; Mata-Contreras, J.; Dubuc, D.; Martin, F. Highly-Sensitive Microwave Sensors Based on Open Complementary Split Ring Resonators OCSRRs for Dielectric Characterization and Solute Concentration Measurement in Liquids. *IEEE Access* **2018**, *6*, 48324–48338. [CrossRef]
25. O'Hara, J.F.; Singh, R.; Brener, I.; Smirnova, E.; Han, J.; Taylor, A.J.; Zhang, W. Thin-film sensing with planar terahertz metamaterials: Sensitivity and limitations. *Opt. Express* **2008**, *16*, 1786–1795. [CrossRef]
26. Al-Naib, I.A.I.; Jansen, C.; Koch, M. Thin-film sensing with planar asymmetric metamaterial resonators. *Appl. Phys. Lett.* **2008**, *93*, 083507. [CrossRef]
27. Abdolrazzaghi, M.; Zarifi, M.H.; Daneshmand, M. Sensitivity enhancement of split ring resonator based liquid sensors. In Proceedings of the 2016 IEEE SENSORS, Orlando, FL, USA, 30 October–3 November 2016; IEEE: Piscataway, NJ, USA, 2016; pp. 1–3.
28. Albishi, A.M.; Ramahi, O.M. Highly Sensitive Microwaves Sensors for Fluid Concentration Measurements. *IEEE Microw. Wirel. Compon. Lett.* **2018**, *28*, 287–289. [CrossRef]
29. Albishi, A.M.; Alshebeili, S.A.; Ramahi, O.M. Three-Dimensional Split-Ring Resonators-Based Sensors for Fluid Detection. *IEEE Sens. J.* **2021**, *21*, 9138–9147. [CrossRef]
30. Yesiloz, G.; Boybay, M.S.; Ren, C.L. Effective Thermo-Capillary Mixing in Droplet Microfluidics Integrated with a Microwave Heater. *Anal. Chem.* **2017**, *89*, 1978–1984. [CrossRef]
31. Wong, D.; Yesiloz, G.; Boybay, M.S.; Ren, C.L. Microwave temperature measurement in microfluidic devices. *Lab Chip* **2016**, *16*, 2192–2197. [CrossRef]
32. Yesiloz, G.; Boybay, M.S.; Ren, C.L. Label-free high-throughput detection and content sensing of individual droplets in microfluidic systems. *Lab Chip* **2015**, *15*, 4008–4019. [CrossRef]
33. Albishi, A.M.; Badawe, M.K.E.; Nayyeri, V.; Ramahi, O.M. Enhancing the Sensitivity of Dielectric Sensors With Multiple Coupled Complementary Split-Ring Resonators. *IEEE Trans. Microw. Theory Tech.* **2020**, *68*, 4340–4347. [CrossRef]
34. Boybay, M.S.; Jiao, A.; Glawdel, T.; Ren, C.L. Microwave sensing and heating of individual droplets in microfluidic devices. *Lab Chip* **2013**, *13*, 3840–3846. [CrossRef]
35. Bonache, J.; Gil, M.; Gil, I.; Garcia-Garcia, J.; Martin, F. On the electrical characteristics of complementary metamaterial resonators. *IEEE Microw. Wirel. Compon. Lett.* **2006**, *16*, 543–545. [CrossRef]
36. ANSYS HFSS Version, 15.0.0. Available online: http://www.ansys.com (accessed on 8 June 2018).
37. Keysight ADS Version, 2016.01. Available online: http://www.keysight.com (accessed on 15 July 2018).
38. Chretiennot, T.; Dubuc, D.; Grenier, K. A Microwave and Microfluidic Planar Resonator for Efficient and Accurate Complex Permittivity Characterization of Aqueous Solutions. *IEEE Trans. Microw. Theory Tech.* **2013**, *61*, 972–978. [CrossRef]
39. Rowe, D.J.; al-Malki, S.; Abduljabar, A.A.; Porch, A.; Barrow, D.A.; Allender, C.J. Improved Split-Ring Resonator for Microfluidic Sensing. *IEEE Trans. Microw. Theory Tech.* **2014**, *62*, 689–699. [CrossRef]

40. Abduljabar, A.A.; Rowe, D.J.; Porch, A.; Barrow, D.A. Novel Microwave Microfluidic Sensor Using a Microstrip Split-Ring Resonator. *IEEE Trans. Microw. Theory Tech.* **2014**, *62*, 679–688. [CrossRef]
41. Ebrahimi, A.; Scott, J.; Ghorbani, K. Ultrahigh-Sensitivity Microwave Sensor for Microfluidic Complex Permittivity Measurement. *IEEE Trans. Microw. Theory Tech.* **2019**, *67*, 4269–4277. [CrossRef]
42. Maryott, A.A.; Smith, E.R. *Table of Dielectric Constants of Pure Liquids;* Technical Report 514; Nat. Bureau Standards Circular: Gaithersburg, MD, USA, 1951.

sensors

MDPI

Article

Enhancement of Refractive Index Sensitivity Using Small Footprint S-Shaped Double-Spiral Resonators for Biosensing

Anh Igarashi [1,*], Maho Abe [2], Shigeki Kuroiwa [3], Keishi Ohashi [3] and Hirohito Yamada [1]

1 Graduate School of Engineering, Tohoku University, Sendai 980-8579, Japan
2 Research Institute of Electrical Communication, Tohoku University, Sendai 980-8577, Japan
3 R&D Group, KOKOROMI Inc., Shinjuku-ku, Tokyo 169-0051, Japan; kuroiwa@kokoromill.com (S.K.)
* Correspondence: truong.hoang.anh.c6@tohoku.ac.jp

Abstract: We demonstrate an S-shaped double-spiral microresonator (DSR) for detecting small volumes of analytes, such as liquids or gases, penetrating a microfluidic channel. Optical-ring resonators have been applied as label-free and high-sensitivity biosensors by using an evanescent field for sensing the refractive index of analytes. Enlarging the ring resonator size is a solution for amplifying the interactions between the evanescent field and biomolecules to obtain a higher refractive index sensitivity of the attached analytes. However, it requires a large platform of a hundred square millimeters, and 99% of the cavity area would not involve evanescent field sensing. In this report, we demonstrate the novel design of a Si-based S-shaped double-spiral resonator on a silicon-on-insulator substrate for which the cavity size was 41.6 μm × 88.4 μm. The proposed resonator footprint was reduced by 680 times compared to a microring resonator with the same cavity area. The fabricated resonator exposed more sensitive optical characteristics for refractive index biosensing thanks to the enhanced contact interface by a long cavity length of DSR structures. High quality factors of 1.8×10^4 were demonstrated for 1.2 mm length DSR structures, which were more than two times higher than the quality factors of microring resonators. A bulk sensitivity of 1410 nm/RIU was calculated for detecting 1 μL IPA solutions inside a 200 μm wide microchannel by using the DSR cavity, which had more than a 10-fold higher sensitivity than the sensitivity of the microring resonators. A DSR device was also used for the detection of 100 ppm acetone gas inside a closed bottle.

Keywords: refractive index; ring resonator; evanescent field sensing; cavity length

Citation: Igarashi, A.; Abe, M.; Kuroiwa, S.; Ohashi, K.; Yamada, H. Enhancement of Refractive Index Sensitivity Using Small Footprint S-Shaped Double-Spiral Resonators for Biosensing. *Sensors* **2023**, *23*, 6177. https://doi.org/10.3390/s23136177

Academic Editor: Evgeny Katz

Received: 2 June 2023
Revised: 3 July 2023
Accepted: 4 July 2023
Published: 5 July 2023

1. Introduction

In recent years, biosensors with several requirements of label-free detection, a fast response, a low cost, a high sensitivity, and a compact size measurement for use in the development of healthcare diagnoses and environmental monitoring have received increasing attention [1–6]. Thus far, enzyme-linked immunosorbent assay (ELISA) detection has been the gold-standard labeled immunoassay for detecting biomarkers, including antibodies, antigens, and proteins, with an ultralow detection limit of about 1 pM [7]. ELISA can provide a high specificity and sensitivity because of specific antigen–antibody reactions. However, the ELISA method needs time and additional costs with a high-level laboratory and well-trained specialists due to the difficulties in label-based measurements. A simple label-free and real-time detection method has become increasingly attractive. Optical biosensors, based on the interaction between the evanescent wave and particles, are being developed as sensitive and label-free quantification measurement systems [5,6,8]. The evanescent field is defined as the radiation close to the interface of two refractive media of different indices, and the radiation intensity decreases exponentially with the distance from the boundary. The biosensing mechanism is based on the sensitive interaction between the field and the attached biomolecules that have different refractive indices in the range of the evanescence field height, or penetration depth [9,10]. The refractive index can be used to determine an object, and it has a correlation with the concentration of the particles inside a substance. The detection of a very small amount of changed refractive index has been a target in the development of label-free biosensors using evanescent-wave-based optical sensing [11–15]. Surface plasmon resonance (SPR) has attracted much interest as a label-free biomolecular detection method with a high sensitivity in real time based on measuring the angle, wavelength, and intensity corresponding to a change in the refractive index of the medium near the metal surface [11,15–18]. However, the SPR-based biosensors remain a challenge in system miniaturization and high-cost settings [19].

Meanwhile, silicon photonics has increasingly attracted attention due to the combination of biosensing together with the development of the fabrication process for photonic-integrated circuits. Several promising candidates, such as Mach–Zehnder interferometers (MZIs), Bragg gratings, and microring resonators (MRRs), can quantitatively recognize biomolecules that have different refractive indices [2,3,12,20,21]. Based on the structure having a high contrast between the refractive index of the core layer of silicon and that of a cladding layer of silicon dioxide, silicon nitride, and the media surrounding the sensor surface, silicon photonic sensors impress with their simple structures, reusable devices, and multiplexed biomolecular detectors [1,22]. Among these methods, a ring-resonator-based method has been adopted as a simple, multiplexed-array-compatible, and sensitive biosensor that is capable of detecting a precise resonance wavelength shift as a product of the interaction between attached biomolecules and the electric field. A ring resonator used with microfluidic channels for detecting a small-volume solution and its integration with a CMOS electronic device as an all-in-one chip is highly attractive [5,22–24]. In addition, optical biosensors using ring resonators have increasingly become more reliable owing to the modification of receptor activities or functionalization of the biorecognition materials on the sensing surface, for example by using aptamer molecules to increase the binding events [8]. By these means, a high-sensitivity and -selectivity optical biosensor can be obtained by two methods: amplifying the output signals by increasing the interaction intensity and density using a modified-structural resonator within the penetration depth, and modifying the sensing surface using optimized surface materials [4,25,26]. There are several impactful applications for cancer diagnoses of multiplexing nanophotonic biosensors, the detection of aptamer thin layers, and the monitoring of the hemoglobin concentration in the human body [13,27–29]. A silicon photonics waveguide-array sensor with polydimethylsiloxane polymer cladding has been reported for applications in the detection of volatile organic compounds [30]. This research paper focuses on the improvement of the sensitivity by the method of amplifying the output signals.

Si microring resonator (MRR) biosensors are a potential candidate to achieve a compact-size, simple-fabrication, and high-sensitivity detection method [8]. The evanescence field, which is formed from the high contrast between the Si core layer of the waveguides and the cladding layers, interacts with different attached biomolecules, which pursue different refractive indices, to elevate the change in resonance wavelengths caused by the microring ring cavity [5,23]. MRRs only detect attached biomolecules within the evanescent field of several nanometers or the side wall of the cavity. This property can produce a high-sensitivity optical biosensor as most of the target molecules bind to the sensing surface. Enlarging the sensing surface or increasing the interaction between the evanescent field and the analyte is a method for amplifying the optical output signals to achieve an ultra-sensitivity [26,31]. However, an enlarged microring resonator is not a suitable approach for sensing a small-volume solution within a microfluidic channel because the enlarged cavity requires a large platform of a hundred square millimeters. Utilizing a narrow microchannel for small-volume detection becomes a chip miniaturization issue. Moreover, a large portion of biomolecules within the height of the evanescent field, or the penetration depth, cannot be involved in the sensing activities. The conventional MRR structure which has a ring resonator with a radius of several hundred micrometers might achieve an impact sensitivity; however, only a small part of the biomolecules in the analyte solution can immobilize on the sensor surface and interact with the evanescent field, while the other biomolecules will be free and will pass over the surface.

In this study, we demonstrate a design of a silicon S-shaped double-spiral resonator (DSR) with a compact platform to increase the refractive index sensitivity for detecting a small amount of the analytes in a narrow microchannel. The cavity comprises two spiral microresonators on a silicon-on-insulator (SOI) substrate, which are connected to each other by an S-shaped channel. Two impact effects of the DSR cavity are demonstrated. Firstly, the DSR structure allows the cavity length to be extended in a compact size. The extended-length cavity increases the attached biomolecules onto the area of contact interface of the resonator to generate more interaction between the electric field and biomolecules within the penetration depth, thus achieving impactful changes in the output signal's quality factor and bulk sensitivity [31,32]. Secondly, the structure of the DSR cavity supports an easy fabrication and the alignment with a narrow microfluidic channel formed over the device. To achieve the goal of measuring a small-volume solution using a microchannel, the ideal microchannel width is approximately between 100 and 200 μm. Therefore, minimizing the width of the spiral resonator is a method to fit the cavity in the narrow microchannel easily by the naked eye. Based on these effects, the research on the optical characteristics and the sensitivity of the DSR cavity is meaningful in the development of evanescence-based biosensors.

To modify the DSR structure, simulations using the finite-difference time-domain (FDTD) method were mainly utilized to analyze the resonance characteristics for optimizing the structural parameters, such as the waveguide width, the spacing distance between turns, the curvature radius, and the gap with a bus waveguide. DSR structures were compared regarding their cavity lengths, optical characteristics, and sensitivities in simulation. Several DSR cavities were experimentally fabricated using microfabrication processes based on the simulated structures. The optical transmission spectra, measured using a custom measurement system, were collected to analyze the resonance characteristics. Detection in IPA with different concentrations from 0.1% to 0.4% was demonstrated with 1 μL solutions within the microfluidic channel to investigate the sensitivity. Finally, we examined the resonance wavelength change in response to the presence of acetone vapor in a closed bottle. In this research, IPA and acetone were selected to prove the biosensing performance of a small amount of the analytes, particularly the volatile organic compounds (VOCs). VOCs exist in the human body, and out-of-range concentrations can cause harmful effects to human health. VOCs are also known as biomarkers to discover diseases at an early stage [33,34]. Monitoring a low concentration of VOCs in air and solution with a simple real-time measurement system has been well studied [35–37]. IPA and acetone are popular VOCs with high vapor pressure and evaporate quickly. A simple experimental setting using the DSR-based device was used to examine the sensitivity for detecting small-volume analytes in gas and liquid environments.

2. Sensing Principle

The S-shaped DSR has two single Archimedean spirals (left and right) and an S-shaped channel. DSR biosensors detect attached biomolecules with different values of refractive index (RI), which are based on the interactions between the evanescent field formed over the waveguide of the resonator and the analytes. The input light from a tunable laser is coupled to a bus waveguide through a taper-edged coupler. A part of the light is subsequently coupled to the DSR cavity when the light wavelengths satisfy the resonance condition of

$$\lambda_{res} = \frac{n_{eff}L}{m} \quad (1)$$

where m is an integer, λ_{res} is the resonance wavelength, n_{eff} is the effective refractive index, and L is the length of the DSR cavity [38]. When the resonance occurs, the resonance wavelengths can be detected as dip wavelengths in the transmission spectrum of the output light. The effective refractive index n_{eff} is determined by the refractive indices of the core material and the cladding layer. At the time that the biomolecules attach to the surface of the DSR cavity, the effective index of the cavity changes owing to the interactions between the electrical field and the biomolecules, introducing a shift in the resonance wavelength $\Delta\lambda_{res}$:

$$\Delta\lambda_{res} = \Delta n_{eff} \times \frac{\lambda_{res}}{n_g} \quad (2)$$

The effective group index of the cavity waveguide (n_g) exhibits a dispersion with the wavelength [2]. The change in refractive index (Δn_{eff}) may be caused by conditions of the surrounding environment such as temperature and analyte concentration. The bulk sensitivity S, one of the parameters for evaluating the biosensing performance, is defined as the change in the resonance wavelength $\Delta\lambda_{res}$ for a unit of the changed refractive index Δn of the cladding layer which is altered by the attached particles [38]:

$$S\left(\frac{nm}{RIU}\right) = \frac{\Delta\lambda_{res}}{\Delta n} \quad (3)$$

To increase the sensitivity, the method of increasing the refractive index change amount has been developed. The utilization of a slot waveguide enables a change in refractive index due to a large light intensity by the overlapping of the leaked light from the slot and the attached analyte [4,26,38].

In this research, the mechanism of sensitivity enhancement by DSR with a long cavity can be explained by the enhanced area of the leak electric field intensity of the DSR waveguide. The DSR structures allow a larger area of the contact interfaces for the analyte to interact with electromagnetic field including the top interface and the side walls of the cavity. By extending the cavity length, it is able to use more contact surface for biomolecules to attach to the DSR cavities; hence, the change in resonance wavelengths increased to achieve an increased sensitivity.

Another important parameter improved by the DSR that impacts bio-sensitivity is the effect of the Q factor increment. The Q factor is an indicator for the peak sharpness required to obtain a precise resonance wavelength shift. The Q factor can be calculated using a function of the cavity perimeter L [39,40]:

$$Q = \frac{2\pi n_{eff}}{\lambda_{res}} L. \quad (4)$$

Here, the light loss after a trip through the cavity is neglected. The Q factor shows a relation with the cavity length. The Q factor also depends on the structure of the cavity. In the biosensing application, cavity designs for Q factors of about 10^4 are preferable for measurement ease and detection sensitivity. If the Q factor is too high, the wavelength tuning becomes difficult, and if the Q factor is too low, the shortened photon lifetime can reduce the analyte sensitivity.

Furthermore, the compact size of DSR designs takes advantage of biosensing a small amount of the analyte. In other words, the proposed DSR aims to design and fabricate a cavity structure for high-sensitivity and small-volume detection and easy fabrication. Firstly, the DSR cavity increases the bulk sensitivity and the Q factor due to the enhanced contact interface area between biomolecules and the evanescent field for a more significant resonance wavelength shift. Because of the extended cavity length, the area of the top and the side wall of the cavity waveguide, where the analyte interacts with the evanescent field to contribute to the resonance wavelength changes, is increased. Secondly, the DSR is designed in a micro-sized footprint to minimize the cavity size, to minimize the width of the cavity, thus reducing the amount of solution in a small fluidic channel for µL-level detection and helping the alignment procedure of the resonator into the microchannel to be easier.

In this research, the DSR cavities with about 1 mm cavity length combined with a thin-width polydimethylsiloxane (PDMS) microfluidic channel for detecting 1 µL of solution over the biosensor chip. The minimum microchannel sizes to contain 1 μL volume inside the microchannel are a length of 20 mm, a height of 200 μm, and a width of about 200 μm to 250 μm. The microchannel length was decided based on the chip size which is easy for handling and the size of the inlet and the outlet. The microchannel height was decided based on the fabrication mold of the microchannel. The cavity width is thus limited to be smaller than 50 μm for easing the alignment of the center PDMS microfluidic channel and the resonator. Hence, a cavity size of 40 μm × 90 μm was set to optimize the DSR design.

3. Simulation Analysis

The DSR cavity was constructed by combining two single Archimedean spirals (left and right) and an S-shaped channel, as shown in Figure 1a. The spirals were designed to have the same spacing distance (D_S) between successive turns to make good use of the cavity size. The right spiral was the vertically and horizontally flipped shape of the left spiral. An S-shaped channel waveguide linked two single spirals, for which θ_S is the angle between the vertical axis and the channel. Figure 1b shows a cross-section of the refractive index distribution of the Si-based DSR waveguide (refractive index, n_{Si}: 3.45) built on a SiO_2 substrate (n_{SiO2}: 1.45), where air functions as the cladding layer (n_{air}: 1). Different structures of the resonators were compared using an FDTD simulation (FullWAVE™ simulation tool, Synopsys) in the characteristic optimization of optical spectra to balance a high Q factor and the cavity length. DSR devices with optimized parameters using the simulation results were fabricated and their optical characteristics were analyzed. To save time in the simulation, a 2D FDTD analysis was used, in which the effective index of the waveguide was calculated using the finite element method (FEM) in a 3D waveguide structure with a width of 450 nm and a height of 240 nm [20].

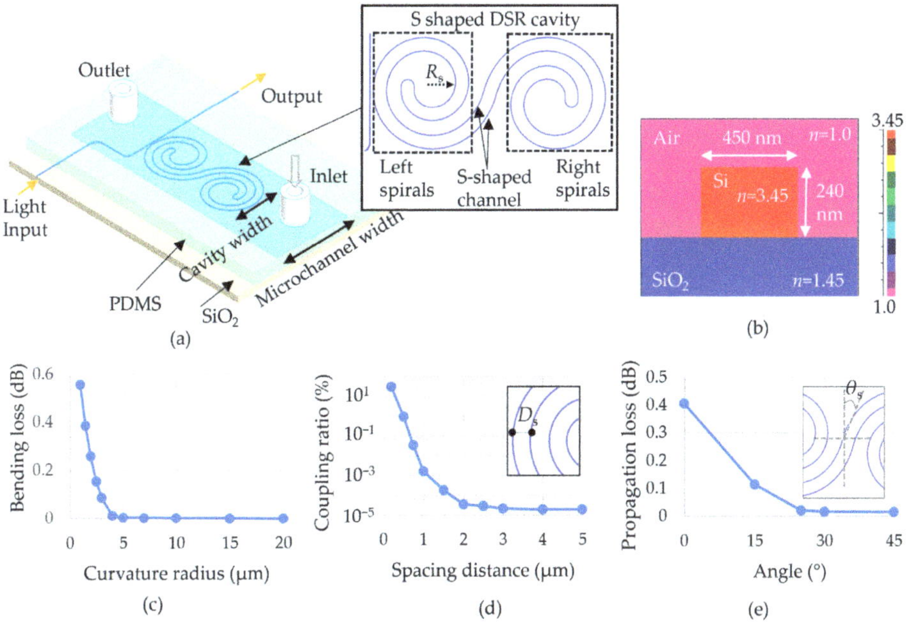

Figure 1. (**a**) Schematic of the double-spiral resonator (DSR) biosensor with a compact cavity size. The DSR is able to detect a small-volume analyte inside the thin-width microchannel. (**b**) Refractive index distribution of a waveguide cross-section. (**c**) Bending loss caused by the curvature radius. (**d**) Coupling ratio when the spacing distance (D_s) changes. D_s is defined as the inset figure. (**e**) Power loss caused by the bending angles at S-shaped channel (θ_s) as defined in the inset figure.

There are three primary factors for designing a small-sized DSR, including the minimum radius R_s of the curvature, the spacing distance D_s, and the shape of an S-shaped channel, to achieve a balance of low propagation loss and a high ratio of sensing surface to cavity size. During the simulation, the radiation loss and material scattering loss were neglected. Firstly, R_s was decided based on the bending loss caused by a bent waveguide's curvature. Figure 1c shows a decreased bending loss depending on the curvature radius with a radius from 1 μm to 20 μm. It was noticed that an R_s with a value larger than 4 μm achieved minimal bending loss, which might provide a high Q factor; therefore, values of $R_s \geq 5$ μm were used for designing the DSR cavity. Secondly, to examine D_s, we simulated the dependence of the coupling ratio, which is the percentage of the light power of one waveguide to another waveguide at a spacing distance D_s. Figure 1d indicates the coupling ratio from one waveguide to another waveguide in a spiral when the spacing distance changed from 0.2 μm to 5 μm. $D_s \geq 3$ was selected to eliminate the light coupling to the other waveguides. Finally, optimizing the S-shaped channel for a low propagation loss was investigated by adjusting the angle θ_S. Figure 1e indicates the propagation loss depending on the angle θ_S from 0 to 45°. The loss proportion was lower than 0.02 dB when $\theta_S \geq$ 25°. Propagation loss occurs when mode-mismatched loss is caused by a bent channel for linking two spirals. Hence, an angle of 25° was utilized to achieve a low loss and minimize the cavity size to about 41 μm × 88 μm.

The transmission spectra for different DSR structures were studied using the FDTD method. The light was coupled to the DSR cavity via a bus waveguide placed at a gap distance apart from the cavity. Here, we designed a stripe-based bus waveguide with a width of 450 nm and a height of 240 nm, as large as the size of the DSR waveguide. The effective index was calculated using a finite-element-method (FEM) simulation. Figure 2a shows the electric field distribution (E_x) of a DSR waveguide for the TE mode. A part of the leaked electric field over the waveguide at the top and side walls of the waveguide contributes to increasing attached biomolecules. A continuous wave is excited from the input of the bus waveguide, and the power intensity was monitored at the output. An example of the simulated transmission spectrum, for which the DSR had a cavity length of 1.2 mm, is shown in Figure 2b. The Q factor was calculated using a function of $\lambda_{res}/\lambda_{res\_FWHM}$, in which λ_{res} is the resonance wavelength or the wavelength at the dip of the resonance, and λ_{res_FWHM} is the wavelength difference of the full width at half the maximum intensity. The free spectral range (FSR) is the distance between two dip wavelengths. Figure 2c shows the Q factors, FSR values, and cavity lengths for different spacing gaps (4, 5, and 6 μm) in the same cavity footprint of 41 μm × 88 μm. The DSR structure with a 5 μm spacing was selected for the fabrication because it could achieve a high Q factor and had a large sensing surface. It was noticed that, in principle, a larger cavity length might result in a higher Q factor; however, in a limited cavity size, the bending loss caused by the curvature radius could increase the bending loss, driving decreased Q factors. This is the reason that the DSR cavity with a 4 μm spacing distance had a smaller Q factor than that of the DSR cavity with a 5 μm spacing distance, even though it had a long cavity. Another reason that we selected the 5 μm spacing distance is that the FSR was long enough for analyzing the data. The FSR is reduced more when the cavity length is longer, which may cause some unrecognition of the changed dip wavelengths for minimal refractive index changes. The DSR with a 5 μm spacing distance and a cavity length of 1.2 mm balanced a large sensing area with a high Q factor.

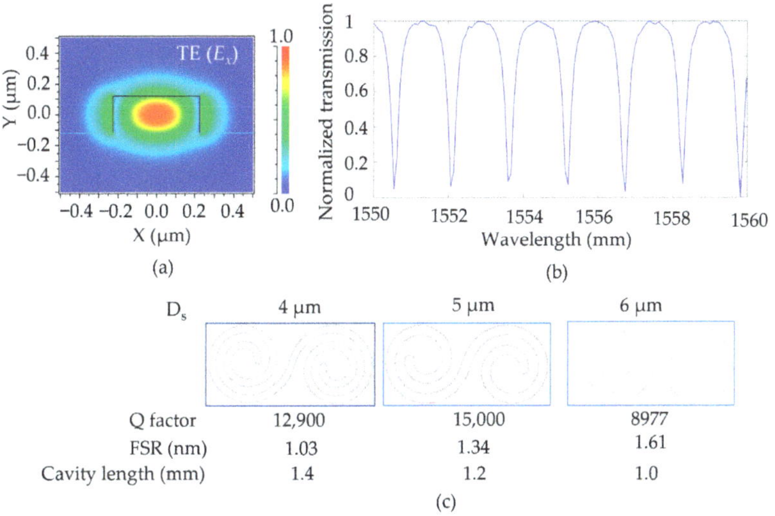

Figure 2. (**a**) Cross-section of the E field of a DSR waveguide with a width of 450 nm and a height of 240 nm in the TE (E_X). (**b**) Simulated transmission spectrum for the DSR cavity with a spacing distance of 5 μm. (**c**) The images, Q factors, FSRs, and cavity lengths of different DSR designs with different spacing distances.

Finally, we designed the C-turn curve at the center of the spiral by calculating the Q factors of different types, as shown in Figure 3a. Here, we compared two types of the C-turns: the A-type, which is an arc with a diameter equal to the spacing distance (5 μm), and the B-type, which includes two arcs of a circle with a radius of 5 μm that join each other at an angle θ_c. Figure 3b shows the column chart of the Q factors (the left axis) and cavity lengths (the right axis) for different types of resonators: MRR cavities with a radius of 20 μm and 189 μm, an A-type C-turn DSR cavity, and B-type C-turn DSR cavities with θ_c = 63°, 70°, or 72°. The 189 μm MRR has the same cavity length with DSR cavities. The 20 μm radius MRR has the same cavity width with the DSR cavities. The DSR designs allow the 10-time-enlarged contact interface in comparison with 20 μm radius MRR for the same cavity width. The DSR platform was reduced 680 times in comparison with 189 μm radius MRR, which has the same cavity length. The DSR structures (light green color) showed Q factors of 1.4–1.9 × 10^5, which is two times higher than that of the MRR cavity with a radius of 20 μm, which had the same cavity width as the DSR structure. This is because the DSR cavity had a longer cavity length. We also found that the Q factors of the DSR structures were greater than that of the MRR cavity with a radius of 189 μm, which had the same cavity length as the DSR. This might be explained by the coupling ratio becoming larger when the light from the bus waveguide is coupled with a larger-radius ring so that the light power is conserved inside the MRR cavity, resulting in a smaller total Q factor. In addition, the B-type DSR expressed a higher Q factor than the A-type DSR. This increased Q factor was the result of less bending loss at the curvature in the B-type C-turn. We also recognized that the angle θ_c affected the Q factor values. The Q factor difference was because different bending curvatures can cause a mismatched mode of the propagating light. Eventually, DSR structures with a spacing gap of 5 μm and an angle θ_c of 72° were chosen for the fabrication.

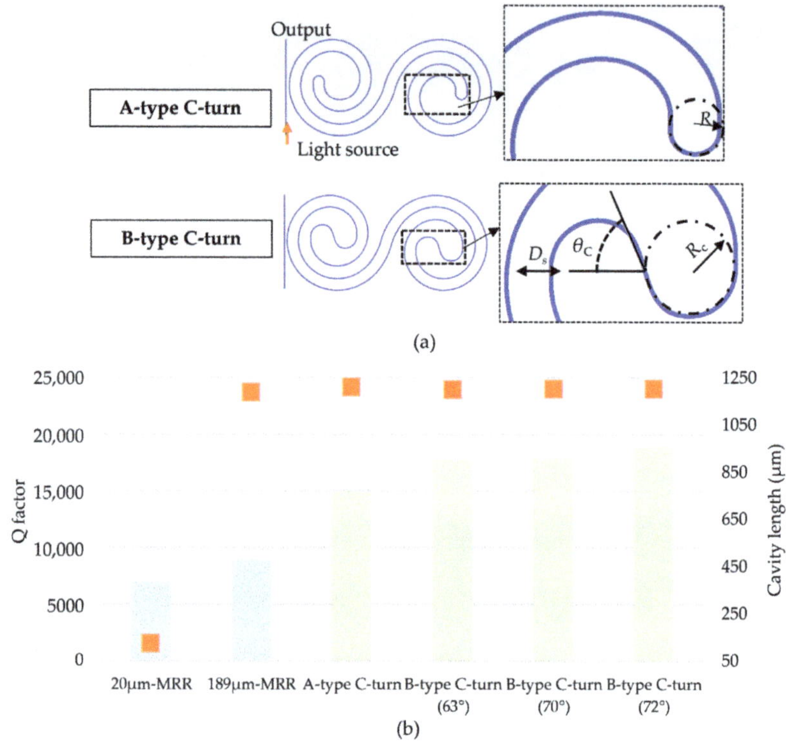

Figure 3. (**a**) Images of two different types of C-turns. The A-type C-turn had an arc diameter (R_c) as large as the spacing distance D_s. The B-type C-turn had two arcs with R_c = 5 μm linked to each other at an angle of θ_C. (**b**) The column chart of Q factors (the left axis) for different cavities of conventional microring resonators (light blue) with R = 20 μm and 189 μm, and DSR cavities (light green) with different C-turn types. The orange square markers (the right axis) are the cavity lengths of the plotted cavities.

Bulk sensitivity was simulated from the changed resonance wavelength when the attached surface or the top cladding surface had a different refractive index. The effective indices of the waveguide were calculated according to the changed refractive index of the top cladding surface. Based on the simulation results, a bulk sensitivity of 1500 nm/RIU was achieved for the DSR structure, while the bulk sensitivity of the 20 μm ring resonator was 150 nm/RIU. These results were compared with the experimental results. Furthermore, a surface sensitivity of 4100 pm/nm, calculated from the resonance wavelength shift for a unit thickness of the attached layer, was calculated. In this simulation, a model of an attached protein layer covering the DSR cavity was utilized [41]. The attached layer's thickness needed to be less than the height of penetration depth. This simulated surface sensitivity was one order of magnitude higher than that of the 20 μm radius MRR. These results highlight the effect of enhancing the area of contact interface on the sensitivity by using DSR structures and the Q factors.

4. Experimental Methods

A 20 μm radius MRR, DSR cavities including the 1 mm length DSR, the 1.2 μm length DSR with an A-type C-turn, and the 1.2 μm length DSR with a B-type C-turn were fabricated on an SOI substrate (20 mm × 20 mm) with a 240 nm thick silicon layer on a 2.5 μm thick buried oxide (BOX) layer. The parallel waveguides at 250 μm apart were designed using

Autodesk AutoCAD 2022 software. Taper-edged couplers were designed for the input and output facets, with the taper size reduced from 3 μm to 450 nm over a length of 30 μm. The length of the waveguides was 8 mm. The fabrication process included four steps of patterning by electron beam lithography (EBL), etching the silicon layer to transfer the waveguide patterns, cleaving the chip, and sticking the microfluidic channel. Firstly, the waveguides were patterned by EBL (ELS-G125S Elionix, Tokyo, Japan) using the 300 nm thick positive resist layer of ZEP520A (ZEON, Tokyo, Japan). Secondly, the pattern was transferred to the silicon layer by a dry etching process using deep reactive-ion etching (deep RIE) (MUC-21 ASE-SRE, Sumitomo Precision Products, Amagasaki, Japan) in CF_4 and SF_6 gas [42]. Thirdly, the device was cleaved at both sides of the input and output facet to achieve an 8 mm × 20 mm chip size. Herewith, the light from a laser source could be transmitted to the waveguide by placing a lensed fiber close to the 3 μm width facet of the taper-edged coupler. The resist layer could be removed in O_2 plasma 100 W for 3 min (PDC210, Yamato Scientific Co., Ltd., Tokyo, Japan). Finally, a 20 mm thick PDMS film with a microfluidic channel was adhered to the sensor plate by heating the device with the PDMS film at 70° for 30 min. The microchannel had a width of 200 μm and a length of 16 mm. The PDMS film was placed at the position which overlapped the cavities. The alignment was achieved without observing under a microscope because the cavity width size was much thinner than the channel width. It was able to inject 1 μL volume solution into the microfluidic channel to contact with the resonator.

The sensing performances of the DSR cavities including the Q factors in air and water, the bulk sensitivities, and the intrinsic limits of detection were experimentally evaluated and compared with the 20 μm radius MRR. To clarify the sensing performances of the DSR cavity for different types of analytes using the microchannel-on-chip, experiments with acetone gas with different concentrations were demonstrated. The optical transmission spectra in various conditions were studied to quantify the sensing performances. The Q factors were calculated by dividing the resonance wavelength from the optical transmission spectrum by its full width at half maximum. The bulk sensitivity was estimated from the linear relationship between the refractive index of solutions with different concentrations and the resonance wavelength shifts. IPA solution (99.7%, Kanto Chemical Co., Tokyo, Japan) was used to measure bulk sensitivities. The solutions with different concentrations from 0.1% to 0.4% (v/v) were made by diluting IPA in deionized water. The intrinsic limit of detection defines a minimum refractive index change for shifting the resonance wavelength by a linewidth.

Figure 4 shows the measurement setup for studying the optical transmission spectra at different conditions. The light (λ: 1.52–1.58 μm) was transmitted from a tunable laser (TSL-200, Santec Holdings Corporation, Komaki, Aichi, Japan). After the light passed through a 3-paddle polarization controller (FPC564, Thorlabs, Inc., Newton, NJ, USA), it was coupled with the taper coupler of the device via a lensed fiber. The output powers for various input wavelengths were collected by a power meter (PM101, Thorlabs, Inc., Newton, NJ, USA) measured at the lensed output fiber. The input wavelength and the output power were plotted with a LabVIEW program. The optical performance stability was evaluated by measuring resonance wavelength shifts for the same cavity over one hour at a room temperature of 25 °C. For measuring the optical characteristics in liquid, the 1 μL volume solutions were injected into the inlet using a micropipette. The measurement for the same solution was repeated every 2 min for 10 min. For measuring the acetone gas, a clean cloth permeated 50 μL of acetone (67-64-1, Kanto Chemical Co., Inc., Tokyo, Japan) was placed at the bottom of a closed 500 mL perfluoxy copolymer (PFA) bottle. A 10 cm long tube connected the PFA bottle to the chip inlet.

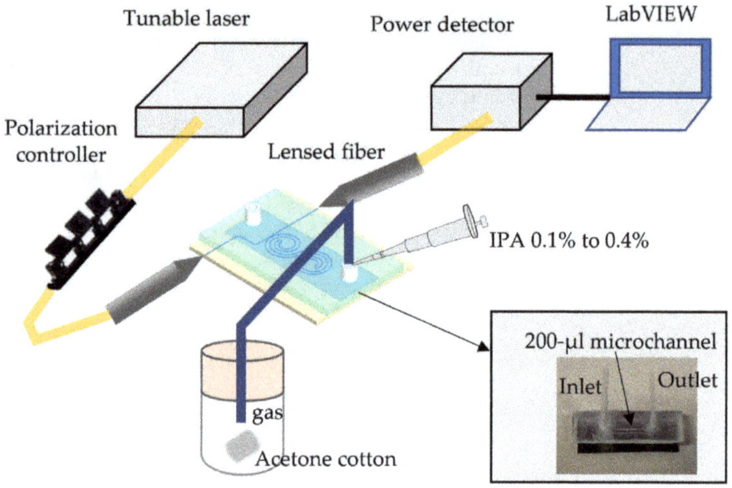

Figure 4. Schematic diagram of custom measurement setup. The PDMS microfluid channel adhered to the Si chip to make the microchannel. A micropipette was used to pump 1 μL of solution into the microchannel for detecting solution bulk sensitivity. In the VOC gas detection, acetone-infused cotton was placed inside a closed PFA bottle connected to the microfluidic inlet by a silicone tube, and the outlet was blocked.

5. Experimental Results and Discussion

Figure 5a shows the different types of cavities of the 20 μm radius MRR, the 1 mm length DSR, the 1.2 μm length DSR with an A-type C-turn, and the 1.2 μm length DSR with a B-type C-turn from left to right. The 20 μm radius MRR had a cavity size of 41 μm × 41 μm. The DSR cavity width was 41.6 μm, and the cavity length was 88.4 μm. The DSR footprint was reduced by 680 times compared to that of the 189 μm MRR, which had the same cavity length. The DSR cavities showed the advantages of reduced exposure-time in comparison with the MRR with the same cavity length due to the small scanning area of the EBL beam in a small platform.

The prototyped waveguide width was 456 ± 6 nm, while the designed width was 450 nm. The gap between the bus waveguide and the resonator was 290 ± 5 nm while the designed gap width was 300 nm. The waveguide size errors might have mostly occurred due to the EBL control of the resist fabrication. The reproducibility of the lithography process depends on the stability of EBL system, operator errors, and environmental conditions, such as the experimental temperature and the humidity, which affect the thickness and the affinity of the resist layer. The fabrication reproducibility using the EBL process can achieve the several-hundred-width waveguide with the error of sub 10 nm. However, the system stability might worsen the patterning process. Because the control of the waveguide width is important in the propagating mode loss, EBL process conditions for system stability such as the beam current and the aperture needed to be modified with a dummy sample before the main chip process. The optical power loss can also result from the roughness of the waveguide side walls and the edge coupler facet caused by the etching process. The deep RIE process with a slow etching rate of 18 nm/min was performed aiming to smooth the side walls. In addition, hydrogen annealing was reported to reduce the side wall roughness [43].

Figure 5b shows the SEM images of two different types of C-turns. The A-type C-turn had an arc diameter equal to the spacing distance, and the arc radius was 2.6 μm. The B-type C-turn had two arcs with a radius of 5.1 μm and a θ_C of 72°. The measured spacing distance was 5 μm, which is within the range required for the simulated spacing distance to prevent the coupling with the next curve. The taper-edged coupler had a length of 30 μm, changing from the waveguide with the size of 456 nm to the inserted facet with a size of 3 μm, as shown in Figure 5c.

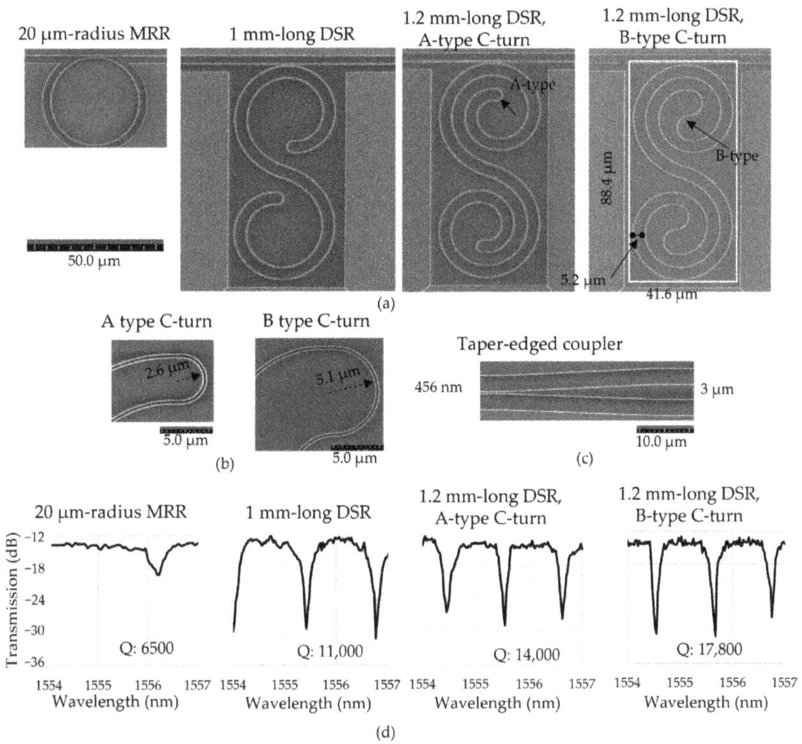

Figure 5. (**a**) SEM images of the 20 μm radius MRR, 1 mm length DSR, 1.2 mm length DSR (A-type C-turn), and 1.2 mm length DSR (B-type C-turn), starting from the left. The DSR cavity size was 41.6 μm × 88.4 μm. (**b**) SEM images of two types of C-turns. (**c**) SEM images of taper-edged coupler. (**d**) Transmission spectra in DI water for fabricated cavities with different cavity lengths.

The transmission spectra of the four types of cavities in air and DI water at a room temperature of 25 °C were collected. The resonance wavelength shift was about ±50 pm. The extinction ratio of the DSRs was 15 dB, while the extinction ratio of the MRR was 5 dB, measured in air. The measured propagation loss from straight waveguides was about −3.5 dB/cm, mainly from the scattering loss due to the roughness of the waveguide. From these resonance properties, the Q factors of the different designs in air and DI water were calculated. The Q factors in the air of the cavities of the 20 μm radius MRR, the 1 mm long DSR, the 1.2 mm long DSR with an A-type C-turn, and the 1.2 μm long DSR with a B-type C-turn were 0.73×10^4, 1.29×10^4, 1.55×10^4, and 1.94×10^4, respectively. Later, the transmission spectra in DI water were monitored using 1 μL of DI water inside the microchannel. Figure 5d shows the transmission spectra taken in DI water for 20 μm radius MRR, the 1 mm long DSR, the 1.2-mm long DSR (A-type C-turn), and the 1.2 mm long DSR (B-type C-turn), starting from the left. The Q factors in the DI water were reduced by 10% compared with those in the air due to the absorption loss in water. In DI water, the 1.2 mm long DSR with a B-type C-turn obtained the highest Q factor value of 1.78×10^4. Consequently, with the same cavity width of 40 μm, the DSRs had Q factors that were two orders of magnitude higher than that of the 20 μm radius MRR in the air and water environments. Furthermore, the Q factor of the 1 mm long DSR cavity was higher than that of the MRR, but lower than that of the 1.2 mm long DSR. These results indicate that DSRs with a long cavity length can experimentally achieve a higher Q factor based on their cavity length. The shape at the C-turn is also a factor leading to an increased Q factor because it can affect the bending loss. The FSRs of the cavities of the 20 μm radius MRR, the 1 mm long DSR, the 1.2 mm long DSR with A-type C-turn, and the 1.2 mm long DSR with a B-type C-turn were 9.6 nm, 1.4 nm, 1.2 nm, and 1.2 nm, respectively. These results are close to the simulated results.

The 1.2 mm long DSR with a B-type C-turn was used to measure the bulk sensitivity in comparison with 20 μm radius MRR. Figure 6a shows the linear relationship between the refractive index (concentrations of 0.1% to 0.4%) of IPA solutions and the resonance wavelength shift of the 1.2 mm long DSR with a B-type C-turn. The refractive indices of IPA solutions were estimated from the linear relationship between the solution concentration and the refractive index [44]. We achieved a bulk sensitivity for the DSR of 1416 nm/RIU. The bulk sensitivity of the 20 μm radius MRR was 120 nm/RIU, as calculated by measuring the IPA solutions with concentrations of 1% to 3%. The selected IPA concentrations were different in the DSR cavity and MRR cavity to ensure the resonance wavelength shifted to be smaller than FSR. The bulk sensitivities of the 1.2 mm long DSR and 20 μm radius MRR in simulation were 1500 nm/RIU and 150 nm/RIU, respectively, as discussed in Section 3. Considering the reasons of the propagating loss caused by the fabrication process, the measurement system errors, and the surface hydrophilic, the experimental bulk sensitivity is very close to the simulation data. This result indicates the bulk sensitivity of the DSR was one order of magnitude larger than that of the 20 μm MRR in the simulation and experimental results as the result of the extended contact interface in the surface and the side wall of the DSR cavity. This result also indicates the highly exact simulation which can reduce much cost and time in fabrication. A standard deviation of the peak shift of ±20 pm was obtained. The peak shift of the same 1 μL IPA solution (concentration of 1%) was studied about ±50 pm, which was the result of the instability in the room temperature, the power and polarization matching of the input light from the tunable laser. Figure 6b shows the transmission spectrum of the 1 μL IPA solution at the times of 0 h, 1 h, 2 h, and 3 h. Therefore, the deviation of the peak shift in the bulk sensitivity measurement was in the range of the measurement condition error. In addition, it was able to keep the small-volume solution inside the microchannel for 3 h; even IPA is a volatile solution. Therefore, the effect of a longer cavity length increasing the bulk sensitivity was obvious in both the experimental results and the simulation, and using a compact cavity footprint combing with the microchannel could reduce the measurement solution.

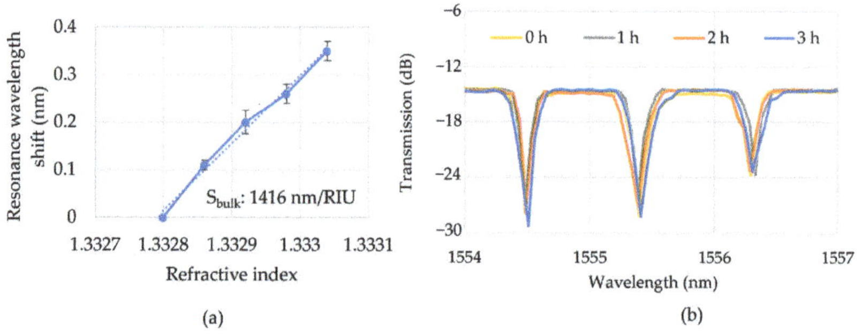

Figure 6. (**a**) Resonant wavelength shift for 1.2 mm length DSR cavity (B-type C-turn) over different refractive indices by using IPA solutions with concentrations of 0.1%, 0.2%, 0.3%, and 0.4%. (**b**) Measured transmission spectrum of IPA solution (0.1%) using DSR cavity at 0 h, 1 h, 2 h, and 3 h.

The 1.2 mm long DSR with a B-type C-turn was also used to detect acetone gas. Figure 7a shows resonance wavelength shifts when the acetone concentrations in the bottle were 100 ppm, 200 ppm, 300 ppm, and 400 ppm. The concentration sensitivity of 120 pm/100 ppm was calculated from the linear relationship between the resonance wavelength shift and the acetone concentration. A standard deviation of ±40 pm was obtained, which might result from the measurement environment. From this result, the detection limit of acetone concentration using the DSR cavity and microchannel fluidic was 100 ppm. Figure 7b shows the results of monitoring the resonance wavelength of acetone gas with a concentration of 100 ppm using a DSR cavity in a sequence of air, acetone gas, and air in real-time detection. The resonance wavelength shifted to 110 pm after the acetone gas appeared inside the bottle. A fast response in a resonance wavelength shift of several seconds was observed. After about 10 min of maintaining the same resonance wavelength shift, the resonance wavelength shift decreased gradually to the shift in the air. The wavelength shifts during the monitoring in the air and acetone conditions were in the range of the temperature shift. It is obvious that the acetone gas particles approached the cavity and interacted with the electric field over the cavity. However, it also vaporized through the PDSM film and attached to the side wall of the channel. Different materials, such as thermoplastic elastomers and soft thermoplastic elastomers, could be used instead of PDMS for the microchannel to obtain slow evaporation [45]. The DSR structure improved the sensitivity and quality factor for small-volume detection. The investigation using IPA solutions and acetone vapor indicated the variety of biosensing applications for measuring in different environments that are possible for label-free biosensing.

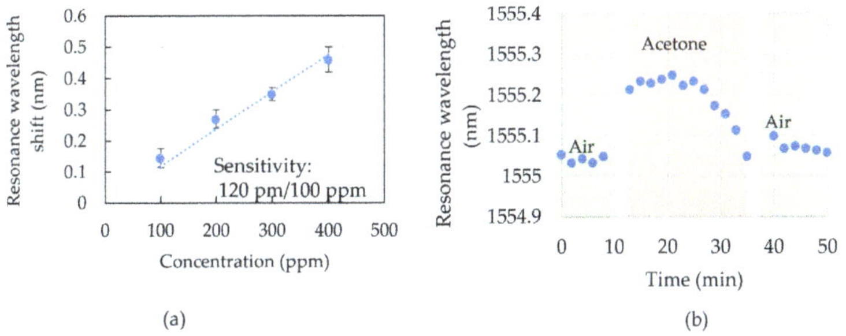

Figure 7. (**a**) Resonant wavelength shift for 1.2 mm length DSR cavity (B-type C-turn) over different acetone concentrations of 100 ppm, 200 pp, 300 ppm, and 400 ppm. (**b**) Resonant wavelength shifts as the environment inside the bottle changed in the sequence of air, acetone, and air. Acetone gas had a concentration of 100 ppm.

6. Challenges and Future Prospects

The S-shaped double-spiral resonator could reduce the analyte solution and enhance the refractive index sensitivity and the Q factor, which is the key characteristic evaluating the biosensing capability of evanescence-based biosensors. There were several challenges in the prototyped device fabrication method and measurement setup discussed. However, these issues might be overcome with the development of microfabrication in lab-on-chip technologies. One of the challenges of the DSR cavity for biosensing is the easy recognition of the wavelength shift due to the FSR length being shortened with the enhanced cavity length. A false reading out of the wavelength shift could occur if the wavelength shift by a large concentration were larger than the FSR. In the experiment, the real-time measurement of the transmission spectrum was studied to prevent the false reading-out. Hence, the DSR cavity length should be considered to optimize the target measurement concentrations. Another challenge of biosensing based on the refractive index changes is the selectivity to distinguish one from multiple attached biomolecules. This difficulty is because the effective refractive index would be defined by the total biomolecules attached to the contact interface to cause the resonance shift. The analyte types should be known as reference data for detection. To increase the selectivity, recent research of a serial microring resonator for detecting multiple analytes simultaneously was reported [46]. Different resonance wavelength MRRs with large FSR were designed based on hybrid plasmonic waveguides to recognize different solutions.

Further work will study the biosensing applications using the proposed DSR sensor. With the advantage of detecting a small amount of analyte inside the microfluidic, DSR can be applied for measuring a label-free detection of a low concentration of a protein analyte binding with target biomarkers by an optimized surface modification process [47,48]. Furthermore, the microchannel-on-sensor was helpful in preventing a small solution from evaporating. Another potential application of the DSR cavity is cell metabolism measurement in real time with high sensitivity because a greater portion of the cell membrane could interact with the evanescent field in a small footprint.

7. Conclusions

We demonstrated a novel structure of the S-shaped double-spiral resonator for an enhanced refractive index sensitivity to detect a small amount of analyte within a narrow-width microchannel. The DSR had a cavity length of about 1 mm and used dense spirals along the flow direction of the microfluidic channel on a compact footprint with a width of 40 μm, which could only contain a single 20 μm radius ring resonator with a cavity length of 125 μm. The DSR cavity with the enlarged cavity length resulted in a tenfold increase of the bulk sensitivity for different-concentration IPA solutions in comparison with the MRR with the same ring diameter of 40 μm due to the increased contact interface between the biomolecules and the waveguide leaky light. The Q factor of the DSR cavity with a cavity length of 1.2 mm, a cavity footprint of 41.6 μm × 88.4 μm, and a C-turn radius of 5.1 μm achieved a Q factor of ~1.9×10^4, more than two times higher than that of the 20 μm radius MRR as the effect of the extended cavity length. A 1 μL solution was able to be kept inside the microfluidic for the measurement in 8 h with the resonance wavelength stability of ±5 pm. Real-time detection of acetone gas using the DSR cavity was achieved with a concentration sensitivity of 180 pm/100 ppm and a concentration detection limit of 100 ppm. The impact increase in the sensitivity, the Q factor, and small-volume detection in a small sensor footprint in real time are the impact advances for the biosensing development of an accurate, high-response, compact, and low-cost detection method of an ideal all-in-one optical biosensor.

Author Contributions: Conceptualization, A.I., H.Y. and K.O.; funding acquisition, A.I.; methodology, A.I.; data analysis, A.I. and M.A.; technical supervision, S.K., K.O. and H.Y.; writing—original draft preparation, A.I.; writing—review and editing, A.I., S.K., K.O. and H.Y. All authors have read and agreed to the published version of the manuscript.

Funding: This work received financial support from the Tohoku University Center for Gender Equality Promotion, Tohoku University, and was partially funded by the Kao Foundation for Arts and Sciences.

Institutional Review Board Statement: Not applicable.

Informed Consent Statement: Not applicable.

Data Availability Statement: The data analyzed during the current study are available from the corresponding author upon reasonable request.

Acknowledgments: The authors would like to acknowledge N. Matsuda for providing measurement systems and all helpful discussions, S. Abe for his technical discussions, and Y. Shang for his technical support. The authors would like to thank the technical supports for device fabrication and evaluation from Micro System Integration Center (μSIC) and Fundamental Technology Center, Research Institute of Electrical Communication, Tohoku University.

Conflicts of Interest: The authors declare no conflict of interest.

Abbreviations

ELISA	Enzyme-linked immunosorbent assay
DSR	Double-spiral microresonator
SPR	Surface plasmon resonance
MZIs	Mach–Zehnder interferometers
CMOS	Complementary metal-oxide semiconductor
SOI	Silicon-on-insulator
MRR	Microring resonator
RI	Refractive index
FDTD	Finite-difference time-domain
IPA	Isopropyl alcohol
PDMS	Polydimethylsiloxane
FEM	Finite element method
FSR	Free spectral range
BOX	Buried oxide
EBL	Electron beam lithography
RIE	Reactive-ion etching
VOCs	Volatile organic compounds

References

1. Washburn, A.L.; Luchansky, M.S.; McClellan, M.S.; Bailey, R.C. Label-free, multiplexed biomolecular analysis using arrays of silicon photonic microring resonators. *Procedia Eng.* **2011**, *25*, 63–66. [CrossRef]
2. Park, M.K.; Yiying, J.Q.; Kee, J.S.; Song, J.; Kao, L.T.H.; Fang, Q.; Guo-Qiang, L.; Netto, V.; Fosse, E.M.L. Silicon micro-ring resonators for label-free aptamer-based biosensing. In Proceedings of the 2012 7th IEEE Conference on Industrial Electronics and Applications (ICIEA), Singapore, 18–20 July 2012; pp. 1599–1602. [CrossRef]
3. Lo, S.M.; Hu, S.; Gaur, G.; Kostoulas, Y.; Weiss, S.M.; Fauchet, P.M. Photonic crystal microring resonator for label-free biosensing. *Opt. Express* **2017**, *25*, 7046–7054. [CrossRef] [PubMed]
4. Luan, E.; Yun, H.; Laplatine, L.; Dattner, Y.; Ratner, D.M.; Cheung, K.C.; Chrostowski, L. Enhanced Sensitivity of Subwavelength Multibox Waveguide Microring Resonator Label-Free Biosensors. *IEEE J. Sel. Top. Quantum Electron.* **2019**, *25*, 1–11. [CrossRef]
5. Rho, D.; Breaux, C.; Kim, S. Label-Free Optical Resonator-Based Biosensors. *Sensors* **2020**, *20*, 5901. [CrossRef]
6. Fan, X. Sensitive surface plasmon resonance label-free biosensor on a fiber end-facet. *Light Sci. Appl.* **2022**, *11*, 325. [CrossRef]
7. Hu, R.; Sou, K.; Takeoka, S. A rapid and highly sensitive biomarker detection platform based on a temperature-responsive liposome-linked immunosorbent assay. *Sci. Rep.* **2020**, *10*, 18086. [CrossRef]
8. Long, F.; Zhu, A.; Shi, H. Recent advances in optical biosensors for environmental monitoring and early warning. *Sensors* **2013**, *13*, 13928–13948. [CrossRef]
9. Puumala, L.S.; Grist, S.M.; Wickremasinghe, K.; Al-Qadasi, M.A.; Chowdhury, S.J.; Liu, Y.; Mitchell, M.; Chrostowski, L.; Shekhar, S.; Cheung, K.C. An Optimization Framework for Silicon Photonic Evanescent-Field Biosensors Using Sub-Wavelength Gratings. *Biosensors* **2022**, *12*, 840. [CrossRef]
10. Ouyang, H.; Striemer, C.C.; Fauchet, P.M. Quantitative analysis of the sensitivity of porous silicon optical biosensors. *Appl. Phys. Lett.* **2006**, *88*, 163108. [CrossRef]
11. Tognazzi, A.; Rocco, D.; Gandolfi, M.; Locatelli, A.; Carletti, L.; De Angelis, C. High Quality Factor Silicon Membrane Metasurface for Intensity-Based Refractive Index Sensing. *Optics* **2021**, *2*, 193–199. [CrossRef]
12. Bekmurzayeva, A.; Ashikbayeva, Z.; Myrkhiyeva, Z.; Nugmanova, A.; Shaimerdenova, M.; Ayupova, T.; Tosi, D. Label-free fiber-optic spherical tip biosensor to enable picomolar-level detection of CD44 protein. *Sci. Rep.* **2021**, *11*, 19583. [CrossRef] [PubMed]
13. Zhou, G.; Yan, S.; Chen, L.; Zhang, X.; Shen, L.; Liu, P.; Cui, Y.; Liu, J.; Li, T.; Ren, Y. A Nano Refractive Index Sensing Structure for Monitoring Hemoglobin Concentration in Human Body. *Nanomaterials* **2022**, *12*, 3784. [CrossRef] [PubMed]
14. Lai, H.; Kuo, T.-N.; Xu, J.-Y.; Hsu, S.-H.; Hsu, Y.-C. Sensitivity Enhancement of Group Refractive Index Biosensor through Ring-Down Interferograms of Microring Resonator. *Micromachines* **2022**, *13*, 922. [CrossRef] [PubMed]
15. Elsharkawi, A.S.A.; Shaban, H.; Gomaa, L.R.; Du, Y.C. A Highly Sensitive Sensing Technique via Surface Plasmons With Tunable Prism Refractive Index. *IEEE Sens. J.* **2023**, *23*, 9917–9924. [CrossRef]
16. Vahala, K.J. Optical microcavities. *Nature* **2003**, *424*, 839–846. [CrossRef]
17. Gosu, R.; Zaheer, S.M. Principle of Surface Plasmon Resonance (OneStep). In *Methods for Fragments Screening Using Surface Plasmon Resonance*; Zaheer, S.M., Gosu, R., Eds.; Springer: Singapore, 2021; pp. 5–7.
18. Homola, J. Surface Plasmon Resonance Sensors for Detection of Chemical and Biological Species. *Chem. Rev.* **2008**, *108*, 462–493. [CrossRef]
19. Steglich, P.; Lecci, G.; Mai, A. Surface Plasmon Resonance (SPR) Spectroscopy and Photonic Integrated Circuit (PIC) Biosensors: A Comparative Review. *Sensors* **2022**, *22*, 2901. [CrossRef]

20. Troia, B.; De Leonardis, F.; Passaro, V.M.N. Cascaded ring resonator and Mach-Zehnder interferometer with a Sagnac loop for Vernier-effect refractive index sensing. *Sens. Actuators B Chem.* **2017**, *240*, 76–89. [CrossRef]
21. Andrew, W.P.; Shaoqi, F.; Hong, C.; Ting, L.; Hui, C.; Xianshu, L. Microring and microdisk resonator-based devices for on-chip optical interconnects, particle manipulation, and biosensing. In *Laser Resonators and Beam Control XIII*; SPIE LASE: San Francisco, CA, USA, 2011; Volume 7913, p. 791313. [CrossRef]
22. Ramirez, J.C.; Grajales García, D.; Maldonado, J.; Fernández-Gavela, A. Current Trends in Photonic Biosensors: Advances towards Multiplexed Integration. *Chemosensors* **2022**, *10*, 398. [CrossRef]
23. Steglich, P.; Hülsemann, M.; Dietzel, B.; Mai, A. Optical Biosensors Based on Silicon-On-Insulator Ring Resonators: A Review. *Molecules* **2019**, *24*, 519. [CrossRef]
24. Sun, Y.; Fan, X. Optical ring resonators for biochemical and chemical sensing. *Anal. Bioanal. Chem.* **2011**, *399*, 205–211. [CrossRef]
25. Leo Tsui, H.C.; Alsalman, O.; Mao, B.; Alodhayb, A.; Albrithen, H.; Knights, A.P.; Halsall, M.P.; Crowe, I.F. Graphene oxide integrated silicon photonics for detection of vapour phase volatile organic compounds. *Sci. Rep.* **2020**, *10*, 9592. [CrossRef] [PubMed]
26. Yan, H.; Huang, L.; Xu, X.; Chakravarty, S.; Tang, N.; Tian, H.; Chen, R.T. Unique surface sensing property and enhanced sensitivity in microring resonator biosensors based on subwavelength grating waveguides. *Opt. Express* **2016**, *24*, 29724–29733. [CrossRef] [PubMed]
27. Ali, L.; Mohammed, M.U.; Khan, M.; Yousuf, A.H.B.; Chowdhury, M.H. High-Quality Optical Ring Resonator-Based Biosensor for Cancer Detection. *IEEE Sens. J.* **2020**, *20*, 1867–1875. [CrossRef]
28. Chen, Y.; Liu, Y.; Shen, X.; Chang, Z.; Tang, L.; Dong, W.F.; Li, M.; He, J.J. Ultrasensitive Detection of Testosterone Using Microring Resonator with Molecularly Imprinted Polymers. *Sensors* **2015**, *15*, 31558–31565. [CrossRef] [PubMed]
29. Huertas, C.S.; Domínguez-Zotes, S.; Lechuga, L.M. Analysis of alternative splicing events for cancer diagnosis using a multiplexing nanophotonic biosensor. *Sci. Rep.* **2017**, *7*, 41368. [CrossRef]
30. Janeiro, R.; Flores, R.; Viegas, J. Silicon photonics waveguide array sensor for selective detection of VOCs at room temperature. *Sci. Rep.* **2019**, *9*, 17099. [CrossRef]
31. Xu, D.X.; Densmore, A.; Delâge, A.; Waldron, P.; McKinnon, R.; Janz, S.; Lapointe, J.; Lopinski, G.; Mischki, T.; Post, E.; et al. Folded cavity SOI microring sensors for high sensitivity and real time measurement of biomolecular binding. *Opt. Express* **2008**, *16*, 15137–15148. [CrossRef]
32. Igarashi, A.; Shang, Y.; Kuroiwa, S.; Ohashi, K.; Yamada, H. A compact-size and ultrasensitive optical biosensor using a double-spiral microresonator. In Proceedings of the 2022 IEEE Sensors, Dallas, TX, USA, 30 October–2 November 2022; pp. 1–4. [CrossRef]
33. Pathak, A.K.; Swargiary, K.; Kongsawang, N.; Jitpratak, P.; Ajchareeyasoontorn, N.; Udomkittivorakul, J.; Viphavakit, C. Recent Advances in Sensing Materials Targeting Clinical Volatile Organic Compound (VOC) Biomarkers: A Review. *Biosensors* **2023**, *13*, 114. [CrossRef]
34. Rath, R.J.; Farajikhah, S.; Oveissi, F.; Dehghani, F.; Naficy, S. Chemiresistive Sensor Arrays for Gas/Volatile Organic Compounds Monitoring: A Review. *Adv. Eng. Mater.* **2023**, *25*, 2200830. [CrossRef]
35. Zhang, C.; Ghosh, A.; Zhang, H.; Shi, S. Langasite-based surface acoustic wave resonator for acetone vapor sensing. *Smart Mater. Struct.* **2019**, *29*, 015039. [CrossRef]
36. Zamboni, R.; Zaltron, A.; Chauvet, M.; Sada, C. Real-time precise microfluidic droplets label-sequencing combined in a velocity detection sensor. *Sci. Rep.* **2021**, *11*, 17987. [CrossRef] [PubMed]
37. Zhou, J.; Husseini, D.A.; Li, J.; Lin, Z.; Sukhishvili, S.; Coté, G.L.; Gutierrez-Osuna, R.; Lin, P.T. Mid-Infrared Serial Microring Resonator Array for Real-Time Detection of Vapor-Phase Volatile Organic Compounds. *Anal. Chem.* **2022**, *94*, 11008–11015. [CrossRef] [PubMed]
38. Taniguchi, T.; Hirowatari, A.; Ikeda, T.; Fukuyama, M.; Amemiya, Y.; Kuroda, A.; Yokoyama, S. Detection of antibody-antigen reaction by silicon nitride slot-ring biosensors using protein G. *Opt. Commun.* **2016**, *365*, 16–23. [CrossRef]
39. Ramón José Pérez, M. Fiber-Optic Ring Resonator Interferometer. In *Interferometry*; Mithun, B., Bruno, U., Eds.; IntechOpen: London, UK, 2018; p. 4.
40. Kazanskiy, N.L.; Khonina, S.N.; Butt, M.A. A Review of Photonic Sensors Based on Ring Resonator Structures: Three Widely Used Platforms and Implications of Sensing Applications. *Micromachines* **2023**, *14*, 1080. [CrossRef]
41. Vörös, J. The density and refractive index of adsorbing protein layers. *Biophys. J.* **2004**, *87*, 553–561. [CrossRef] [PubMed]
42. Abe, S.; Hara, H.; Masuda, S.; Yamada, H. Fabrication of vertical-taper structures for silicon photonic devices by using local-thickness-thinning process. *Jpn. J. Appl. Phys.* **2022**, *61*, SK1005. [CrossRef]
43. Bellegarde, C.; Pargon, E.; Sciancalepore, C.; Petit-Etienne, C.; Hugues, V.; Robin-Brosse, D.; Hartmann, J.M.; Lyan, P. Improvement of Sidewall Roughness of Submicron SOI Waveguides by Hydrogen Plasma and Annealing. In Proceedings of the 2018 IEEE Photonics Conference (IPC), Reston, VA, USA, 30 September–4 October 2018; pp. 1–4. [CrossRef]
44. Chu, K.-Y.; Thompson, A.R. Densities and Refractive Indices of Alcohol-Water Solutions of n-Propyl, Isopropyl, and Methyl Alcohols. *J. Chem. Eng. Data* **1962**, *7*, 358–360. [CrossRef]
45. Roy, E.; Galas, J.-C.; Veres, T. Thermoplastic elastomers for microfluidics: Towards a high-throughput fabrication method of multilayered microfluidic devices. *Lab Chip* **2011**, *11*, 3193–3196. [CrossRef]

46. Butt, M.A.; Khonina, S.N.; Kazanskiy, N.L. A serially cascaded micro-ring resonator for simultaneous detection of multiple analytes. *Laser Phys.* **2019**, *29*, 046208. [CrossRef]
47. Hayashi, H.; Sakamoto, N.; Hideshima, S.; Harada, Y.; Tsuna, M.; Kuroiwa, S.; Ohashi, K.; Momma, T.; Osaka, T. Tetrameric jacalin as a receptor for field effect transistor biosensor to detect secretory IgA in human sweat. *J. Electroanal. Chem.* **2020**, *873*, 114371. [CrossRef]
48. Kuroiwa, S.; Hayashi, H.; Toyama, R.; Kaneko, N.; Horii, K.; Ohashi, K.; Momma, T.; Osaka, T. Potassium-regulated Immobilization of Cortisol Aptamer for Field-effect Transistor Biosensor to Detect Changes in Charge Distribution with Aptamer Transformation. *Chem. Lett.* **2021**, *50*, 892–895. [CrossRef]

sensors MDPI

Article

Using Machine Learning Algorithms to Determine the Post-COVID State of a Person by Their Rhythmogram

Sergey V. Stasenko [1,*], Andrey V. Kovalchuk [2], Evgeny V. Eremin [3], Olga V. Drugova [4], Natalya V. Zarechnova [5], Maria M. Tsirkova [6], Sergey A. Permyakov [3], Sergey B. Parin [3] and Sofia A. Polevaya [3]

1 Neurotechnology Department, Institute of Biology and Biomedicine, Lobachevsky State University of Nizhny Novgorod, 603022 Nizhny Novgorod, Russia
2 Laboratory of Autowave Processes, Institute of Applied Physics, Russian Academy of Sciences, 603950 Nizhny Novgorod, Russia; aka.xzib1t@gmail.com
3 Faculty of Social Sciences, Lobachevsky State University of Nizhny Novgorod, 603022 Nizhny Novgorod, Russia; eugenevc@gmail.com (E.V.E.); permyakov@fsn.unn.ru (S.A.P.); parins@mail.ru (S.B.P.); sofia.polevaia@fsn.unn.ru (S.A.P.)
4 Department of Medical Biophysics, Privolzhsky Research Medical University, 603005 Nizhny Novgorod, Russia; olgadrugova@gmail.com
5 GBUZ NO "Nizhny Novgorod Regional Clinical Oncological Dispensary", 603126 Nizhny Novgorod, Russia; nvzar@mail.ru
6 Clinical Hospital No. 2, Privolzhsky District Medical Center, 603032 Nizhny Novgorod, Russia; cirkova_mariya@mail.ru
* Correspondence: stasenko@neuro.nnov.ru

Abstract: This study introduces a novel method for detecting the post-COVID state using ECG data. By leveraging a convolutional neural network, we identify "cardiospikes" present in the ECG data of individuals who have experienced a COVID-19 infection. With a test sample, we achieve an 87 percent accuracy in detecting these cardiospikes. Importantly, our research demonstrates that these observed cardiospikes are not artifacts of hardware–software signal distortions, but rather possess an inherent nature, indicating their potential as markers for COVID-specific modes of heart rhythm regulation. Additionally, we conduct blood parameter measurements on recovered COVID-19 patients and construct corresponding profiles. These findings contribute to the field of remote screening using mobile devices and heart rate telemetry for diagnosing and monitoring COVID-19.

Keywords: machine learning algorithms; electrocardiogram; post-COVID state; COVID-19; data analysis

Citation: Stasenko, S.V.; Kovalchuk, A.V.; Eremin, E.V.; Drugova, O.V.; Zarechnova, N.V.; Tsirkova, M.M.; Permyakov, S.A.; Parin, S.B.; Polevaya, S.A. Using Machine Learning Algorithms to Determine the Post-COVID State of a Person by Their Rhythmogram. *Sensors* **2023**, *23*, 5272. https://doi.org/10.3390/s23115272

Academic Editor: Nicole Jaffrezic-Renault

Received: 25 April 2023
Revised: 19 May 2023
Accepted: 30 May 2023
Published: 1 June 2023

1. Introduction

Currently, several viruses have had significant socio-economic and medical impacts, including SARS, A/H1N1, H5N1/H7N9, MERS, Ebola, and the most widespread and impactful of them all, SARS-CoV-2 [1,2]. SARS-CoV-2 is responsible for causing COVID-19, a potentially severe acute respiratory infection that reached pandemic status in 2020. COVID-19 can manifest as mild or severe, affecting various organs [3,4].

Common symptoms of COVID-19 include fever, fatigue, dry cough, loss of smell, and loss of taste [5,6]. Most infected individuals experience mild or asymptomatic cases with spontaneous recovery [2,7]. However, a portion of cases can progress to severe forms requiring oxygen therapy, with some patients becoming critically ill [8]. COVID-19 can trigger intense inflammation, known as a cytokine storm, leading to fatal pneumonia and acute respiratory distress syndrome [9]. The disease can also have long-term systemic effects on cardiovascular health and other complications, such as acute heart failure, renal failure, septic shock, and cognitive impairment [10,11].

SARS-CoV-2 can induce neurotropic manifestations and damage the central nervous system, causing immune cell infections, encephalitis, encephalopathy, and demyelination [12]. Cognitive problems associated with COVID-19 include difficulty concentrating, motor activity reduction, impaired coordination, loss of smell/taste, decreased sensitivity, and cognitive decline [13,14]. Post-COVID syndrome, experienced by a percentage of patients, presents a wide range of symptoms that can persist for several weeks or longer, including weakness, shortness of breath, headaches, joint pains, cognitive impairment, gastrointestinal issues, and more [15].

The causes of post-COVID syndrome are still not fully understood, and various hypotheses are being explored, including direct damage to cells and organs, persistence of the virus in the body, and autoimmune reactions [16–30].

These findings highlight the wide-ranging impacts of SARS-CoV-2 and COVID-19 on human health, necessitating ongoing research and care for affected individuals.

Post-COVID syndrome remains challenging to determine, given the involvement of various organ systems and disorders [31]. Mass spectrometric analysis of blood samples has been used to evaluate inflammatory reactions by assessing 96 proteins [32]. Rapid diagnostic methods with high throughput are crucial for timely treatment and rehabilitation of COVID-19 patients [33–43].

Studies have shown that abnormal ECGs and elevated troponin levels at admission are associated with major adverse events in COVID-19 patients [33]. ECG abnormalities are independent of pulmonary infection severity, and reflect various cardiovascular complications [36]. ST segment alteration and signs of left ventricular hypertrophy are associated with worse prognosis, while an abnormal T wave or the presence of ST segment elevation/depression can indicate COVID-19 patient mortality [34,37].

Deep learning models, such as ECGConvnet, have shown promise in distinguishing COVID-19 from other cardiovascular diseases with high accuracy [38]. One-dimensional convolutional neural networks (1D-CNN) have been used to automatically detect COVID-19 using ECG signals [39].

Heart rate sensors on wearable devices have been utilized to identify potential early indicators of COVID-19 symptoms, such as elevated resting heart rate and altered HR/step measurements [40]. HRV analysis using smartphone cameras and wrist-worn devices has shown differences in HRV indicators for some patients during different phases of COVID-19 [41].

In patients with hypoxic respiratory failure, a study found a significant drop in SDNN (a measure of heart rate variability), accompanied by a substantial increase in CRP levels within 72 h, indicating systemic inflammation [42,43].

Cardiac rhythmography, initially limited to stationary registration conditions, has evolved with advancements in telemetric systems, enabling assessment during various natural activities [44–47].

Convolutional neural networks (CNNs) have emerged as powerful tools for ECG analysis, demonstrating high accuracy in automating heartbeat classification and arrhythmia detection [48,49]. These CNN-based models have shown superior performance, comparable to cardiologist-level expertise, in accurately identifying and categorizing arrhythmias [48,49]. Deep learning approaches, including CNNs, have also shown potential in automating complex cardiac analyses beyond ECG interpretation [50,51].

Remote assessment of human functional state during various activities has been made possible through different methods and technologies [52–56].

In the context of COVID-19, remote screening using mobile devices and heart rate telemetry has shown promise in diagnosing and monitoring the disease. Studies have observed "cardiospikes" in the rhythmogram records of COVID-19 patients, potentially serving as markers for the disease [57].

In this article, we propose a new method for detecting the post-COVID state using ECG data. The method is based on the detection of "cardiospikes" observed in the ECG data of patients who have had a COVID-19 infection. We utilize a convolutional neural network, which has proven itself as an effective method for analyzing ECG data. The detection accuracy in the test sample was 87 percent. Furthermore, we demonstrate that the observed cardiospikes are not the result of hardware–software signal distortions, but have an endogenous nature, making them potential markers for COVID-specific modes of heart rhythm regulation. Additionally, we measured the blood parameters of patients who had recovered from COVID-19 infection and constructed corresponding profiles.

The structure of this work is as follows. In Section 2, we provide a detailed explanation of the ECG data recording method, including the number of records and their organization. We also outline the algorithms used for data analysis and neural network training. Moving on to Section 3, we present the key findings of our study, which include the visual representation of a "cardiospike" the verification process for identifying these spikes, the blood parameter profiles of individuals who have recovered from a COVID-19 infection, and the outcomes of training a neural network to detect cardiospikes in ECG recordings. In Section 4, we delve into a discussion of our results, and propose potential avenues for further research. Finally, we summarize the study's outcomes in Section 5.

2. Methods

A schematic diagram of the cardiac rhythmogram recording and analysis system is depicted in Figure 1.

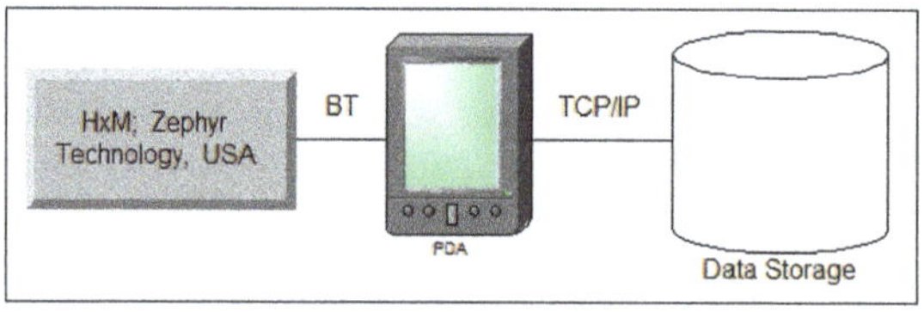

Figure 1. Illustration depicting the setup for recording and analyzing the cardiac rhythmogram system.

Event-related heart rate telemetry technology offers several advantages, including mobility (allowing subjects to move freely while reliably recording the signal from a significant distance), uninterrupted recording, autonomous measurement, resistance to external interference, and real-time data acquisition and processing.

To achieve performance goals, we used the ZephyrTM HxMTM Smart-Zephyr BIO PACH BH3-M1 (HxM, Zephyr Technology, Annapolis, MD, USA) sensor platform, including a microprocessor; radio signal receiver/transmitter; and ECG, acceleration, and distance sensors.

Data were transmitted to a smartphone/PC via Bluetooth SPP-2.4 GHz channel at 1-second intervals (Figure 1). Each packet includes sensor platform ID, last 15 R-R intervals, and time relative to recording start. Data is pre-processed on the Android smartphone, then transmitted to a server via GSM channels. Experimental data can be exported in TXT/CSV formats for further processing.

2.1. *Data Collection*

The neural network algorithm for post-COVID state recognition based on ECG data was trained using a database from a clinical hospital. The database includes ECG records from 970 COVID-19 patients and over 1000 records from healthy subjects. Patients with extrasystole were excluded, but "cardiospike" anomalies were identified in COVID-19 patients' ECG data and used to train the algorithm.

All ECG data, including the training data, are available on the COGNITOM Web platform (cogni-nn.ru) as a zip archive containing a CSV file. The CSV file includes 5 columns with the identifier, RR interval value, markup for spikes, and measurement time for each COVID-19 record. This study was conducted according to the guidelines of the Declaration of Helsinki and approved by the Ethics Committee of Lobachevsky University (Protocol number 3 from 8 April 2021).

2.2. Method for Determining Blood Parameters

The diagnostic blood tests were conducted at the Department of Clinical and Laboratory Diagnostics of the POMC. A biochemical blood test was performed using the "Konelab-20" automatic biochemical analyzer, manufactured by Thermo Fisher Scientific (Waltham, MA, USA). The parameters measured included ALT (U/l), AST (U/l), albumin (g/l), bilirubin unbound (indirect) (μmol/l), total bilirubin (μmol/l), bilirubin bound (direct) (μmol/l), glucose (venous) (mmol/l), blood creatinine (μmol/l), lactate dehydrogenase (LDH) (U/l), urea (mmol/l), C-reactive protein (mg/l), and ferritin (mcg/l).

Hematological studies were performed using the ABX MIKROS 60 hematological analyzer. The measured indicators included absolute basophil content ($10^9/l$), hematocrit (%), hemoglobin (g/l), leukocytes ($10^9/l$), lymphocytes (%), absolute lymphocyte content ($10^9/l$), monocytes (%), absolute monocyte content ($10^9/l$), absolute neutrophil content ($10^9/l$), band neutrophils (%), segmented neutrophils (%), erythrocyte sedimentation rate (mm/h), mean erythrocyte hemoglobin (pg), mean platelet volume (fl), mean erythrocyte volume (fl), mean erythrocyte hemoglobin concentration (g/l), platelet crit (%), platelets ($10^9/l$), erythrocyte distribution width (coefficient of variation) (%), platelet distribution width (%), absolute eosinophil content ($10^9/l$), and erythrocytes ($10^{12}/l$).

Hemostasis was assessed using coagulation screening indicators, which included activated partial thromboplastin time (APTT), prothrombin time (PT), fibrinogen concentration (according to Clauss), international normalized ratio (INR), and soluble fibrin monomer complexes (SFMK).

The acid–base composition of venous blood was assessed using the Rapidlab 865 blood gas analyzer, manufactured by Bayer Diagnostics, in terms of pH, electrolytes, metabolites, and oximetry. The measured indicators included absolute oxygen content O2CT (ml/dl), anion difference (mmol/l), hematocrit (%), glucose (mmol/l), deoxyhemoglobin (%), excess BE in blood (mmol/l), potassium (mmol/l), ionized calcium (mmol/l), lactate (mmol/l), sodium (mmol/l), total hemoglobin (g/l), oxyhemoglobin (%), negative logarithm of hydrogen ion concentration, oxygen partial pressure (mmHg), oxygen partial pressure at 50% saturation (mmHg), carbon dioxide partial pressure (mmHg), standard blood bicarbonate (mmol/l), functional oxygen saturation (%), and chloride level (mmol/l).

2.3. Machine Learning Algorithms

To solve the problem of segmenting the flow of RR intervals, a dilated convolution network was chosen. Convolutional networks have a higher performance and learning speed than recurrent networks (RNNs), which are commonly used in time series detection and segmentation problems. Convolutional networks, on the other hand, require more convolutions to achieve a sufficient receptive field, i.e., capture long enough time sequences. To eliminate this shortcoming, stretched convolutions were used. An example for 3 layers is shown in Figure 2.

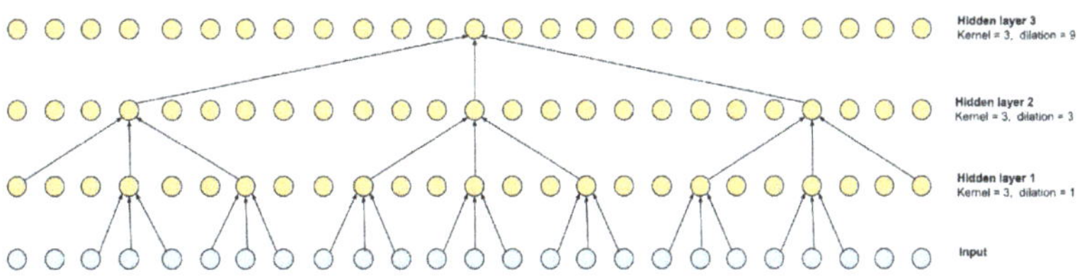

Figure 2. Dilated convolution with kernel = 3 and its receptive field.

The stretch for each subsequent layer can be defined as $d = k^{n-1}$, where n is the number of the convolutional layer. As can be seen from the scheme shown in Figure 2, the receptive field is 27 and is defined as $r = (k-1)\sum_{i=1}^{L} k^{i-1} + 1$, where L is the number of layers. Of course, such a multilayer scheme can be equivalent to 1 convolutional layer with a kernel of 27, but the calculation speed of the multilayer version is higher.

As shown in [58], the speed of the convolutional layer can also be increased by dividing the convolution into horizontal channel-by-channel convolutions and vertical inter-channel mixing. This scheme is borrowed for the base block of the RR interval segmenter. The residual block (see Figure 3) is a variant of the MobileNetv3 base block adapted for time series. It retains the basic principle of expanding the feature space from size (T, C) to (T, H) (expansion layer), per-channel convolution with gaps, non-linearity based on GELU [59], projection of the input (T, H) based on the Squeeze-and-Excitation approach [60], as well as a layer of inter-channel mixing to the original size (T, C) (compression layer), to the output of which the input (residual) is mixed. To speed up convergence, parameterized dimension reset (Ts, S) (skip) was used, which allows more deep layers to be involved in learning at the earliest stage [61]. It is worth noting that the temporal dimension of the skip output is truncated on the left and right $Ts = T - 2P$, where P is the size of the padding. The value of P is determined by the truncation of the receptive field of the extreme elements, caused by the addition of convolutions with boundary values to preserve the dimension, as well as the maximum duration of the desired object.

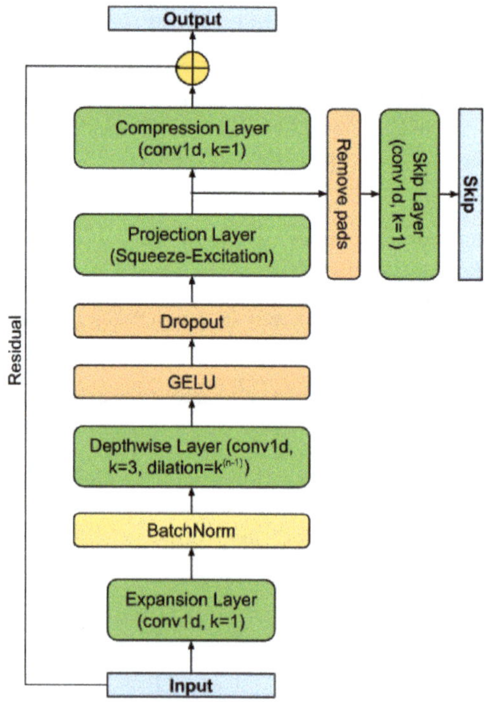

Figure 3. The residual block.

Figure 4 shows the schematic of the RR interval detector. It consists of F filters, each with L base blocks connected in series. The skip branches of each of the layers are summed up and fed into the final segmenting unit. The diagram of this block is shown in Figure 5. The block converts the dimension input (Ts, S) into the output (Ts, M), where M is the number of classes (in the work $M = 1$).

To train the model, we used labeled records of RR intervals, which are sequences of delays between adjacent QRS impulses on a cardiogram. The records are divided into segments of length T with an overlap of $2P$; the response of the Ts model corresponds to the central part of the segment, which is $T = P + Ts + P$. On the records of RR intervals, the maximum values of the desired spike were previously marked, which is the goal of the study; these values unambiguously describe the position and size of the spike. The loss function was chosen based on the well-known focal loss [62]. The AdamW function [63] was used as an optimizer.

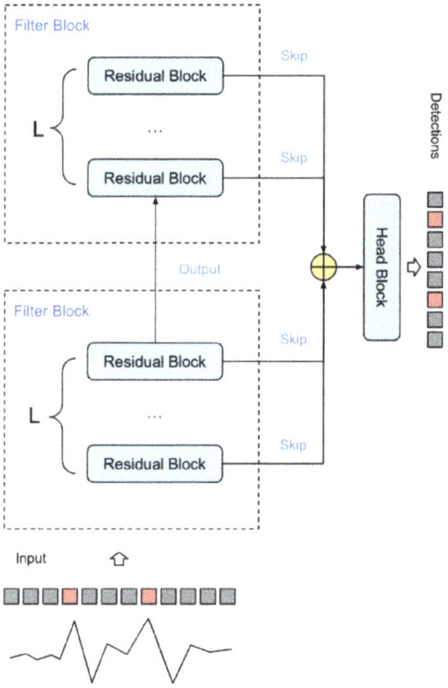

Figure 4. Spike detector circuit.

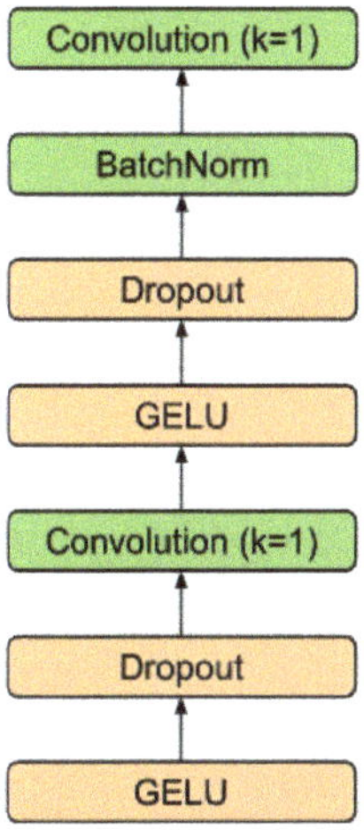

Figure 5. Head block.

3. Results

3.1. Registration and Collection of ECG Data

The heart, acting as a phase sensor, provides information about an organism's functional state through rhythmograms, which are optimal signals for displaying physiological systems' operating modes. Rhythmograms are used in various fields, including sports medicine and psychophysiology, to objectively assess functional state. In COVID-19 patients, rhythmograms are registered and processed on the Cognite web platform alongside other examinations. An experimental database was created, consisting of records from 970 patients and a total of 110,330 samples. This database revealed the rigidity [64,65] of RR intervals and the presence of cardiospikes (Figure 6), which are low-amplitude anomalies observed in patients diagnosed with COVID-19. These findings were obtained from the ZephyrSmart sensor platform.

The spike pattern in the data differentiates two consecutive jumps, based on the RR number, from the average value. Specifically, a longer RR is followed by a shorter one, followed by a slight relaxation, as depicted in Figure 7.

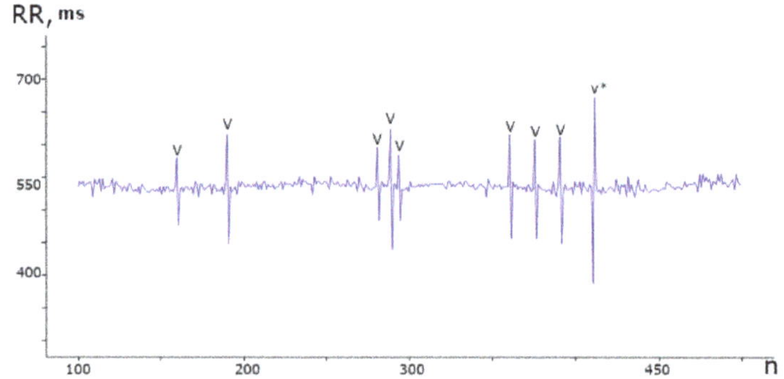

Figure 6. Segment of the rhythmogram of a patient with COVID-19 comprising 400 counts, displaying 9 instances of spike anomalies. These anomalies exhibit a distinct repetitive pattern and varying amplitudes within the range of ±100 ms from the mean value.

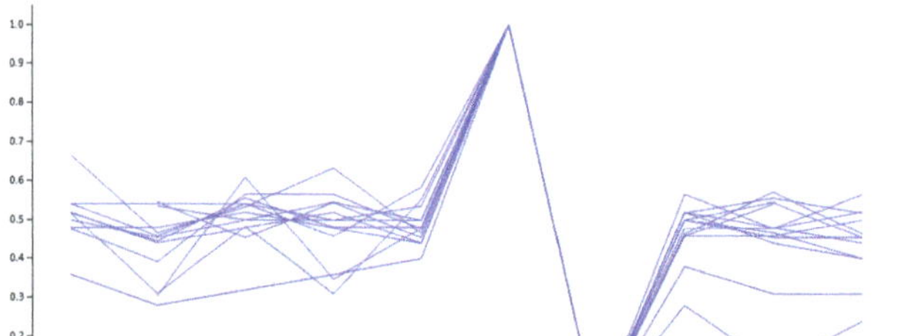

Figure 7. An illustration of a collection of normalized spikes from diverse subjects who have been diagnosed with COVID-19.

3.2. Verification of Cardiospikes

An additional investigation was conducted to rule out the possibility that cardiospikes are linked to idiosyncrasies in signal recording and processing within the sensory platform.

The accuracy of ZephyrSmart data was verified by comparing it with data obtained from a professional electrocardiograph, Poly-Spectrum-8 (Neurosoft). Data logging was synchronized between ZephyrSmart and Poly-Spectrum-8 based on the start time. Upon comparing the rhythmograms obtained from the two devices, it was observed that cardiospikes present in the ZephyrSmart rhythmogram were not replicated in the Poly-Spectrum-8 rhythmogram, despite a high overall correlation between the two signals (Figure 8). To simplify visual comparisons, the data are displayed on a single scale.

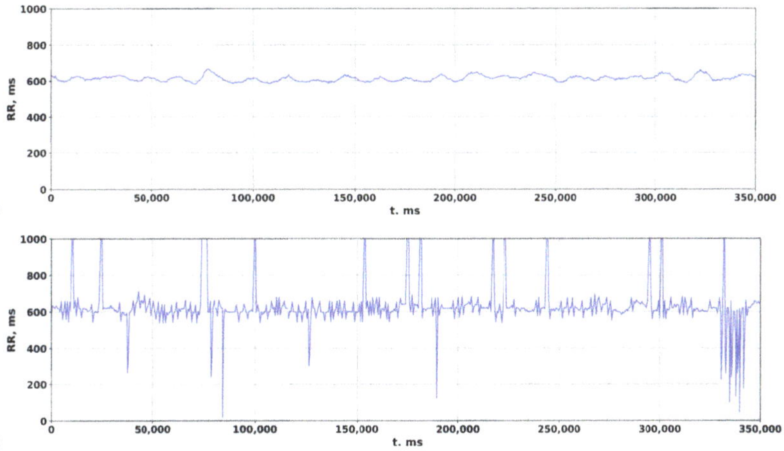

(a)
(b)

Figure 8. Comparing rhythmograms obtained from various devices in a patient with COVID-19 over a long-term period: (**a**) Poly-Spectrum-8; (**b**) Zephyr.

Based on this observation, the authors hypothesized that the algorithm for detecting QRS complexes, which includes filtering, preprocessing, and postprocessing methods, may affect the final rhythmogram signal by removing spike episodes. This hypothesis aligns with the statistical principle of heart rate variability analysis, which focuses on studying the totality of RR intervals rather than individual events. To test this hypothesis, various established algorithms for detecting R peaks in the original ECG signal were investigated.

We used algorithms from open sources [66,67]:

- Pan–Tompkins;
- Hamilton;
- Two-Moving-Average.

The rhythmogram results of processing the ECG of a patient with COVID-19 are depicted in Figure 9. In the standard package, Poly-Spectrum-8, the Two-Moving-Average algorithm is utilized to calculate rhythmograms from the ECG, which normalizes the signal and reduces deviations from the mean values (Figure 9c). However, when the Pan–Tompkins and Hamilton algorithms are used to calculate rhythmograms from the same ECG, distinctive patterns of RR intervals appear in the rhythmogram (Figure 9a,b), resembling cardiospikes observed in ZephyrSmart (refer to Figure 6).

These findings suggest that cardiospikes are not a result of hardware–software signal distortions, but rather have an endogenous nature, and may serve as markers for COVID-specific modes of heart rhythm regulation.

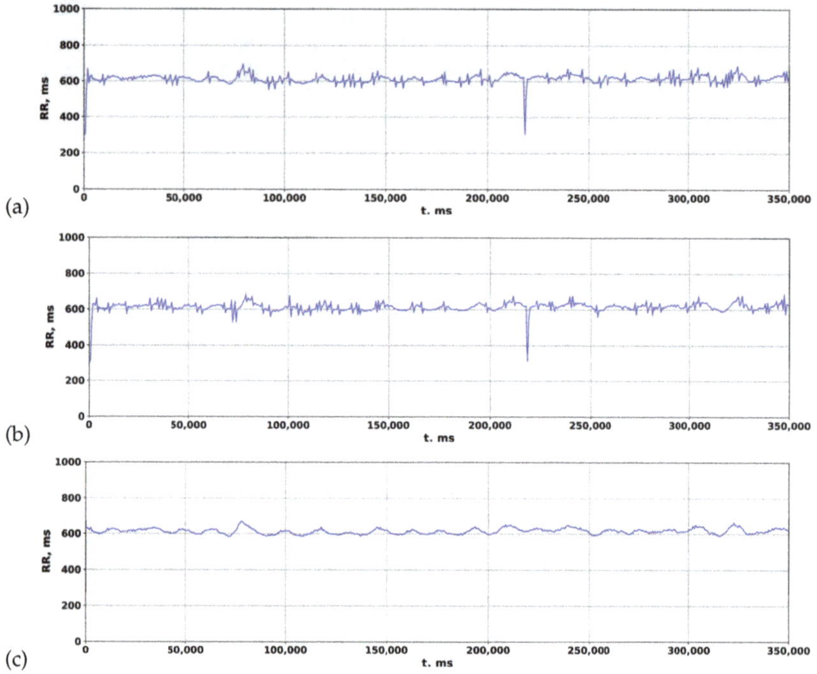

Figure 9. Comparison of rhythmograms in a patient with COVID-19 obtained using different algorithms: (**a**) Pan–Tompkins; (**b**) Hamilton; (**c**) Two-Moving-Average.

3.3. Measurement of Blood Parameters

Along with the registration of RR intervals in patients with COVID-19, blood parameters were measured. To search for possible reasons for the appearance of cardiospikes, 11 patients were selected with sufficiently high frequencies of cardiospike appearance in the cardiointervalogram. We studied the space of parameters from four modules of indicators reflecting the state of the blood: biochemistry (Figure 10); hematological studies (Figure 11); hemostasis (Figure 12); acid–base composition of venous blood (Figure 13). In each module, patients were found with deviations from the normal values of the parameters under consideration. The results of the analysis are shown in the Figures 10–13. The "biochemistry" module is distinguished by the greatest excess of normal values in terms of ALT, ferritin (0.64), C-reactive protein (0.64), and glucose (0.55). The module "hematological studies" shows the largest deviations of the following indicators: segmented neutrophils (0.73), lymphocytes (0.82), and ESR (0.55). In the module "hemostasis", values exceed the norm for fibrinogen (0.75) and RFMC test (1). The greatest number of significant parameters occurred in the module "acid–base composition of venous blood": glucose (1), deoxyhemoglobin (1), excess BE in the blood (1), lactate (1), sodium (1), oxyhemoglobin (1), functional oxygen saturation (1), partial pressure of oxygen and carbon dioxide (0.83), anion difference (0.83), standard blood bicarbonate (0.67).

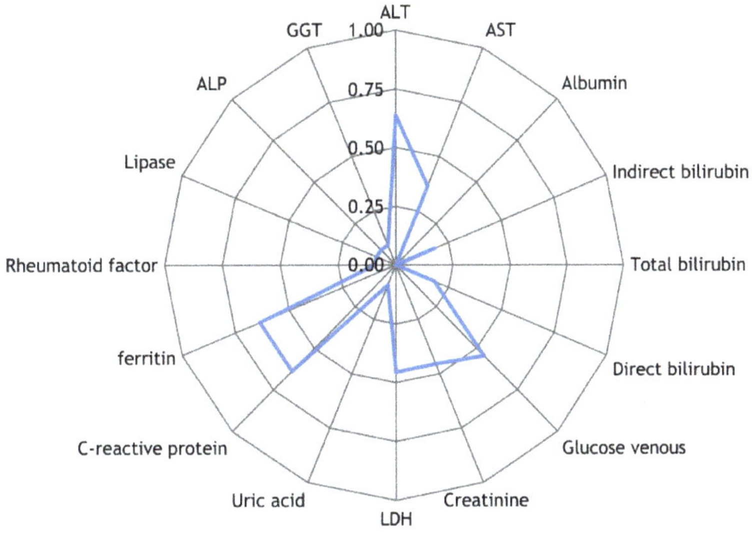

Figure 10. The space of parameters from the biochemistry module of indicators reflecting the state of the blood.

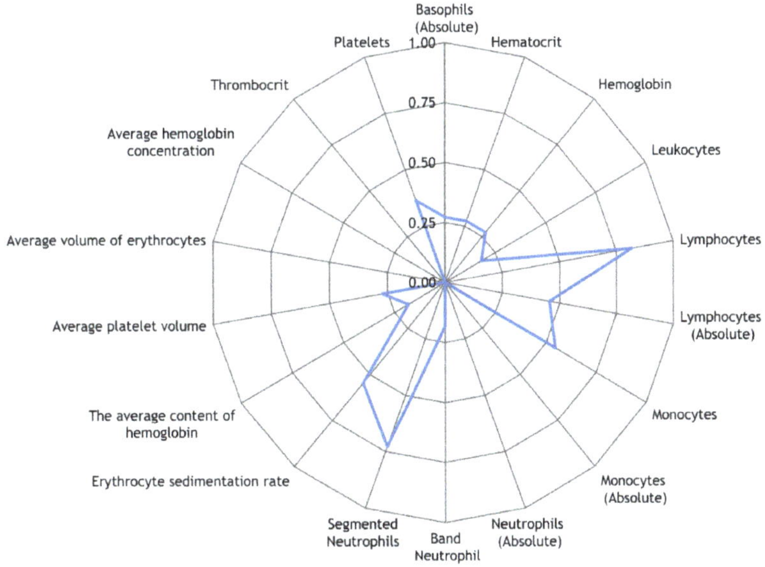

Figure 11. The space of parameters from the hematological studies module of indicators reflecting the state of the blood.

The sample included patients with a volume of lung lesions from 44 to 92% according to the CT results. Most of them had ground glass lesions; two descriptions showed a change in transparency according to the type of fibrotic changes.

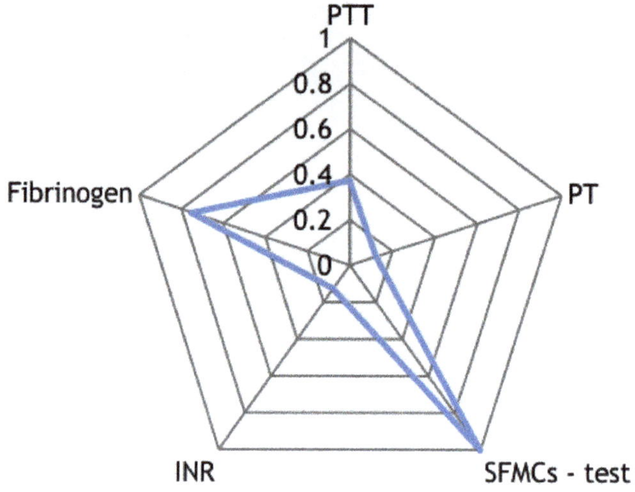

Figure 12. The space of parameters from the hemostasis module of indicators reflecting the state of the blood.

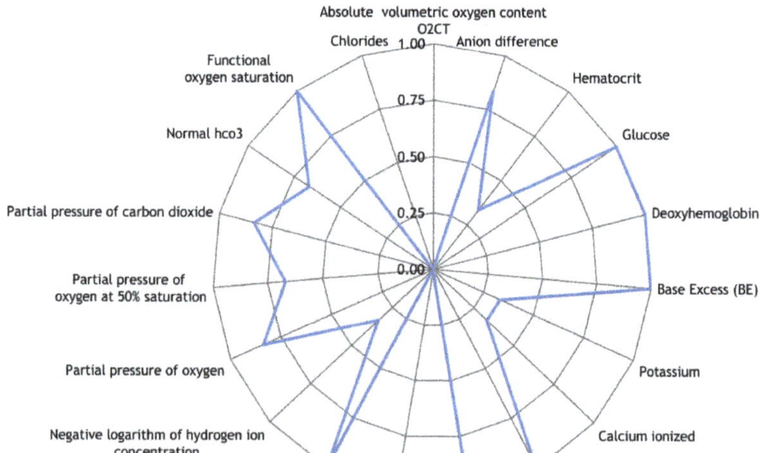

Figure 13. The space of parameters from the acid–base composition of venous blood module of indicators reflecting the state of the blood.

3.4. Data Analysis

An experimental database of records of 970 patients with a total size of 110,330 samples was collected. Of these, 2041 samples were marked as spikes. In all experiments, fixed parameters were used: sample length $T = 32$, padding $P = 4$, layers per block $L = 4$, and number of blocks $F = 2$. The choice of parameters (H, C, S) was based on the analysis of cross-validation experiments with the division of the database into 10 subsets. The metrics for evaluating the result were AP (average precision), F_1-score [68], and $F_{1/2}$-score. These metrics were computed from precision–recall curves, and were chosen as most closely reflecting the quality of the final model, despite the fact that each of them is insufficient at assessing the quality. As an estimate, among the 10 curves of the cross-validation sample, the worst value for the metrics was chosen.

Based on the measurements in Table 1, the group of model parameters (H, C, S) $(40, 24, 72)$ was selected as the main parameters. The experiment results for the optimal values of the parameters are presented in Figure 14a,c,e, which display the values of the metrics depending on the gradient descent iteration.

Table 1. Dependence of metrics on model parameters during cross-validation.

	H (Fixed $C = 24$, $S = 72$)			C (Fixed $H = 40$, $S = 72$)				S (Fixed $H = 40$, $C = 24$)		
Value	**32**	**40**	**56**	**16**	**24**	**32**	**40**	**56**	**72**	**96**
F_1	0.86	0.886	**0.89**	0.869	**0.886**	0.886	0.876	0.872	0.886	**0.89**
$F_{1/2}$	0.85	0.87	**0.891**	0.857	0.878	**0.879**	0.857	0.857	**0.87**	0.867
AP	0.93	0.943	**0.944**	0.932	**0.943**	0.934	0.93	0.929	0.943	**0.946**

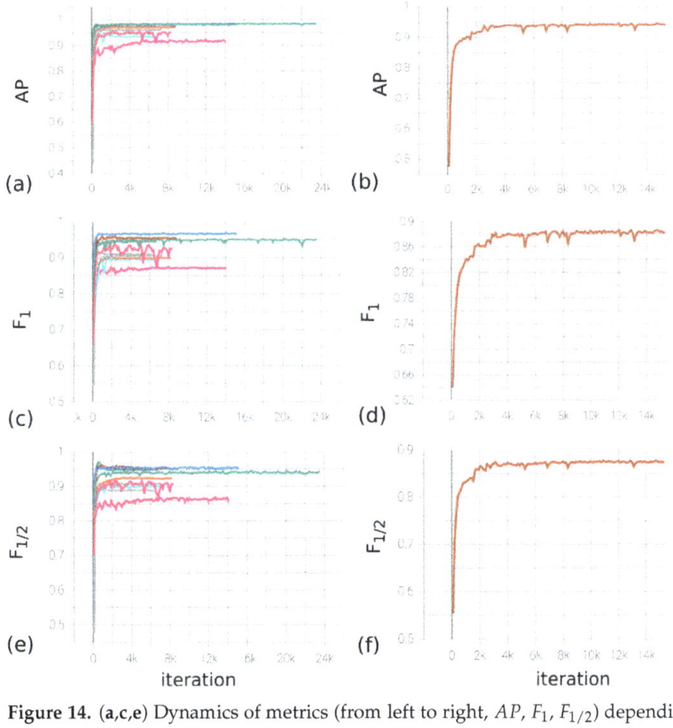

Figure 14. (**a**,**c**,**e**) Dynamics of metrics (from left to right, AP, F_1, $F_{1/2}$) depending on the iteration number during cross-validation for (H, C, S) = (40, 24, 72). (**b**,**d**,**f**) Dynamics of metrics (from left to right, AP, F_1, $F_{1/2}$) depending on the iteration number on the full base for (H, C, S) = (40, 24, 72).

The LR (learning rate) parameter of the optimizer was chosen to be 0.01, with the possibility of decreasing by a factor of 10 every 10 epochs. The decrease in LR occurred when the loss function for the validation set reached a plateau. The final model was trained on the division of the base into training/validation/test in the ratio 1861/180/50, respectively. Figure 14b,d,f contain the dependence of metrics on iteration. Based on the experiments, the F_1-score metric on the test data indicates that the model's accuracy is at least 0.89. Furthermore, the estimates for the model accuracy are 0.94 and 0.87, respectively.

4. Discussion

It is well known that disturbances in the work of the cardiovascular system and its regulation lead not only to the appearance of various heart rhythm distortions, but also to changes in the basic characteristics of the heart rate, such as heart rate variability (HRV). HRV analysis is rarely used to diagnose specific diseases. The only HRV analysis proposed by the American Heart Association for diagnostic use is the observation that low 24 h HRV in diabetes mellitus is an early sign of diabetic neuropathy [69].

The main goal of analyzing a patient's HRV is to assess their functional state in order to evaluate the effectiveness of treatment, assess the severity of the disease, predict the risk of sudden death or dangerous complications in various diseases, and so on. The interest in HRV is primarily due to the property of the heart rhythm over time, where rigidity of a patient's heart rhythm indicates a deterioration in their condition. R. Baevsky explains this property of the rhythm with the concept of the relationship between the adaptive capabilities of the human body and its HRV: low HRV reflects the poor adaptability of the cardiovascular system to random or permanent effects on the body [70].

In this regard, the following requirement is usually put forward for the quantitative parameters of HRV: when the patient's condition worsens, the statistical parameters of HRV should decrease, and when the condition improves, they should increase. Since the 1960s, HRV has been studied at rest for short (up to 5 min) time intervals, while meeting strict requirements for rhythm stationarity [71].

In this case, two main characteristics of the rhythm are used: the cardiorhythmogram of rest and the magnitude of the accompanying arrhythmia. As significant parameters of HRV, both the characteristics of the histogram of the distribution of RR intervals and derivatives of these characteristics are used, including the parameters of variational pulsometry, as well as the spectral characteristics of the sequence of RR intervals. Numerous studies have shown that the deterioration of the patient's functional state correlates with both a decrease in RR at rest and a decrease in the severity of the accompanying arrhythmia. Thus, RR at rest of 700 ms and above significantly increases the risk of cardiac death in various cardiovascular diseases, and rigidity of the resting rhythm, regardless of the value of RR, significantly increases the risk of cardiac death in myocardial infarction [72].

Since the 1980s, commercial Holter electrocardiogram (ECG) monitoring systems have been available. This made it possible to start active study of HRV over long (mainly daily) periods of time. The study of HRV in patients with various cardiovascular and other pathologies has been carried out, and many works have been published on the analysis of HRV over long periods of time. In the vast majority of these works, quantitative parameters of HRV were used, which are used in the analysis of a stationary rhythm over short periods of time [47].

It is obvious that the use of these parameters for long periods of time was effective only in the analysis of stationary (rigid or close to rigid) rhythms characterizing a very poor functional state of a person [73]. At the same time, periodic local (3–6 intervals) fluctuations in the duration of RR intervals remained outside the focus of researchers' interests for a long time. A positive example of an in-depth clinical study of the significance of local fluctuations in the duration of RR intervals is the work of [74].

In a study by Asarcikli [75], autonomic function in the post-COVID period was examined using HRV analysis. Comparing 24 h ECG recordings of post-COVID patients with healthy controls revealed significant differences in HRV scores. Time and frequency domain indices, including SDNN 24 h, RMSSD, high frequency, and low-frequency/high-frequency ratio, were higher in post-COVID patients. The prevalence of SDNN > 60 ms and RMSSD > 40 ms was also higher among post-COVID patients. Logistic regression models supported the presence of a parasympathetic overtone in post-COVID patients, independent of covariates. These findings suggest autonomic imbalance as a potential explanation for persistent orthostatic symptoms in the post-COVID period.

The processing of medical data is essential in terms of making subsequent diagnoses and prescribing treatment or rehabilitation measures. Existing clinical systems for analyzing ECG data are based on the Two-Moving-Average algorithm, which, as our study shows, smooths the signal in such a way that significant information for diagnosing the patient's condition disappears. However, by employing alternative algorithms, such as Pan–Tompkins and Hamilton, in the calculation of rhythmograms from ECG data, it becomes possible to acquire a more comprehensive data description.

The progress in ECG data analysis methods and their integration into clinical data analysis systems has the potential to extend the existing diagnostic capabilities for diseases, leading to significant enhancements in patients' quality of life. Furthermore, it can facilitate the development of mobile systems for diagnosing patients' conditions, which is a crucial undertaking in the prevention and treatment of diseases with high rates of transmission and systemic impacts on the body.

5. Conclusions

In this study, it was observed that patients with COVID-19 showed a distinct change in heart rate characterized by a transient increase and subsequent compensatory decrease in RR intervals. To explain the reason for such dynamics, the blood parameters of patients who had COVID-19 infection were analyzed. Based on the data obtained, it is suggested that these RR interval dynamics could be associated with profound disorders of blood coagulation and viscosity, leading to mechanical stress on the myocardium, as well as an imbalance in the acid–base balance and damage to the lung tissue. Cardiospikes, which are associated with cardiac arrhythmias in patients who had a COVID-19 infection, were found on the recorded RR intervals.

The confirmation of telemetry data using a standard electrocardiograph revealed the persistence of this phenomenon across all recording devices, emphasizing the importance of the algorithm for determining RR intervals. Importantly, the study shows that the observed spikes were not artifacts of hardware–software signal distortions, but were innate in nature, indicating their potential as markers for COVID-specific rhythm regulation regimens.

A new method for determining the post-COVID status from ECG data has been presented. By utilizing a convolutional neural network, cardiospikes present in ECG data from individuals who had a COVID-19 infection were successfully identified. The method achieved an 87 percent accuracy in detecting these cardiospikes in the test sample. These results contribute to the development of remote screening using mobile devices and heart rate telemetry for the diagnosis and monitoring of COVID-19. Further development of the study could aim at improving the accuracy of detecting cardiospikes, as well as exploring the possibility of detecting extrasystoles using the proposed method.

It should be noted that the proposed method for detecting cardiospikes was tested on data collected using the ZephyrSmart sensor platform, which allows the recording of a pure ECG signal. Other existing commercial analogues used in medical applications have built-in RR interval detection algorithms, such as the Two-Moving-Average. Our study demonstrates that such algorithms can strongly smooth the final signal by removing information about cardiospikes.

The developed method for detecting a post-COVID state based on the presence of cardiospikes in ECG recordings can serve as a valuable tool for conducting more detailed studies of the condition of patients during long-term follow-up and at various stages of the disease. This will contribute to the development of effective methods and approaches for the treatment of COVID-19.

Author Contributions: Conceptualization, S.B.P. and S.A.P. (Sofia A. Polevaya); methodology, S.B.P. and S.A.P. (Sofia A. Polevaya); software, E.V.E., S.V.S. and A.V.K.; validation, S.B.P. and S.A.P. (Sofia A. Polevaya); formal analysis, E.V.E., S.V.S. and A.V.K.; investigation, S.V.S., A.V.K., E.V.E., O.V.D., N.V.Z., M.M.T., S.A.P. (Sergey A. Permyakov), S.B.P. and S.A.P. (Sofia A. Polevaya); resources, S.V.S., A.V.K., O.V.D., N.V.Z., M.M.T., S.A.P. (Sergey A. Permyakov), S.B.P. and S.A.P. (Sofia A. Polevaya); data curation, O.V.D., N.V.Z., M.M.T., S.A.P. (Sergey A. Permyakov), S.B.P. and S.A.P. (Sofia A. Polevaya); writing—original draft preparation, S.V.S., O.V.D. and A.V.K.; writing—review and editing, S.V.S., O.V.D., A.V.K. and S.A.P. (Sofia A. Polevaya); visualization, S.V.S., O.V.D. and A.V.K.;

supervision, S.V.S., A.V.K. and S.A.P. (Sofia A. Polevaya); project administration, S.V.S. and S.A.P. (Sofia A. Polevaya); funding acquisition, S.A.P. (Sofia A. Polevaya). All authors have read and agreed to the published version of the manuscript.

Funding: This work was supported by the Russian Science Foundation (Project No. 22-18-20075).

Institutional Review Board Statement: This study was conducted according to the guidelines of the Declaration of Helsinki and approved by the Ethics Committee of Lobachevsky University (Protocol №3 from 8 April 2021).

Informed Consent Statement: Informed consent has been obtained from all subjects involved in the study.

Data Availability Statement: The data that support the findings of this study are available from the corresponding author upon reasonable request.

Acknowledgments: We are grateful to the reviewers for their constructive comments and valuable suggestions that have helped to improve the quality of the paper.

Conflicts of Interest: The authors declare no conflict of interest.

References

1. Yuki, K.; Fujiogi, M.; Koutsogiannaki, S. COVID-19 pathophysiology: A review. *Clin. Immunol.* **2020**, *215*, 108427. [CrossRef] [PubMed]
2. Heymann, D.; Shindo, N. COVID-19: What is next for public health? *Lancet* **2020**, *395*, 542–545. [CrossRef]
3. Mudatsir, M.; Fajar, J.; Wulari, L.; Soegiarto, G.; Ilmawan, M.; Purnamasari, Y.; Mahdi, B.; Jayanto, G.; Suhendra, S.; Setianingsih, Y. Others Predictors of COVID-19 severity: A systematic review and meta-analysis. *F1000Research* **2021**, *9*, 1107. [CrossRef] [PubMed]
4. Trypsteen, W.; Van Cleemput, J.; Snippenberg, W.; Gerlo, S.; Vandekerckhove, L. On the whereabouts of SARS-CoV-2 in the human body: A systematic review. *PLoS Pathog.* **2020**, *16*, e1009037. [CrossRef] [PubMed]
5. Saniasiaya, J.; Islam, M.; Abdullah, B. Prevalence and characteristics of taste disorders in cases of COVID-19: A meta-analysis of 29,349 patients. *Otolaryngol. Neck Surg.* **2021**, *165*, 33–42. [CrossRef] [PubMed]
6. Agyeman, A.; Chin, K.L.; Ersdorfer, C.; Liew, D.; Ofori-Asenso, R. Smell and taste dysfunction in patients with COVID-19: A systematic review and meta-analysis. *Mayo Clin. Proc.* **2020**, *95*, 1621–1631. [CrossRef]
7. Oran, D.; Topol, E. The proportion of SARS-CoV-2 infections that are asymptomatic: A systematic review. *Ann. Intern. Med.* **2021**, *174*, 655–662. [CrossRef]
8. World Health Organization. *Coronavirus Disease 2019 (COVID-19): Situation Report*; World Health Organization: Geneva, Switzerland , 2020; Volume 46.
9. Rahmati, M.; Moosavi, M. Cytokine-targeted therapy in severely ill COVID-19 patients: Options and cautions. *Mortality* **2020**, *4*, 179–180. [CrossRef]
10. Beeching, N.; Fletcher, T.; Fowler, R. Complications. Coronavirus disease 2019 (COVID-19). *BMJ Best Pract.* **2020** .
11. Miskowiak, K.; Johnsen, S.; Sattler, S.; Nielsen, S.; Kunalan, K.; Rungby, J.; Lapperre, T.; Porsberg, C. Cognitive impairments four months after COVID-19 hospital discharge: Pattern, severity and association with illness variables. *Eur. Neuropsychopharmacol.* **2021**, *46*, 39–48 . [CrossRef]
12. Beaud, V.; Crottaz-Herbette, S.; Dunet, V.; Vaucher, J.; Bernard-Valnet, R.; Du Pasquier, R.; Bart, P.; Clarke, S. Pattern of cognitive deficits in severe COVID-19. *J. Neurol. Neurosurg. Psychiatry* **2021**, *92*, 567–568 . [CrossRef]
13. Devita, M.; Bordignon, A.; Sergi, G.; Coin, A. The psychological and cognitive impact of COVID-19 on individuals with neurocognitive impairments: Research topics and remote intervention proposals. *Aging Clin. Exp. Res.* **2021**, *33*, 733–736 . [CrossRef] [PubMed]
14. Rogers, J.; Chesney, E.; Oliver, D.; Pollak, T.; McGuire, P.; Fusar-Poli, P.; Zandi, M.; Lewis, G.; David, A. Psychiatric and neuropsychiatric presentations associated with severe coronavirus infections: A systematic review and meta-analysis with comparison to the COVID-19 pandemic. *Lancet Psychiatry* **2020**, *7*, 611–627. [CrossRef] [PubMed]
15. Sudre, C.; Murray, B.; Varsavsky, T.; Graham, M.; Penfold, R.; Bowyer, R.; Pujol, J.; Klaser, K.; Antonelli, M.; Canas, L.; et al. Attributes and predictors of long COVID. *Nat. Med.* **2021**, *27*, 626–631. [CrossRef] [PubMed]
16. Juan Jose, J.; Huda, M. The effects of COVID-19 on hypothalamus: Is it another face of SARS-CoV-2 that may potentially control the level of COVID-19 severity. *Int. J. Clin. Stud. Med. Case Rep.* **2020**, *7*, 5. [CrossRef]
17. Stasenko, S.; Hramov, A.; Kazantsev, V. Loss of neuron network coherence induced by virus-infected astrocytes: A model study. *Sci. Rep.* **2023**, *13*, 6401. [CrossRef]
18. Lu, Y.; Li, X.; Geng, D.; Mei, N.; Wu, P.; Huang, C.; Jia, T.; Zhao, Y.; Wang, D.; Xiao, A.; et al. Cerebral micro-structural changes in COVID-19 patients–an MRI-based 3-month follow-up study. *EClinicalMedicine* **2020**, *25*, 100484. [CrossRef]
19. Hajra, A.; Mathai, S.; Ball, S.; Bandyopadhyay, D.; Veyseh, M.; Chakraborty, S.; Lavie, C.; Aronow, W. Management of thrombotic complications in COVID-19: An update. *Drugs* **2020**, *80*, 1553–1562. [CrossRef]

20. De Melo, G.; Lazarini, F.; Levallois, S.; Hautefort, C.; Michel, V.; Larrous, F.; Verillaud, B.; Aparicio, C.; Wagner, S.; Gheusi, G.; et al. COVID-19-associated olfactory dysfunction reveals SARS-CoV-2 neuroinvasion and persistence in the olfactory system. *BioRxiv* **2020**. [CrossRef]
21. Choi, B.; Choudhary, M.; Regan, J.; Sparks, J.; Padera, R.; Qiu, X.; Solomon, I.; Kuo, H.; Boucau, J.; Bowman, K.; et al. Persistence and evolution of SARS-CoV-2 in an immunocompromised host. *N. Engl. J. Med.* **2020**, *383*, 2291–2293. [CrossRef]
22. Hu, F.; Chen, F.; Ou, Z.; Fan, Q.; Tan, X.; Wang, Y.; Pan, Y.; Ke, B.; Li, L.; Guan, Y.; et al. A compromised specific humoral immune response against the SARS-CoV-2 receptor-binding domain is related to viral persistence and periodic shedding in the gastrointestinal tract. *Cell. Mol. Immunol.* **2020**, *17*, 1119–1125. [CrossRef]
23. Varga, Z.; Flammer, A.; Steiger, P.; Haberecker, M.; Andermatt, R.; Zinkernagel, A.; Mehra, M.; Schuepbach, R.; Ruschitzka, F.; Moch, H. Endothelial cell infection and endotheliitis in COVID-19. *Lancet* **2020**, *395*, 1417–1418. [CrossRef] [PubMed]
24. Booz, G.; Altara, R.; Eid, A.; Wehbe, Z.; Fares, S.; Zaraket, H.; Habeichi, N.; Zouein, F. Macrophage responses associated with COVID-19: A pharmacological perspective. *Eur. J. Pharmacol.* **2020**, *887*, 173547. [CrossRef] [PubMed]
25. Da Silva, S.; Ju, E.; Meng, W.; Paniz Mondolfi, A.; Dacic, S.; Green, A.; Bryce, C.; Grimes, Z.; Fowkes, M.; Sordillo, E.; et al. Broad SARS-CoV-2 cell tropism and immunopathology in lung tissues from fatal COVID-19. *J. Infect. Di* **2021**, jiab195.
26. Banerjee, A.; Nasir, J.; Budylowski, P.; Yip, L.; Aftanas, P.; Christie, N.; Ghalami, A.; Baid, K.; Raphenya, A.; Hirota, J.; et al. Isolation, sequence, infectivity, and replication kinetics of severe acute respiratory syndrome coronavirus 2. *Emerg. Infect. Dis.* **2020**, *26*, 2054. [CrossRef]
27. Wang, E.; Mao, T.; Klein, J.; Dai, Y.; Huck, J.; Jaycox, J.; Liu, F.; Zhou, T.; Israelow, B.; Wong, P.; et al. Diverse functional autoantibodies in patients with COVID-19. *Nature* **2021**, *595*, 283–288. [CrossRef] [PubMed]
28. Zuo, Y.; Estes, S.; Ali, R.G.; Hi, A.; Yalavarthi, S.; Shi, H.; Sule, G.; Gockman, K.; Madison, J.; Zuo, M.; et al. Prothrombotic autoantibodies in serum from patients hospitalized with COVID-19. *Sci. Transl. Med.* **2020**, *12*, eabd3876. [CrossRef]
29. Garvin, M.; Alvarez, C.; Miller, J.; Prates, E.; Walker, A.; Amos, B.; Mast, A.; Justice, A.; Aronow, B.; Jacobson, D. A mechanistic model and therapeutic interventions for COVID-19 involving a RAS-mediated bradykinin storm. *Elife* **2020**, *9*, e59177. [CrossRef]
30. Afrin, L.; Weinstock, L.; Molderings, G. COVID-19 hyperinflammation and post-COVID-19 illness may be rooted in mast cell activation syndrome. *Int. J. Infect. Dis.* **2020**, *100*, 327–332. [CrossRef] [PubMed]
31. Nalbian, A.; Desai, A.; Wan, E. Post-COVID-19 condition. *Annu. Rev. Med.* **2023**, *74*, 55–64. [CrossRef]
32. Doykov, I.; Hällqvist, J.; Gilmour, K.; Grjean, L.; Mills, K.; Heywood, W. 'The long tail of COVID-19'-The detection of a prolonged inflammatory response after a SARS-CoV-2 infection in asymptomatic and mildly affected patients. *F1000Research* **2020**, *9*. [CrossRef]
33. Bergamaschi, L.; D'Angelo, E.; Paolisso, P.; Toniolo, S.; Fabrizio, M.; Angeli, F.; Donati, F.; Magnani, I.; Rinaldi, A.; Bartoli, L.; et al. The value of ECG changes in risk stratification of COVID-19 patients. *Ann. Noninvasive Electrocardiol.* **2021**, *26*, e12815. [CrossRef]
34. Wang, Y.; Chen, L.; Wang, J.; He, X.; Huang, F.; Chen, J.; Yang, X. Electrocardiogram analysis of patients with different types of COVID-19 . *Ann. Noninvasive Electrocardiol.* **2020**, *25*, e12806. [CrossRef]
35. Chorin, E.; Dai, M.; Kogan, E.; Wadhwani, L.; Shulman, E.; Nadeau-Routhier, C.; Knotts, R.; Bar-Cohen, R.; Barbhaiya, C.; Aizer, A.; et al. Electrocardiographic risk stratification in COVID-19 patients. *Front. Cardiovasc. Med.* **2021**, *8*, 636073. [CrossRef] [PubMed]
36. Angeli, F.; Spanevello, A.; De Ponti, R.; Visca, D.; Marazzato, J.; Palmiotto, G.; Feci, D.; Reboldi, G.; Fabbri, L.; Verdecchia, P. Electrocardiographic features of patients with COVID-19 pneumonia. *Eur. J. Intern. Med.* **2020**, *78*, 101–106. [CrossRef] [PubMed]
37. Mehraeen, E.; Alinaghi, S.; Nowroozi, A.; Dadras, O.; Alilou, S.; Shobeiri, P.; Behnezhad, F.; Karimi, A. A systematic review of ECG findings in patients with COVID-19. *Indian Heart J.* **2020**, *72*, 500–507. [CrossRef] [PubMed]
38. Bassiouni, M.; Hegazy, I.; Rizk, N.; El-Dahshan, E.; Salem, A. Automated detection of COVID-19 using deep learning approaches with paper-based ecg reports. *Circuits Syst. Signal Process.* **2022**, *41*, 5535–5577. [CrossRef]
39. Nguyen, T.; Pham, H.; Le K.; Nguyen, A.; Thanh, T.; Do, C. Detecting COVID-19 from digitized ECG printouts using 1D convolutional neural networks. *PLoS ONE* **2022**, *17*, e0277081. [CrossRef]
40. Mishra, T.; Wang, M.; Metwally, A.; Bogu, G.; Brooks, A.; Bahmani, A.; Alavi, A.; Celli, A.; Higgs, E.; Dagan-Rosenfeld, O.; et al. Early detection of COVID-19 using a smartwatch. *MedRxiv* **2020**. [CrossRef]
41. Ponomarev, A.; Tyapochkin, K.; Surkova, E.; Smorodnikova, E.; Pravdin, P. Heart rate variability as a prospective predictor of early COVID-19 symptoms. *MedRxiv* **2021**. [CrossRef]
42. Hasty, F.; Garcia, G.; Davila, H.; Wittels, S.; Hendricks, S.; Chong, S. Heart rate variability as a possible predictive marker for acute inflammatory response in COVID-19 patients. *Mil. Med.* **2021**, *186*, e34–e38. [CrossRef]
43. Smilowitz, N.; Kunichoff, D.; Garshick, M.; Shah, B.; Pillinger, M.; Hochman, J.; Berger, J. C-reactive protein and clinical outcomes in patients with COVID-19. *Eur. Heart J.* **2021**, *42*, 2270–2279. [CrossRef]
44. Parin, V.V.; Baevsky, P.M. *Introduction to Medical Cybernetics*; University of Pennsylvania: Philadelphia, PA, USA, 1966; p. 220.
45. Kaznacheev, V.P.; Baevsky, R.M.; Berseneva, A.P.; Domakhina, G.M.; Polyakov, Y.V. On some features of the adaptation of the organism in connection with the profession and age. *Labor Hyg. Occup. Dis.* **1978**, *2*, 21–26.
46. Grigoriev, A.I.; Bayevsky, R.M. *The Concept of Health and the Problem of the Norm in Space Medicine*; Slovo: Moscow, Russia, 2001; p. 95.
47. Shlyk, N.I.; Sapozhnikova, E.N.; Kirillova, E.N.; Semenov, V.G. Typological features of the functional state of regulatory systems in schoolchildren and young athletes (according to the analysis of heart rate variability). *Hum. Physiol.* **2009**, *35*, 85–93. [CrossRef]

48. Acharya, U.; Oh, S.; Hagiwara, Y.; Tan, J.; Adam, M.; Gertych, A.; Tan, R.S. A deep convolutional neural network model to classify heartbeats. *Comput. Biol. Med.* **2017**, *89*, 389–396. [CrossRef]
49. Hannun, A.; Rajpurkar, P.; Haghpanahi, M.; Tison, G.; Bourn, C.; Turakhia, M.; Ng, A. Cardiologist-level arrhythmia detection and classification in ambulatory electrocardiograms using a deep neural network. *Nat. Med.* **2019**, *25*, 65–69. [CrossRef]
50. Zhang, J.; Gajjala, S.; Agrawal, P.; Tison, G.; Hallock, L.; Beussink-Nelson, L.; Lassen, M.; Fan, E.; Aras, M.; Jordan, C.; et al. Fully automated echocardiogram interpretation in clinical practice: Feasibility and diagnostic accuracy. *Circulation* **2018**, *138*, 1623–1635. [CrossRef] [PubMed]
51. Rajpurkar, P.; Hannun, A.; Haghpanahi, M.; Bourn, C.; Ng, A. Cardiologist-level arrhythmia detection with convolutional neural networks. *arXiv* **2017**, arXiv:1707.01836.
52. Nekrasova, M.M.; Karatushina, D.I.; Parin, S.B.; Polevaya, S.A. Application of information technologies for assessment of professional risks for high-altitude assemblers during periodic medical examination. *Med. Alm.* **2011**, *3*, 26–31.
53. Polevaya, S.A.; Parin, S.B.; Runova, E.V.; Nekrasova, M.M.; Fedotova, I.V.; Bakhchina, A.V.; Kovalchuk, A.V.; Shishalov, I.S. Telemetric and information technologies for monitoring of the functional state of athletes. *Mod. Technol. Med.* **2012**, *4*, 94–98.
54. Runova, E.V.; Grigorieva, V.N.; Bakhchina, A.V.; Parin, S.B.; Shishalov, I.S.; Kozhevnikov, I.S.; Nekrasova, M.M.; Karatushina, D.I.; Grigorieva, D.I.; Polevaya, S.A. Vegetative correlates of arbitrary mappings emotional stress. *Int. J. Psychophysiol.* **2013**, *4*, 69–77.
55. Chernigovskaya, T.; Parin, S.; Parina, I.; Konina, A.; Urikh, D.; Yachmonina, Y.; Chernova, M.; Polevaya, S.A. Simultaneous interpreting and stress: Pilot experiment. *Int. J. Psychophysiol.* **2016**, *108*, 165. [CrossRef]
56. Chernigovskaya, T.V.; Parina, I.S.; Alekseeva, S.V.; Konina, A.A.; Urich, D.K.; Yachmonina, Y.O.; Parin, S.B. Simultaneous interpreting and stress: Pilot experiment. *Mod. Technol. Med.* **2019**, *11*, 132–140. [CrossRef]
57. Polevaya, S.; Eremin, E.; Bulanov, N.; Bakhchina, A.; Kovalchuk, A.; Parin, S. Event-related telemetry of heart rate for personalized remote monitoring of cognitive functions and stress under conditions of everyday activity. *Mod. Technol. Med.* **2019**, *11*, 109–114. [CrossRef]
58. Li, Z.; Liu, F.; Yang, W.; Peng, S.; Zhou, J. A survey of convolutional neural networks: Analysis, applications, and prospects. *IEEE Trans. Neural Netw. Learn. Syst.* **2021**, *33*, 6999–7019 . [CrossRef]
59. Hendrycks, D.; Gimpel, K. Gaussian error linear units (gelus). *arXiv* **2016**, arXiv:1606.08415.
60. Jin, X.; Xie, Y.; Wei, X.; Zhao, B.; Chen, Z.; Tan, X. Delving deep into spatial pooling for squeeze-and-excitation networks. *Pattern Recognit.* **2022**, *121*, 108159. [CrossRef]
61. Shafiq, M.; Gu, Z. Deep residual learning for image recognition: A survey. *Appl. Sci.* **2022**, *12*, 8972. [CrossRef]
62. Zaidi, S.; Ansari, M.; Aslam, A.; Kanwal, N.; Asghar, M.; Lee, B. A survey of modern deep learning based object detection models. *Digit. Signal Process.* **2022**, *126*, 103514. [CrossRef]
63. Loshchilov, I. & Hutter, F. Decoupled weight decay regularization. *arXiv* **2017**, arXiv:1711.05101.
64. Shirshov, Y.A.; Govorin, A.I. Vegetative disorders in patients with influenza A (H1N1). *Sib. Med J.* **2011**, *5*, 41–44.
65. Zufarov, A.A. Indicators of heart rate variability in acute respiratory syndrome in children. *Young Sci.* **2020**, *3*, 98–103.
66. Nepi, D.; Sbrollini, A.; Agostinelli, A.; Maranesi, E.; Morettini, M.; Di Nardo, F.; Fioretti, S.; Pierleoni, P.; Pernini, L.; Valenti, S.; et al. Validation of the heart-rate signal provided by the Zephyr bioharness 3.0. In Proceedings of the 2016 Computing in Cardiology Conference (CinC), Vancouver, BC, Canada, 11–14 September 2016; pp. 361–364.
67. Howell, L.; Porr, B. *Popular ECG R Peak Detectors Written in Python*; Zenodo: Geneva, Switzerland, 2019.
68. Manning, C. *An Introduction to Information Retrieval*; Cambridge University Press: Cambridge, UK, 2009.
69. Electrophysiology, T. Heart rate variability: Standards of measurement, physiological interpretation, and clinical use. *Circulation* **1996**, *93*, 1043–1065. [CrossRef]
70. Baevskiy, R.; Ivanov, G.; Chireykin, L.; Gavrilushkin, A.; Dovgalevskiy, P.; Kukushkin, Y.A.; Mironova, T.; Prilutskiy, D.; Semenov, Y.N.; Fedorov, V.; et al. Analiz variabel'nosti serdechnogo ritma pri ispol'zovanii razlichnykh elektrokardiograficheskikh sistem: Metod. rekomendatsii [Analysis of Heart Rate Variability Using Various Electrocardiographic Systems: Guidelines]. *Vestn. Aritmologii* **2001**, *24*, 66–85.
71. Parin, V.; Baevski, R. *Introduction to Medical Cybernetics: By VV Parin and RM Bayevskiy*; National Aeronautics: Washington, DC, USA, 1967.
72. Wolf, M.; Varigos, G.; Hunt, D.; Sloman, J. Sinus arrhythmia in acute myocardial infarction. *Med. J. Aust.* **1978**, *2*, 52–53. [CrossRef] [PubMed]
73. Sobolev, A. Methods for analyzing heart rate variability over long periods of time. *Metod. Anal. Vari* **2009** .
74. Sobolev, A.; Ryabykina, G.; Kozhemyakina, E. Specificity of the effect of double fractures of the rhythmogram on the daily variability of the sinus rhythm in patients with pulmonary and arterial hypertension. *Syst. Hypertens.* **2021**, *18*, 43–49. [CrossRef]
75. Asarcikli, L.; Hayiroglu, M.; Osken, A.; Keskin, K.; Kolak, Z.; Aksu, T. Heart rate variability and cardiac autonomic functions in post-COVID period. *J. Interv. Card. Electrophysiol.* **2022**, *63*, 715–721. [CrossRef]

sensors

MDPI

Communication

Novel Siloxane Derivatives as Membrane Precursors for Lactate Oxidase Immobilization

Darya V. Vokhmyanina *, Olesya E. Sharapova, Ksenia E. Buryanovataya and Arkady A. Karyakin

Chemistry Faculty of M.V. Lomonosov, Moscow State University, 119991 Moscow, Russia
* Correspondence: vokhmyaninadv@my.msu.ru

Abstract: We report new enzyme-containing siloxane membranes for biosensor elaboration. Lactate oxidase immobilization from water–organic mixtures with a high concentration of organic solvent (90%) leads to advanced lactate biosensors. The use of the new alkoxysilane monomers—(3-aminopropyl)trimethoxysilane (APTMS) and trimethoxy[3-(methylamino)propyl]silane (MAPS)—as the base for enzyme-containing membrane construction resulted in a biosensor with up to a two times higher sensitivity (0.5 $A \cdot M^{-1} \cdot cm^{-2}$) compared to the biosensor based on (3-aminopropyl)triethoxysilane (APTES) we reported previously. The validity of the elaborated lactate biosensor for blood serum analysis was shown using standard human serum samples. The developed lactate biosensors were validated through analysis of human blood serum.

Keywords: lactate biosensor; Prussian blue; siloxane; immobilization from water–organic mixtures

1. Introduction

Lactate is considered as a marker for glycolysis, the anaerobic glucose metabolism, which makes it a useful metabolite for both clinical diagnostics and sports medicine. The lactate dynamics in blood was shown to be a predictor of death from shock in 1964 [1] and can be used with this aim for various hypoxia-caused diseases [2,3], including COVID-19 [4]. Sports medicine requires monitoring of blood lactate for both training and evaluating the so-called "lactate threshold" indicating the sportsperson's physical training level [5]. Lactate is also a fermentation byproduct and can be used as a marker for food naturalness [6].

All these applications assume the analysis of complex objects of a biological nature such as blood or food samples. Thus, lactate represents a relevant target for biosensorics implying usage of highly selective biomolecules for analyte recognition. Starting from the early 1980s, lactate oxidase (LOx) became the terminal enzyme for lactate biosensors' elaboration [7,8]. The first lactate biosensors utilized oxygen sensing by the Clark electrode [7]. However, this approach suffered from the influence of the oxygen concentration in the sample on the result. Later research concentrated on using hydrogen peroxide (the byproduct of the enzyme-catalyzed reaction) detection [6,9]. For blood analysis, the most advanced approach seems to be the use of the hydrogen-peroxide low-potential reduction reaction provided by the advantageous catalyst—Prussian blue [10–12].

Lactate oxidase (LOx) is one of the less stable oxidases; so, the immobilization protocol is of great importance. For optimum stability and bioreaction efficiencies, the preferred host matrix must be one that isolates the biomolecule, protecting it from self-aggregation, while providing essentially the same local aqueous microenvironment as in the biological media [13]. The use of one of the most known matrixes for enzyme immobilization—negatively charged Nafion—obviously should dramatically reduce the dynamic range of the resulting biosensor because the analyte (lactate) is also negatively charged. Sol–gel membranes offer a better way to immobilize LOx within their porous matrix due to the simple sol–gel processing conditions and the possibility of tailoring [14,15]. This approach is unique because immobilization is based on the siloxane polymer growing

Citation: Vokhmyanina, D.V.; Sharapova, O.E.; Buryanovataya, K.E.; Karyakin, A.A. Novel Siloxane Derivatives as Membrane Precursors for Lactate Oxidase Immobilization. *Sensors* **2023**, *23*, 4014. https://doi.org/10.3390/s23084014

Academic Editor: Evgeny Katz

Received: 16 March 2023
Revised: 12 April 2023
Accepted: 14 April 2023
Published: 15 April 2023

around the biomolecule. The entrapped enzyme remains accessible for analytes because of the porous nature of the sol–gel network [16]. ORMOSILS (organically modified silane precursors) showed promising results in preserving the native activity of biomolecules compared to inorganic sol–gel glasses. The introduction of various functional groups such as amino, glycidoxy, vinyl, etc., into alkoxide monomers leads to organically modified sol–gel membranes. ORMOSILS provide a versatile way to prepare modified sol–gel materials. The intrinsic properties of sol–gel matrixes (e.g., porosity, surface area, polarity, and rigidity) are highly dependent on the progress of hydrolysis and condensation reactions as well as the choice of monomers, water to monomer molar ratios, solvents, etc. [14,17].

Uniform gel membranes should be deposited from diluted alkoxysilane solutions (<3–5%). Since the optimal amount of water is the one required for hydrolysis of alkoxysilane, the H_2O content in the membrane-casting solution of trialkoxysilanes should be less than 9–15%. Thus, water–organic mixtures with a high content of organic solvent must be used for the enzyme immobilization in a siloxane gel membrane. Using a previously reported immobilization protocol from water–organic mixtures with a high content of organic solvents [10] made it possible to obtain a reusable biosensor on the base of lactate oxidase [11].

Thus, with the appropriate use of ORMOSILS together with the advanced protocol of enzyme immobilization from water–organic mixtures, one can alter the ultimate physicochemical properties of the sensing material produced and may elaborate new advantageous biosensors with improved analytical figures of merit for clinical applications. In this regard, the search for new derivatives of alkoxysilanes as membrane-forming components for the immobilization of lactate oxidase seems necessary to elaborate advanced biosensors. We have investigated siloxane monomers with various substituents in order to obtain the advantageous analytical performance of lactate biosensors. From the seven examined siloxanes, MAPS, MTES, and ETES were never reported as membrane-forming agents for enzyme immobilization, and VTMS was used for glucose oxidase immobilization [18], Supporting Information], VTES was reported as a membrane-forming agent for LOx immobilization [10], but no analytical characteristics were published. APTMS was used to develop the lactate biosensor with LOx immobilized on a layer of siloxane gel [19]. The analytical performances for this biosensor were a linear range of 5×10^{-5}–5×10^{-3} M and a detection limit of 1×10^{-5} M. APTES is the most widely used from the examined precursors, and the biosensor based on APTES was prepared as described previously [11]. The use of a membrane based on a new siloxane monomer (MAPS) for the immobilization of lactate oxidase made it possible to obtain a biosensor with twice the sensitivity compared to a biosensor based on the most widely used siloxane in biosensors, APTES.

2. Materials and Methods

2.1. Reagents and Objects of Analysis

Experiments were carried out with Milli-Q water (18.2 MΩ·cm). Inorganic salts, hydrogen peroxide (30% solution), potassium lactate (60% solution), (3-aminopropyl)triethoxysilane (APTES) (99%), (3-aminopropyl)trimethoxysilane (APTMS) (97%), trimethoxy[3-(methylamino) propyl]silane (MAPS) (95%), vinyltrimethoxysilane (VTMS) (97%), triethoxyvinylsilane (VTES) (97%), triethoxymethylsilane (MTES) (99%), triethoxy(ethyl)silane (ETES) (96%), and organic solvents were obtained from Sigma-Aldrich (Burlington, MA, USA) or Reachim (Moscow, Russia) at the highest purity and used as received.

Lactate oxidase (LOx, EC1.1.3.2) from Pediococcus species (Sorachim, Lausanne, Switzerland) was used in the form of a lyophilized protein with a declared activity of 32.8 U/mg. Standardized human serum samples were obtained from Spinreact (Girona, Spain).

Planar three-electrode hydrogen peroxide sensors (i.e., a Prussian-blue-modified carbon working electrode, a carbon counter electrode, and a Ag/AgCl reference electrode) were purchased from Rusens LTD (Moscow, Russia). Sensor performance characteristics in batch-regime mode showed a sensitivity of 0.7 ± 0.1 A·M^{-1}·cm^{-2} and a lower detection limit of 5×10^{-7} M.

2.2. Biosensor Preparation

Lactate-oxidase-containing membrane-casting mixtures were made by suspending an aqueous enzyme solution (10 mg/mL) in isopropanol containing siloxane (APTES, APTMS, MAPS, VTMS, VTES, MTES, or ETES). Siloxane solutions in isopropyl alcohol were prepared from commercial stock solution immediately before use and were used for no more than 6 h. The final concentrations in the water–isopropanol mixture were lactate oxidase 1 mg/mL, siloxane 0.1–3 $_{vol}$%, and water 10 $_{vol}$%. The mixture (2 μL) was drop cast onto a rough screen-printed Prussian-blue-modified carbon working electrode (Rusens LTD, Moscow, Russia) straightaway after preparation and dried in a refrigerator (4 °C) for 12 h. The enzyme-containing membrane was formed on the electrode surface after solvent evaporation and the polycondensation process. The resulted biosensors were stored in a dry state in a sealed envelope at 4 °C between the measurements.

2.3. Electrochemical Measurements

Electrochemical investigations were carried out using a PalmSens 4 potentiostat (PalmSens BV, Houten, The Netherlands). All the applied potentials mentioned in the paper refer to the internal Ag pseudo-reference electrode (potential of 0.25 V versus an SHE). The response of the biosensors towards the lactate was evaluated by chronoamperometry in batch mode. All the measurements were performed in 0.05 M phosphate buffer solution with 0.1 M KCl at pH 6.0 and at an applied potential of 0.0 mV. The calibration curves of lactate were separately obtained with three different biosensors (using each biosensor for all the concentration values tested). The operational stability of the elaborated lactate biosensor was investigated in 0.25 mM of lactate in batch mode upon stirring. The time of 50% loss of the initial signal was used to characterize the operational stability. The residual sensitivity was determined by comparing the analytical characteristics obtained immediately after the biosensor construction and during 12 months of storage in a dry state in a sealed envelope at 4 °C.

2.4. Control Serum Analysis

Standardized human serum samples with normal and pathologic analyte concentrations were prepared as described in the product instructions by reconstituting lyophilized human serum with 5 mL of distilled water. The prepared human serum samples were diluted 50 times by phosphate buffer solution prior to analysis. The lactate amperometric detection was carried out in the flow injection mode using the calibration curve obtained with standard solutions in a range of 0.01–0.1 mM.

3. Results and Discussion

A lactate biosensor was elaborated by lactate oxidase immobilization in a siloxane-based membrane on the top of the Prussian-blue-modified working electrode surface of the screen-printed three-electrode structures. To obtain uniform and stable membranes, the sol–gel procedure had a low water content of 10% in the water–isopropanol mixture. The latter was chosen in accordance with the optimum of lactate oxidase surviving in the water–organic mixtures known from previous works [10].

Siloxane-based membranes are promising materials for lactate oxidase immobilization due to the absence of a negative charge, which leads to the absence of electrostatic barriers for substrate diffusion to the immobilized enzyme [20]. As the structure of the monomer affects the polycondensation process rate and the enzyme environment in the resulting membrane, various siloxanes were investigated as matrices for lactate oxidase immobilization (Table 1). All the data in Table 1 and below are our original data, unless labeled otherwise with the corresponding reference.

Table 1. The analytical performance of lactate biosensors based on different siloxanes.

Membrane Forming Agent	Sensitivity, $A \cdot M^{-1} \cdot cm^{-2}$	LOD, M	Linear Range, μM
APTES, 1.5 $_{vol}$%	0.28 ± 0.03	9×10^{-7}	1–100
APTMS, 0.1 $_{vol}$%	0.31 ± 0.04	5×10^{-5}	50–500
MAPS, 1.0 $_{vol}$%	0.5 ± 0.02	5×10^{-7}	1–1000
VTMS, 1.0 $_{vol}$%	0.26 ± 0.05	1×10^{-6}	5–500
VTES, 0.5 $_{vol}$%	0.44 ± 0.05	9×10^{-7}	5–100
MTES, 2.0 $_{vol}$%	0.13 ± 0.08	1×10^{-6}	1–1000
ETES, 1.5 $_{vol}$%	0.092 ± 0.008	1×10^{-6}	5–1000

The content of the siloxanes in the water–organic mixture used for enzyme immobilization was optimized in the range of 0.1–3 $_{vol}$% to achieve the highest sensitivity of the resulting lactate-sensitive electrode; the optimal amounts are presented in Table 1. The triethoxysiloxane and trimethoxysiloxane containing the same substituent, vinyl, which is not involved in hydrolysis, were used for enzyme immobilization (VTES and MTES in Table 1). Even if the monomers were distinguished only by the ester groups (methoxy- or ethoxy-), the resulting biosensors had different analytical performances. The optimal monomer concentration in the membrane-casting mixture also varied depending on the ester group. This may be due to the different hydrolysis reaction rates when using ethoxy- or methoxy- derivatives [21] leading to different membrane density and enzyme microenvironments. The structure of non-hydrolyzed group also had an impact on the biosensor characteristics. Biosensors based on derivatives with short alkyl groups (methyl- or ethyl-) were characterized by relatively low sensitivity (Figure S1). The most sensitive biosensors were obtained using 3-aminopropyl-siloxanes: APTES, APTMS, and MAPS. The biosensors based on VTES also showed good sensitivity (Figure S2).

The level of L-lactate in the blood normally ranges from 0.5 to 2.2 mmol/L [22]. During intense physical activity, this index can reach 12–25 mmol/L [23]. The standard protocol for clinical lactate analyzers demands a fiftyfold sample dilution. Thus, clinical diagnostics require a lactate biosensor with a linear range from 10 to 500 μM. Using MAPS, VTMS, ETES, and MTES provided biosensors with the relevant characteristics (see Table 1). Cyclic voltammograms of the lactate biosensors demonstrated a catalytic shape in the lactate solutions with concentrations up to 1 mM (Figure S3).

Figure 1a displays an example of the amperometric responses of the elaborated biosensor to various lactate injections in batch mode. The biosensor was made by lactate oxidase immobilization in the sol of MAPS (1% content in the mixture) over the Prussian-blue-modified electrode. There was no signal decrease during the measurement even at high lactate levels. The calibration curve for the membrane composition based on data for five different biosensors is shown in Figure 1b. A linear response was observed in a wide lactate concentration range of 1–1000 μM. The sensitivity evaluated as the slope of the calibration graph was of 0.5 $A \cdot M^{-1} \cdot cm^{-2}$, which is almost two times higher than for the biosensors based on APTES made by the technique elaborated previously [11].

The analytical performances of the siloxane-based biosensors depended on the siloxane content in the casting mixture as shown in Figure 2. An increase in the siloxane concentration should lead to an increase in the membrane density and hinder the substrate diffusion to the enzyme, as seen for (3-aminopropyl)trimethoxysilane-based membranes [11]. At the same time, biosensors based on MAPS showed optimum sensitivity with siloxane concentrations in the range of 1–1.5% possibly due to the optimal enzyme environment in such membranes.

The relative selectivity of the developed biosensor to various interferences is of great interest. In general, the selectivity of the Prussian-blue-based biosensors, which operated due to hydrogen peroxide reduction, to the so called reductants (ascorbate, urate, paracetamol), was provided by the low operation potential [10]. The selectivity of the biosensors based on lactate oxidase to saccharides and hydroxy acids was due to the high selectivity of the enzyme [24].

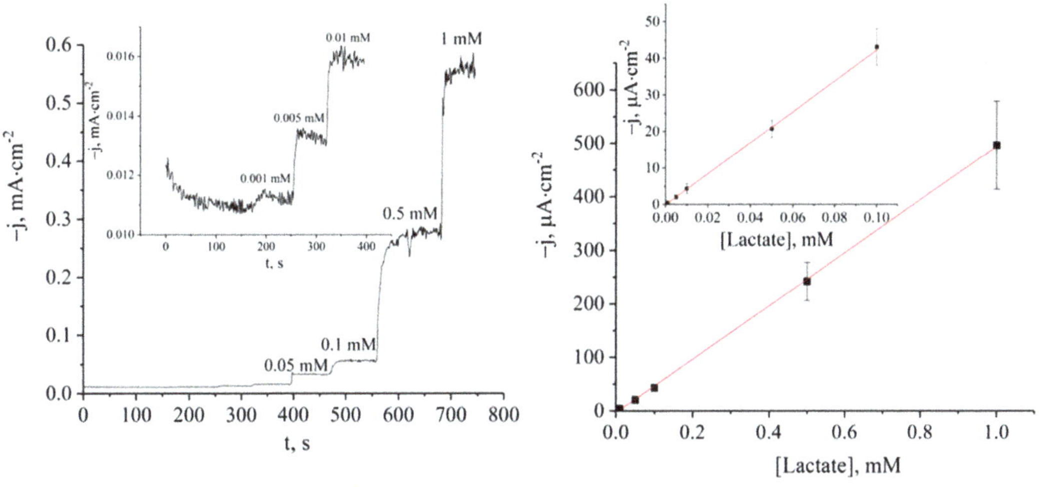

(a) (b)

Figure 1. (**a**) The calibration curve for the lactate biosensor with a 1 $_{vol}$% of MAPS in an enzyme-containing membrane solution; (**b**) calibration graph for the lactate biosensor in the batch mode (E = 0.00 V, phosphate buffer, pH 6.0). The zoomed initial parts of the correspondent dependencies are shown in the insets.

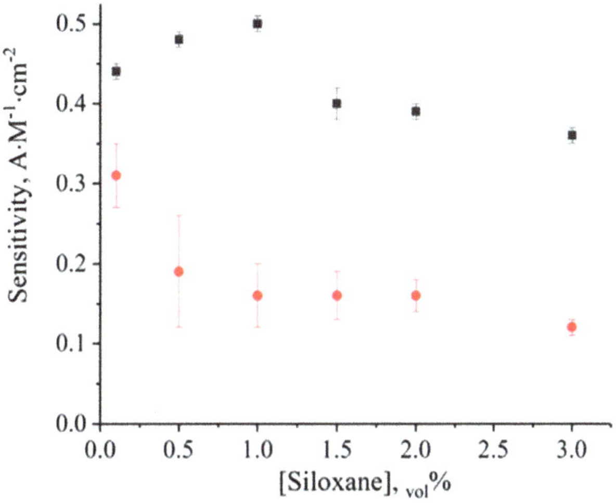

Figure 2. Dependence of the lactate biosensor sensitivity on the siloxane concentration in the membrane solution. Siloxanes: MAPS (■) and APTMS (•).

The storage stability is also of great importance. Figure 3 shows that no significant sensitivity loss was observed during one year of storage in a refrigerator.

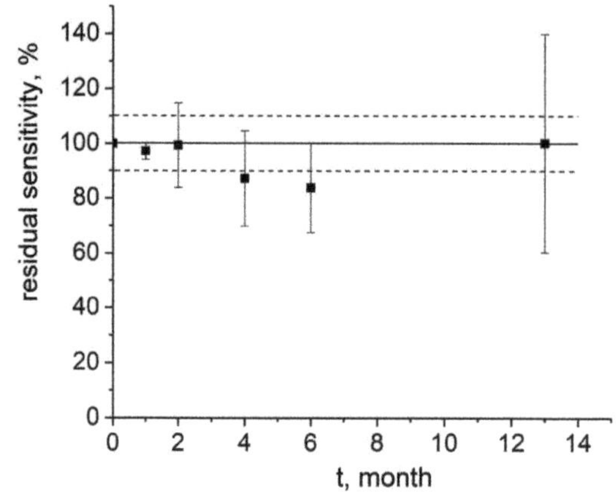

Figure 3. Residual sensitivity of the lactate biosensor after storage in the refrigerator (4 °C).

A potential limitation of the developed biosensor is the half-inactivation time (operational stability), the same as for the APTES-based biosensor: 4 h in 0.25 mM lactate upon stirring (Figure S4). It can be improved by using stabilized layers of Prussian blue as suggested in [25] or by using Nafion together with siloxane as in [26] during subsequent studies.

The biosensor was validated in the course of analysis of standardized human serum with normal and pathologic lactate concentrations. The data obtained using the FIA system equipped with the elaborated lactate biosensor were in good agreement with the levels of lactate shown in the sample passport data (Table 2). Hence, the elaborated biosensor is valid for lactate detection in blood serum.

Table 2. Standardized human serum analysis.

Sample	Measured Data, mM	Passport Data, mM
Normal human serum	2.1 ± 0.2	1.6 ± 0.3
Pathological human serum	3.20 ± 0.03	3.2 ± 0.6

Thus, a new membrane-forming compound was found, allowing elaboration of advantageous biosensors for lactate detection, which can be used for laboratory serum analysis for both clinical and sports medicine.

4. Conclusions

The use of water–organic mixtures with a high concentration of organic compounds makes it possible to use a wide range of ORMASILS as a membrane-forming component. Both hydrolyzable (methoxy- or ethoxy-) and non-hydrolyzable substituents affect the analytical performance of the obtained biosensors. The choice of the monomer used and its concentration in the membrane-casting solution can lead to biosensors with attractive characteristics. Thus, it was shown that a lactate biosensor based on 1% MAPS had a sensitivity of 0.5 $A \cdot M^{-1} \cdot cm^{-2}$ with a detection limit of 0.5 μM. The biosensor exhibited an appropriate stability and an excellent selectivity and may find an application in clinical analysis and food quality control. The use of the proposed approach with different siloxane monomers may lead to further improvement of biosensor characteristics. Moreover, interesting results can be obtained using polysiloxanes, as well as siloxane copolymers with different structures. Using molecularly imprinted polymers on the basis of different polysiloxanes in addition to the enzyme immobilization technique might provide a more appropriate enzyme microenvironment. Moreover, the proposed approach may be suitable for the immobilization of other enzymes, as was shown for GOx and the APTES-based membranes [18], especially if the enzyme substrate is negatively charged.

Supplementary Materials: The following supporting information can be downloaded at: https://www.mdpi.com/article/10.3390/s23084014/s1, Figure S1: Dependence of the lactate biosensor sensitivity on the siloxane concentration in the membrane solution. Siloxanes: ETES (■) and MTES (•); Figure S2: Dependence of the lactate biosensor sensitivity on the siloxane concentration in the membrane solution. Siloxanes: VTMS (■) and VTES (•); Figure S3: Cyclic voltammogram for the lactate biosensor based on 1% MAPS in phosphate buffer solution with different lactate concentrations (2 mV/s); Figure S4: The operational stability of the lactate biosensors based on 1% MAPS; (E = 0.00 V, 0.25 mM lactate, phosphate buffer, pH 6.0, upon stirring).

Author Contributions: Conceptualization, D.V.V.; methodology, D.V.V. and A.A.K.; investigation, D.V.V., O.E.S. and K.E.B.; resources, A.A.K.; data curation, D.V.V.; writing—original draft preparation, D.V.V.; writing—review and editing, A.A.K.; visualization, D.V.V.; supervision, A.A.K.; project administration, D.V.V.; funding acquisition, A.A.K. All authors have read and agreed to the published version of the manuscript.

Funding: The financial support of the Russian Science Foundation (RSF) through grant no. 19-13-00131 (https://rscf.ru/en/project/19-13-00131/, accessed on 15 March 2023) is greatly acknowledged.

Institutional Review Board Statement: Not applicable.

Informed Consent Statement: Not applicable.

Data Availability Statement: Not applicable.

Conflicts of Interest: The authors declare no conflict of interest.

References

1. Broder, G.; Weil, M.H. Excess lactate: An index of reversibility of shock in human patients. *Science* **1964**, *143*, 1457–1459. [CrossRef] [PubMed]
2. Schuster, H.P. Prognostic value of blood lactate in critically ill patients. *Resuscitation* **1984**, *11*, 141–146. [CrossRef] [PubMed]
3. Hameed, S.M.; Aird, W.C.; Cohn, S.M. Oxygen delivery. *Crit. Care Med.* **2003**, *31*, S658–S667. [CrossRef]
4. Yadigaroğlu, M.; Çömez, V.V.; Gültekin, Y.E.; Ceylan, Y.; Yanık, H.T.; Yadigaroğlu, N.Ö.; Yücel, M.; Güzel, M. Can lactate levels and lactate kinetics predict mortality in patients with COVID-19 with using qCSI scoring system? *Am. J. Emerg. Med.* **2023**, *66*, 45–52. [CrossRef] [PubMed]
5. Jacobs, I. Blood Lactate. *Sports Med.* **1986**, *3*, 10–25. [CrossRef] [PubMed]
6. Bravo, I.; Revenga-Parra, M.; Pariente, F.; Lorenzo, E. Reagent-less and robust biosensor for direct determination of lactate in food samples. *Sensors* **2017**, *17*, 144. [CrossRef]
7. Mizutani, F.; Sasaki, K.; Shimura, Y. Sequential determination of L-lactate and lactate dehydrogenase with immobilized enzyme electrode. *Anal. Chem.* **1983**, *55*, 35–38. [CrossRef]
8. Mascini, M.; Moscone, D.; Palleschi, G. A lactate electrode with lactate oxidase immobilized on nylon net for blood serum samples in flow systems. *Anal. Chim. Acta* **1984**, *157*, 45–51. [CrossRef]
9. Romero, M.R.; Garay, F.; Baruzzi, A.M. Design and optimization of a lactate amperometric biosensor based on lactate oxidase cross-linked with polymeric matrixes. *Sens. Actuator B Chem.* **2008**, *131*, 590–595. [CrossRef]
10. Yashina, E.I.; Borisova, A.V.; Karyakina, E.E.; Shchegolikhina, O.I.; Vagin, M.Y.; Sakharov, D.A.; Tonevitsky, A.G.; Karyakin, A.A. Sol-Gel Immobilization of Lactate Oxidase from Organic Solvent: Toward the Advanced Lactate Biosensor. *Anal. Chem.* **2010**, *82*, 1601–1604. [CrossRef]
11. Pribil, M.M.; Cortés-Salazar, F.; Andreyev, E.A.; Lesch, A.; Karyakina, E.E.; Voronin, O.G.; Girault, H.H.; Karyakin, A.A. Rapid optimization of a lactate biosensor design using soft probes scanning electrochemical microscopy. *J. Electroanal. Chem.* **2014**, *731*, 112–118. [CrossRef]
12. Garjonyte, R.; Yigzaw, Y.; Meskys, R.; Malinauskas, A.; Gorton, L. Prussian Blueand lactate oxidase-based amperometric biosensor for lactic acid. *Sens. Actuator B Chem.* **2001**, *79*, 33–38. [CrossRef]
13. Gupta, R.; Chaudhury, N.K. Entrapment of biomolecules in sol–gel matrix for applications in biosensors: Problems and future prospects. *Biosens. Bioelectron.* **2007**, *22*, 2387–2399. [CrossRef] [PubMed]
14. Dave, B.C.; Dunn, B.; Valentine, J.S.; Zink, J.I. Sol-gel encapsulation methods for biosensors. *Anal. Chem.* **1994**, *66*, 1120A–1127A. [CrossRef]
15. Lev, O.; Tsionsky, L.; Rabinovich, L.; Glezer, V.; Sampath, S.; Pankratov, I.; Gun, J. Organically modified sol-gel sensors. *Anal. Chem.* **1995**, *67*, 22A–30A. [CrossRef]

16. Flora, K.K.; Brennan, J.D. Effect of Matrix Aging on the Behavior of Human Serum Albumin Entrapped in a Tetraethyl Orthosilicate-Derived Glass. *Chem. Mater.* **2001**, *13*, 4170–4179. [CrossRef]
17. Winter, R.; Hua, D.W.; Song, X.; Mantulin, W.; Jonas, J. Structural and dynamical properties of the sol-gel transition. *J. Phys. Chem.* **1990**, *94*, 2706–2713. [CrossRef]
18. Karpova, E.V.; Shcherbacheva, E.V.; Galushin, A.A.; Vokhmyanina, D.V.; Karyakina, E.E.; Karyakin, A.A. Non-invasive diabetes monitoring through continuous analysis of sweat using flow-through glucose biosensor. *Anal. Chem.* **2019**, *91*, 3778–3783. [CrossRef]
19. Gomes, S.P.; Odložilíková, M.; Almeida, M.G.; Araújo, A.N.; Couto, C.M.; Montenegro, M.C.B. Application of lactate amperometric sol–gel biosensor to sequential injection determination of l-lactate. *J. Pharm. Biomed. Anal.* **2007**, *43*, 1376–1381. [CrossRef]
20. Nikitina, V.N.; Daboss, E.V.; Vokhmyanina, D.V.; Solovyev, I.D.; Andreev, E.A.; Komkova, M.A.; Karyakin, A.A. The widest linear range of glucose test strips based on various mediators and membranes for whole blood analysis. *J. Electroanal. Chem.* **2023**, 117445. [CrossRef]
21. Bernards, T.N.M.; van Bommel, M.J.; Boonstra, A.H. Hydrolysis-condensation processes of the tetra-alkoxysilanes TPOS, TEOS and TMOS in some alcoholic solvents. *J. Non-Cryst. Solids.* **1991**, *134*, 1–13. [CrossRef]
22. Pundir, C.S.; Narwal, V.; Batra, B. Determination of lactic acid with special emphasis on biosensing methods: A review. *Biosens. Bioelectron.* **2016**, *86*, 777–790. [CrossRef] [PubMed]
23. Batra, B.; Narwal, V.; Pundir, C.S. An amperometric lactate biosensor based on lactate dehydrogenase immobilized onto graphene oxide nanoparticlesmodified pencil graphite electrode. *Eng. Life Sci.* **2016**, *16*, 786–794. [CrossRef]
24. Vokhmyanina, D.; Daboss, E.; Sharapova, O.; Mogilnikova, M.; Karyakin, A. Single Printing Step Prussian Blue Bulk-Modified Transducers for Oxidase-Based Biosensors. *Biosensors* **2023**, *13*, 250. [CrossRef]
25. Karpova, E.V.; Karyakina, E.E.; Karyakin, A.A. Accessing Stability of Oxidase-Based Biosensors via Stabilizing the Advanced H_2O_2 Transducer. *J. Electrochem. Soc.* **2017**, *164*, B3056. [CrossRef]
26. Vokhmyanina, D.V.; Andreeva, K.D.; Komkova, M.A.; Karyakina, E.E.; Karyakin, A.A. 'Artificial peroxidase' nanozyme—Enzyme based lactate biosensor. *Talanta* **2020**, *208*, 120393. [CrossRef]

sensors

MDPI

Article

Psychophysiological Parameters Predict the Performance of Naive Subjects in Sport Shooting Training

Artem Badarin [1,2,*], Vladimir Antipov [1], Vadim Grubov [1], Nikita Grigorev [3], Andrey Savosenkov [3], Anna Udoratina [3], Susanna Gordleeva [1], Semen Kurkin [1], Victor Kazantsev [1] and Alexander Hramov [1]

[1] Baltic Center of Neurotechnology and Artificial Intelligence, Immanuel Kant Baltic Federal University, Kaliningrad 236041, Russia

[2] Neuroscience and Cognitive Technology Laboratory, Innopolis University, Kazan 420500, Russia

[3] Neurodynamics and Cognitive Technology Laboratory, Lobachevsky State University of Nizhny Novgorod, Nizhny Novgorod 603022, Russia

* Correspondence: badarin.a.a@mail.ru

Abstract: In this study, we investigated the neural and behavioral mechanisms associated with precision visual-motor control during the learning of sport shooting. We developed an experimental paradigm adapted for naïve individuals and a multisensory experimental paradigm. We showed that in the proposed experimental paradigms, subjects trained well and significantly increased their accuracy. We also identified several psycho-physiological parameters that were associated with shooting outcomes, including EEG biomarkers. In particular, we observed an increase in head-averaged delta and right temporal alpha EEG power before missing shots, as well as a negative correlation between theta-band energies in the frontal and central brain regions and shooting success. Our findings suggest that the multimodal analysis approach has the potential to be highly informative in studying the complex processes involved in visual-motor control learning and may be useful for optimizing training processes.

Keywords: EEG; training; brain; sport shooting; biomarkers

Citation: Badarin, A.; Antipov, V.; Grubov, V.; Grigorev, N.; Savosenkov, A.; Udoratina, A.; Gordleeva, S.; Kurkin, S.; Kazantsev, V.; Hramov, A. Psychophysiological Parameters Predict the Performance of Naive Subjects in Sport Shooting Training. *Sensors* **2023**, *23*, 3160. https://doi.org/10.3390/s23063160

Academic Editor: Evgeny Katz

Received: 20 February 2023
Revised: 10 March 2023
Accepted: 13 March 2023
Published: 16 March 2023

1. Introduction

Sport shooting represents a complex sensorimotor process requiring a high level of visuospatial work. Shooting sports demand athletes maintain a good psychological state [1], stress control ability, and the ability to efficiently allocate cognitive resources (e.g., attention) during the shooting and aiming period [2]. As a consequence, training in sport shooting is a non-trivial challenge that often requires an individualized approach, especially in a sport with such high achievements [3]. The identification of the psychological and psychophysiological profile of a successful shooter is associated with superior performance, and the building of a training strategy focused on achieving the quickest achievement of this state could help in solving this problem.

The development of modern, compact, and mobile devices for multimodal monitoring of human physiological parameters and the rapid progress in neuroimaging technologies makes it possible to monitor the current state of an athlete concurrently with their behavioral performance to form representations of a successful profile. Currently, research in this direction is mainly focused on identifying biomarkers of the cardiovascular and respiratory systems operation [2], gaze behavior [4], as well as EEG biomarkers of successful shooters [3]; this research is generally based on the comparison of novice shooters with professional athletes [4–6]. However, with this approach, it is impossible to obtain information about the "trajectory" of the transformation from a novice shooter to a professional. A promising experimental paradigm from this point of view is the paradigm aimed at comparing successful and unsuccessful attempts at sport shooting training sessions in a naïve group. This approach makes it possible to identify what distinguishes successful attempts in novice athletes at the level of physiological parameters and EEG characteristics and to investigate the effect of training in detail. Recently, a trend for research in this direction has emerged. Note the study [7] that revealed EEG and kinematic biomarkers of precision motor control and changes in the neurophysiological substrates in naïve participants that may underlie motor learning during simulated marksmanship in immersive virtual reality.

However, many issues still remain unexplored. In particular, it is unknown exactly how and which physiological parameters and EEG characteristics change during sports shooting training; for example, which parameters correlate with shooting success and can thus claim to be biomarkers that are components of a professional athlete's profile. Moreover, most studies generally examine the dynamics of one or two physiological parameters during shooting training (e.g., a study [2] utilized synchronized monitoring of EEG and electrocardiogram (ECG) to understand the mechanism of dual activation of the brain and heart in pistol athletes during shooting performances). At the same time, a deeper understanding of the relationship between physiological and psychological processes and training success can only be achieved by simultaneously considering as many physiological parameters as possible. Biomarkers of successful sport shooting should be searched not at the level of operation of individual subsystems of the human body but at the level of their joint operation and interaction; therefore, it is necessary to use multimodal registration of physiological parameters to solve this problem [2].

The present study takes a step toward solving the problems formulated. Here, we analyze multimodal data of subjects (EEG, ECG, electrooculogram (EOG), respiration activity (R), and fatigue tests) naïve to sport shooting training and study correlations between the psychophysiological parameters and shooting performance of the subjects. The special aspect of this study is the analysis of changes in fatigue levels during training and its effect on shooting success.

From a fundamental point of view, sport provides an ideal model for understanding neural adaptations associated with intensive training over time. We believe that the increased knowledge of links between physiological parameters, brain activity, and behavior characteristics will help to improve the effect of sport shooting training and thus enhance sports performance.

2. Materials and Methods

2.1. Participants

Experimental study included 21 healthy volunteers (all male, age 19–25, with an average age of 21 and a standard deviation of $\sim$1.5 , right-handed). All subjects had no diseases that affected sight or locomotor functions. A healthy lifestyle was advised for the subjects prior to the experiment, which included sufficient night rest, no alcohol or drug consumption, and moderate physical activity. All subjects were volunteers; they were informed about the details of the study prior to participation, were able to ask related questions, and after that, provided informed consent. All participants were naïve to sport shooting, so before the experiment, a trained coach explained to them the basic principles and safety regulations. This study was conducted according to the guidelines of the Declaration of Helsinki and approved by the Ethics Committee of Lobachevsky University (Protocol №3 from 8 April 2021).

2.2. Experimental Setup

During the experiment, we recorded multimodal data from a subject: EEG, EOG, ECG, respiration activity (R). The placement of all sensors is shown in Figure 1A. All these signals were recorded by a wearable EEG recorder "Encephalan-EEGR-19/26" (Medicom MTD, Russia). The sampling rate for all types of data was 250 Hz. For EEG recording, we used 31 Ag/AgCl electrodes placed on the scalp according to the international scheme "10-10" (Figure 1A, grey circles). Other biological signals, besides EEG, were acquired through additional POLY channels of "Encephalan". To record EOG, we used 2 electrodes ("EOG+" and "EOG-") above and below the right eye (Figure 1A, green circles). The resulting EOG signal was calculated as the difference between these two signals. The right eye was chosen as it is usually the one used while aiming the shot. To record ECG, we placed 1 electrode on the subject's back near the left scapula (Figure 1A, blue circle). Respiration activity was collected via a belt-shaped sensor wrapped around the subject's chest (Figure 1A, white stripe). The stretching and contraction of the belt are associated with the expansion and compression of the thorax during respiration.

When choosing the sensors' placement, we tried not to restrict the subject's movement and, at the same time, tried to minimize the influence of this movement on the recorded signals. The "Encephalan" device was placed on the small of the back with a special belt, and all wires from the device to the sensors were tightly packed together and fixed on the back. The "Encephalan" was connected to the PC through Bluetooth, so this connection provided no additional restriction on the subject's movement.

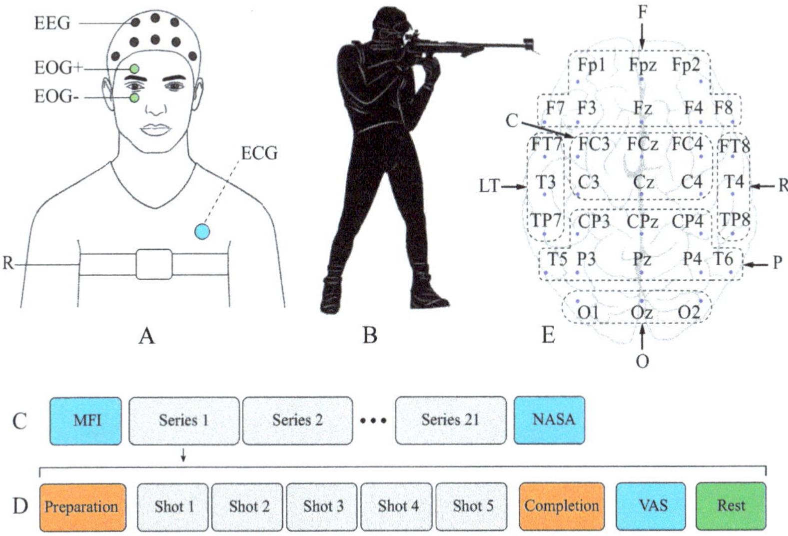

Figure 1. (**A**) Experimental setup with sensors: EEG (grey circles), EOG (green circles), ECG (blue circle), respiration (white stripe); (**B**) Shooting stance; (**C**) General design of the experimental session; (**D**) Design of individual series; (**E**) Scheme of EEG electrodes placement "10-10". Chosen areas of EEG signals averaging are shown with dotted frames: frontal (F), central (C), parietal (P), occipital (O), left temporal (LT), and right temporal (RT).

2.3. Experimental Procedure

The shooting was performed from an upright position, as illustrated in Figure 1B. For the experiment, we chose an air rifle with characteristics close to the real rifle used by sportsmen in biathlons. The rifle's dimensions are 1010/270/85 mm (length/height/width), and its weight is ∼4 kg. The rifle uses a 4.5 mm caliber with a 5-round magazine and open sights. Since this was an air rifle, the recoil was not significant. Protective gear included shooting glasses but not headphones. The subjects shot at 5 separate targets at a distance of 10 m. The targets mimicked the ones used in biathlons at a distance of 50 m, so the targets in the experiment were properly scaled in size. The subject had visual and audial feedback after each shot—the successfully struck target changed color and provided distinct sound.

The experimental session included 21 series of shootings with Multidimensional Fatigue Inventory (MFI-20) [8], and the NASA Task Load Index (NASA-TLX) [9] tests before the first and after the last series correspondingly (see Figure 1C). The first series was treated as a test, so these results were excluded from further analysis. Each series included the following steps (see Figure 1D):

- **Preparation**—the subject received the rifle loaded with 5 bullets from the assistant and assumed shooting stance;
- **Shooting**—the subject performed 5 shots at 5 targets in any order;
- **Completion**—the subject quit shooting stance and handed the rifle back to the assistant for reloading;
- **VAS**—the subject passed a visual analog scale (VAS) test [10] for fatigue estimation;
- **Rest**—the subject rested for 60 s before the next series.

To assess changes in some behavioral and physiological characteristics throughout the experiment, we turned 20 series of shootings into 4 blocks. This was done by averaging results of 5 consecutive series, i.e., 1–5, 6–10, 11–15, 16–20.

MFI-20 is the test aimed at assessing a subject's fatigue through self-report. This test includes 20 questions covering 5 dimensions of fatigue: Physical, Mental, and General Fatigue, as well as Reduced Activity and Motivation. NASA-TLX is another instrument to measure fatigue, but in this case, task-induced fatigue. The test includes several scales and their paired comparisons that help to assess 6 factors: Physical, Mental, and Temporal Demand, as well as Effort, Frustration, and Performance. VAS is used to subjectively measure the fatigue of the subject in his current state. Self-report is performed with the help of a continuous scale, on which the subject chooses the value of his current fatigue. The scale varies between "the lowest" and "the highest fatigue". For all fatigue-assessment tests, we used a tablet computer.

We considered several factors during statistical analysis:

- "block"—reflects the course of the experiment, includes blocks 1–4;
- "phase"—reflects the subject's type of activity in the experiment, including rest and shooting;
- "result"—reflects successfulness on each shot, including hits and misses.

2.4. Data Processing

The goals of preprocessing procedure were the following: for EEG data—to obtain clear signals without noises and artifacts for further time-frequency analysis, for respiration, EOG, and ECG—to obtain signals clear enough for extracting desired features such as blink rate or heart rate.

For EEG preprocessing, we used Fieldtrip toolbox for MATLAB [11]. EEG signals were filtered with a band-pass filter (cut-off frequencies—1 and 70 Hz) and 50 Hz notch filter in preparation for further time-frequency analysis.

To remove eye- and heart-related activity artifacts from EEG, we used a method based on Independent Component Analysis (ICA). For this, we applied *ft_componentanalysis* with the method *runica*. We decomposed EEG data into a set of independent components, searched components with artifacts, removed them, and then restored EEG signals with the remaining components. To ensure data quality, we performed additional visual data analysis with *ft_rejectvisual*. We rejected trials of data and/or EEG channels with severe artifacts remaining after the ICA-based procedure. Most of these artifacts were related to the subject's active movement. We removed "bad" trials from the dataset, while for "bad" channels, we performed a repairing procedure with *ft_channelrepair*.

We performed a time-frequency analysis of EEG signals using continuous wavelet transform (CWT) with Morlet mother wavelet function [12]. We considered wavelet power (WP) as $W_n(f,t)$, where $n = 1, 2, \ldots, N$ is the number of EEG channel ($N = 31$ for the considered dataset), f and t are the frequency and time point. WP is one of the common CWT-based characteristics to describe the time-frequency structure of a signal [13].

To reduce the data dimensionality, we considered averaged CWT spectra. Firstly, we averaged WP over several areas in the cortex: frontal (F), central (C), parietal (P), occipital (O), left temporal (LT), and right temporal (RT) (see Figure 1E). Secondly, we averaged WP over commonly used frequency bands: delta (1–4 Hz), theta (4–8 Hz), alpha (8–13 Hz), and beta (13–30 Hz). In our research, we considered a 2-s time interval just before the subject pulled the trigger. So we additionally averaged WP over this time interval.

We used the NeuroKit2 software package to process signals obtained from the respiratory sensor. NeuroKit2 is an open-source Python package designed to process neurophysiological signals [14]. For primary processing and filtering of the incoming signal, we used a linear detrending method with subsequent application of a low-pass fifth-order IIR Butterworth filter at the frequency of 2 Hz. The procedure is based on the zero-crossing algorithm with the amplitude threshold described in [15]. Then, we determined peaks (beginning of exhalation) and valleys (beginning of inhalation) using different sets of parameters described in [15]. Next, we determined the breathing phase defined between "1" for inspiration (inhalation) and "0" for expiration (exhalation). Then, we calculated the instantaneous frequency of the signal (in "1/min") from a series of peaks. It is calculated as "60/period", where the period is the time between peaks. To interpolate the frequency over the entire duration of the signal, the monotone cubic interpolation method was used. We also calculated the average values of frequencies at different stages of the experiment. For this purpose, the instantaneous respiration rate was calculated for each session at the moments of shooting and rest; further, the obtained rate values were averaged and added up for each subject.

We analyzed EOG to detect eye movement and blinking using the methods of the software package MNE [16], which turned out to be the most effective for this problem. We used a default set of parameters for this method. Additionally, we obtained the values of the signal peaks, which correspond to the moments of the subject's blinks. Next, we calculated the blink rate (in minutes) from the series of peaks as "60/period". Monotone cubic interpolation method was used to interpolate the frequency for the entire duration of the signal. Then, the average values of blink rates at the moments of shooting and rest were obtained for each subject.

To process the ECG signal, we filtered the data using high-pass and low-pass filters in the 1–6 Hz range. Further, R-peaks, which are distinguished by high amplitude and frequency, were selected from the prepared signal. We calculated heart rate as the inverse of the R-R interval ($1/t_{R-R}$). All heart rate values for each individual step were averaged for each subject.

We have considered different time window scales for the analysis of heart rate, respiration rate, and blink rate. To find a difference between stages of the experiment (rest vs. shooting), we averaged heart rate, respiration rate, and blink rate in windows length equal to respective stages. The time length of windows for the resting stage is 60 s, but windows for the shooting stage have different lengths (average length of 22.5 s) because of different rates of shooting across the subjects and shooting stages. Additionally, we analyzed the influence of instantaneous (right at the moment of shot) RR on shooting results.

The main effects at the group level were evaluated via Repeated Measures Analysis of Variance (RM ANOVA). We considered "block", "result", "phase", and cortical area as within-subject factors in those statistical tests where the influence of these factors was considered. The post hoc analysis used either paired samples *t*-test or Wilcoxon signed-rank test, depending on the samples' normality. Normality was tested via the Shapiro–Wilk test. The group-level correlation analysis between all pairs of characteristic changes during the experiment, such as heart rate, respiration rate, characteristic of the brain activity, hit rate, and subjective fatigue, was performed using repeated measures correlation. Correlations between subjective tests (MFI-20, NASA-TLX) and shooting accuracy were searched using Spearman's rank correlation coefficient. We used several open-source statistical packages in Python, such as Pingouin, SciPy, statsmodels, and a package called JASP for statistical analysis and results visualization.

3. Results

3.1. The Behavioral Data Analysis

The results of the assessing subject's state before the experimental task with the MFI-20 test are shown in Figure 2A. The median values are low (less than 8 out of a possible

20) across all scales of MFI-20, which confirms that none of the subjects has asthenia of any type.

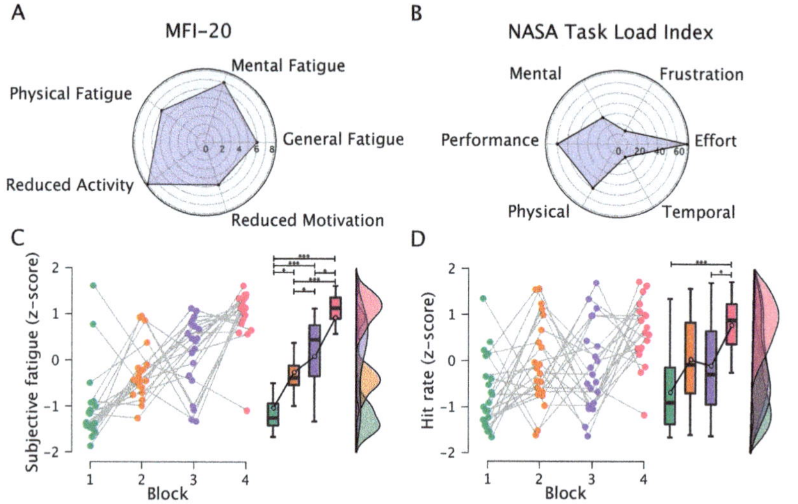

Figure 2. Results of the behavioral data analysis: (**A**) Median values for the scales of MFI-20 in the group of subjects; (**B**) Median values for the scales of NASA-TLX in the group of subjects; (**C**) Subjective fatigue (z-score); (**D**) Hit rate (z-score). Dots correspond to individual subjects, while box and whisker plots show values averaged over the blocks of the experiment. The symbol * denotes statistical significance in post hoc analysis using t-test with Holm's correction for multiple comparisons (*—$p < 0.05$, ***—$p < 0.001$).

To assess the task-induced load, we used a NASA-TLX test, and the results are shown in Figure 2B. We found that the experimental task induces low temporal and mental loads, while the main load is caused by the effort to preserve a certain level of performance.

The results of the change in fatigue level during the task assessed with VAS after each series of shootings are shown in Figure 2C. We considered z-scored results of VAS for a more universal data presentation. We found a significant increase in fatigue from block to block, and post hoc analysis showed significant differences between all blocks of the experiment. However, absolute values for the induced increase in fatigue (i.e., the difference between fatigue at the beginning and at the end of the experiment) are close to 30 out of 100 (maximal value in the scale). We suggest that this result indicates a low overall increase in fatigue during the experiment.

We used the hit rate as a parameter for evaluating the success of performance. The subjects coped well with the task: ∼65% of the shots hit the target on average. We analyzed changes in hit rate over the course of the experiment and found a significant increase in hit rate (RM ANOVA: $p < 0.001$). Post hoc analysis showed significant differences between the first and fourth blocks, as well as between the third and fourth blocks.

3.2. The Physiological Data Analysis

We analyzed changes in physiological characteristics during the experimental task, both in the resting and shooting phases.

3.2.1. Heart Rate

We did not find significant changes in the heart rate during the experiment, as well as no significant differences between heart rates at rest and shooting phases. However, we found an interaction effect between factors "block" and "phase" ($p = 0.000531$). Post hoc analysis showed that there are significant differences in heart rate between blocks 1–3 and 1–4 in the rest phase (see Figure 3A). Additionally, we considered heart rate variability as another characteristic of heart activity but did not find significant changes.

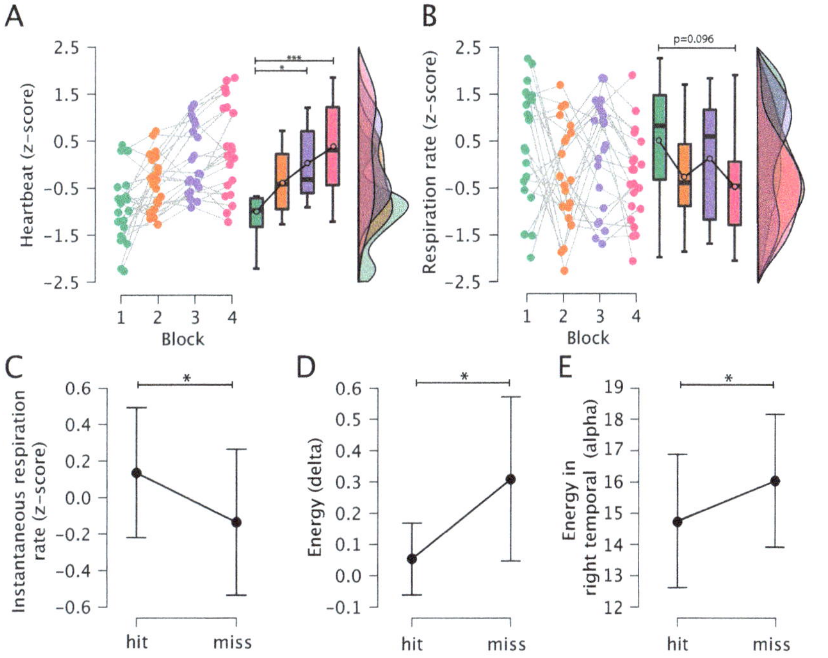

Figure 3. Results of the physiological data analysis: (**A**) Heartbeat at the rest phase (z-score); (**B**) Respiration rate during the shooting phase (z-score); (**C**) Instantaneous respiration rate corresponding to the shot time for hit and miss; (**D**) Average energy value in delta range before the shot for hits and misses; (**E**) Average energy value in the right temporal lobe in the alpha range before the shot for hits and misses. Dots correspond to individuals subjects while box and whisker plots show values averaged over the blocks of the experiment. The symbol * denotes statistical significance in post hoc analysis using t-test with Holm's correction for multiple comparisons (*—$p < 0.05$, ***—$p < 0.001$).

3.2.2. Respiration Rate

Then, we analyzed the dynamics of respiration rate and found the interaction effect between "block" and "phase" ($p = 0.045$) factors, while no changes were detected in respiration rate during the experiment and between the phases. In the post hoc analysis, we found a decrease in respiration rate during the shooting phase, but the statistical significance of these changes is near the accepted threshold (see Figure 3B). Further, we studied the effect of instantaneous respiration rate on shooting success and found a significant difference in the instantaneous respiration rate between misses and hits ($p = 0.043$, see Figure 3C).

3.2.3. Blinking Rate

We have not found significant changes in the blinking rate during the experiment or any relationship between the blinking rate and the hit rate.

3.2.4. Brain Electrical Activity

We analyzed changes in the electrical activity of the brain directly before each shot, both for the "block" and "result" factors. We did not find significant changes during the experiment. However, for energy in the delta range, we found a main effect of shooting results ($p = 0.042$) and cortex areas ($p = 0.013$) (see Figure 3D).

For energy in the alpha range, we did not reveal the main effects. Nevertheless, we found an interaction effect between cortex areas and shooting results ($p = 0.016$). In the post hoc analysis, we found significant changes in the right temporal lobe in the alpha range ($p = 0.049507$); however, a p-value was not adjusted for multiple comparisons.

Finally, we revealed that energies in the delta range and the alpha range in the right temporal lobe were significantly less before a hit compared to a miss.

3.3. Correlation Analysis

To identify the relationships between the characteristics under study, we performed a correlation analysis. The results of correlation analysis are shown in Table 1.

Table 1. The results of correlation analysis.

	Hit Rate	Subjective Fatigue
Heart rate (resting phase)	-	r = 0.42, $p = 0.006$
Respiration rate (resting phase)	r = 0.33 , $p = 0.03$	-
Respiration rate (shooting phase)	r = −0.35, $p = 0.02$	-
Energy (theta; frontal)	r = −0.33, $p = 0.0073$	-
Energy (theta; central)	r = −0.33, $p = 0.0076$	-
NASA-TLX	$\rho = -0.532$, $p = 0.013$	-

We discovered that changes in subjective fatigue positively correlate with average heart rate in the rest phase ($r = 0.42$). Simultaneously, the hit rate correlates with the following parameters: respiration rate in the resting phase ($r = 0.33$), respiration rate in the shooting phase ($r = -0.35$), and energies before the shot in the theta range in the frontal and central regions ($r = -0.33$ and $r = -0.33$, respectively).

Additionally, we identified the correlation between the NASA-TLX and the hit rate ($\rho = -0.532$).

4. Discussion

We analyzed multimodal psychophysiological data (EEG, ECG, EOG, respiration activity, and fatigue) to explore the neural and behavioral mechanisms underlying precision visual-motor control learning during sports shooting tasks. We systematically studied the relationship between physiological parameters, brain activity, and shooting performance over the course of learning to identify biomarkers that can be used to infer complex motor behavior. As expected, naive subjects significantly increased their hit rate during practice. Participants became, on average, $\approx$30% more accurate at shooting targets. Analysis of the physiological activity showed that performance improvements during the course of learning were accompanied by an increase in subjective fatigue and heart rate, wherein the average breathing rate remained unchanged. Respiration rate during shooting negatively correlates with marksmanship performance. We also found that the instantaneous respiration rate before a hit is higher than before a miss. Note that the work [17] did not reveal the influence of the instantaneous respiration rate on the shooting results. However, in study [18], the authors showed that respiration rate is related to the mental load, with high and medium load characterized by a significantly higher rate. In this regard, we hypothesize that our results may reflect a connection between mental load, instantaneous respiration rate, and shooting results. We suggest that in the case of a hit, the subjects were more deeply immersed and concentrated on the task and, accordingly, experienced a higher mental load than in the case of a miss.

Analysis of brain activity reveals several markers associated with shooting success. We found that average energy values in the delta range and the alpha range in the right temporal lobe were significantly less before a successful shot than before a miss. We identified that the hit rate negatively correlates with energies in the theta range in the frontal and central regions during the aiming period before shot execution.

The sport marksmanship task used in this study is one of the most convenient examples of tasks that can be used for investigating neurophysiological mechanisms underlying precise visual–motor coordination in a complex naturalistic context. Usually, studies addressing visual–motor integration by analyzing noninvasive recordings of cortical activity, such as EEG, involve laboratory tasks with minimal mobility to reduce artifact-producing muscle activity. Real-world tasks in natural environments require unrestricted full-body movements arising from full engagement of perception, decision-making, error recognition, and motor control. The shooting task is a controlled, easily replicated natural exercise that is particularly useful for investigating psychophysiological markers of visual-motor skill learning because it produces discrete measures of performance, which can be compared with electrophysiological activity recorded in real-time.

One of the main goals of this study was to identify EEG biomarkers of visual-motor skill learning during sport shooting tasks. Biomarkers are often referred to as quantitative indicators of a biological organism's state and can be used to describe behavior-related psychophysiological processes. In recent years, the relationship of biomarkers with certain skills has been actively investigated [19–21]. The identified associations of skills with biomarkers are a promising tool for training process optimization.

In this study, we analyzed the relationship between EEG power in different frequency bands during the aiming period and shooting performance. We found that novices demonstrated delta and right temporal alpha EEG power increase before missing shots. Our results are in line with other studies reporting an overall reduction of alpha activity for experienced shooters [6]. This effect is interpreted as a greater engagement of task-relevant attentional processes. Janelle et al. [4] showed that shooting task expertise interacted with hemispheric activation levels. They demonstrated stronger alpha activity in the left hemisphere accompanied by its reduction in the right hemisphere for experts as compared to novices during the preparatory period before shot execution. Since shooting places high demands on visuospatial processing, the elevation of alpha power in the left temporal area may indicate a decrease of non-relevant to task cognitive activity (cognitive thinking, self-talk, or language analysis) and show that marksmen focused their attention on the visuospatial work dominated by right-brain areas [22–24].

Our results show the existence of a negative correlation between theta-band energies in the frontal and central brain regions during the preparation period and shooting success. Frontal midline (Fm) theta activation has often been observed in tasks that required consistent attention to a stimulus [25]. Recent studies reported Fm theta power as an indicator of sustained [26,27] and internalized [28,29] attention found in the preparation period in motor performance. Fm theta activity is linked to various kinds of attentional or working memory processes, such as working memory [30–32], learning [33], concentration [34], and action monitoring [35]. Sauseng et al. [26] associated Fm theta power with the number of cognitive resources allocated to attentional processes during a complex finger movement task learning. They clearly showed that Fm theta increased with increasing mental efforts and task demands. The results of our study are in line with these findings, demonstrating weaker theta activation with increasing correct acquisitions of the task and experience by the novice. Sport shooting task highly demands focused attention and precision visual-motor control. Shooting learning requires from the naive subjects a lot of cognitive resources and mental engagement. Therefore, the shooting training process is accompanied by a high level of mental effort reflected by increasing theta energy in the frontal and central brain regions. This explanation is confirmed by comparing the perceived workload level evaluated by NASA-TLX with the hit rate (see Table 1). Subjects with high hit rates reported greater confidence by feeling less workload level (lower levels of stress and pressure). In line with these results, Borghini et al. [36] demonstrated that the variation of the EEG power spectra in frontal areas in the theta band could be used as a measure for the training improvements of novices in flight simulation tasks. Their results showed that behavioral and task performance improvement was accompanied by a significant decrease in the theta band power over the frontal areas. Interestingly, the comparison of the time course of Fm theta during the aiming period in rifle shooting between experts and novices reveals that the theta power increased during the aiming process before the shot only for experts but not for novices [5]. The authors assume that elite marksmen are better able to allocate cortical resources in time while novices are unable to focus attention exactly on the shooting time point.

Note that this study has several limitations. First is the small number of participants (21). The second limitation is that only males participated in this study. Another limitation is using only EEG for the brain activity analysis since EEG has low spatial resolution compared to other techniques such as fMRI (functional magnetic resonance imaging). The last limitation is especially significant in the case of a possible investigation of visual-motor connection. For instance, in a recent paper [37], the usage of fMRI allowed researchers to discover a disrupted visual-motor connection in psychiatric disorders. In this study, however, fMRI is very difficult to use without substantial changes in the experimental paradigm.

5. Conclusions

In conclusion, our study sheds light on the neural and behavioral mechanisms underlying precision visual-motor control learning during sport shooting. We found that performance improvements were accompanied by an increase in subjective fatigue and heart rate and that the respiration rate before a hit was higher than before a miss, potentially reflecting a connection between the mental load and shooting results. Additionally, we identified several EEG biomarkers of visual-motor skill learning, including head-averaged delta and right temporal alpha EEG power increase before missing shots and a negative correlation between theta-band energies in the frontal and central brain regions and shooting success. The results of this study highlight the importance of considering both neural and behavioral factors in precision visual-motor control learning and the potential for using psychophysiological parameters to improve shooting performance. These findings provide valuable insights into the neurophysiological mechanisms underlying visual-motor skill learning and have potential implications for the optimization of training processes.

Author Contributions: Conceptualization, A.B., S.G., V.K. and A.H.; Data curation, A.B. and V.G.; Formal analysis, A.B., V.A., V.G. and S.K.; Funding acquisition, A.B., V.K. and A.H.; Investigation, V.A., N.G., A.S., A.U., S.K. and A.H.; Methodology, A.B., S.G., V.K. and A.H.; Project administration, V.K. and A.H.; Resources, N.G., A.S. and A.U.; Software, V.A.; Supervision, V.K. and A.H.; Validation, V.A., V.G., N.G., A.S., A.U. and S.K.; Visualization, A.B. and S.K.; Writing—original draft, A.B., V.G., S.G., S.K. and A.H.; Writing—review and editing, A.B., S.G. and A.H. All authors have read and agreed to the published version of the manuscript.

Funding: This work was supported by the federal academic leadership program "Priority 2030" of the Ministry of Science and Higher Education of the Russian Federation in part of data collection. A.B. received support from the President Grant (MK-2142.2022.1.2) in part of conceptualization and methodology. A.H. thanks the Presidential Program to Support Leading Scientific Schools of the Russian Federation (grant NSH-589.2022.1.2).

Institutional Review Board Statement: This study was conducted according to the guidelines of the Declaration of Helsinki, and approved by the Ethics Committee of Lobachevsky University (Protocol №3 from 8 April 2021).

Informed Consent Statement: Informed consent has been obtained from all subjects involved in the study.

Conflicts of Interest: The authors declare no conflict of interest. The funders had no role in the design of the study; in the collection, analyses, or interpretation of data; in the writing of the manuscript, or in the decision to publish the results.

Abbreviations

The following abbreviations are used in this manuscript:

EEG	Electroencephalogram
CWT	Continuous wavelet transform
WP	Wavelet power
EOG	Electrooculogramm
ECG	Electrocardiogram
MFI	Multidimensional Fatigue Inventory
NASA-TLX	NASA task load index
VAS	Visual analog scale

References

1. Harris, D.J.; Allen, K.L.; Vine, S.J.; Wilson, M.R. A systematic review and meta-analysis of the relationship between flow states and performance. In *International Review of Sport and Exercise Psychology*; Taylor and Francis: London, UK, 2021; pp. 1–29.
2. Wang, K.; Li, Y.; Liu, H.; Zhang, T.; Luo, J. Relationship between Pistol Players' Psychophysiological State and Shot Performance: Activation Effect of EEG and HRV. *Scand. J. Med. Sci. Sport.* **2022**, *33*, 84–98. [CrossRef]
3. Fang, Q.; Fang, C.; Li, L.; Song, Y. Impact of sport training on adaptations in neural functioning and behavioral performance: A scoping review with meta-analysis on EEG research. *J. Exerc. Sci. Fit.* **2022**, 20, 206–215. [CrossRef]
4. Janelle, C.M.; Hillman, C.H.; Apparies, R.J.; Murray, N.P.; Meili, L.; Fallon, E.A.; Hatfield, B.D. Expertise differences in cortical activation and gaze behavior during rifle shooting. *J. Sport Exerc. Psychol.* **2000**, *22*, 167–182. [CrossRef]
5. Doppelmayr, M.; Finkenzeller, T.; Sauseng, P. Frontal midline theta in the pre-shot phase of rifle shooting: Differences between experts and novices. *Neuropsychologia* **2008**, *46*, 1463–1467. [CrossRef]
6. Hunt, C.A.; Rietschel, J.C.; Hatfield, B.D.; Iso-Ahola, S.E. A psychophysiological profile of winners and losers in sport competition. *Sport. Exerc. Perform. Psychol.* **2013**, *2*, 220. [CrossRef]
7. Liu, S.; Clements, J.M.; Kirsch, E.P.; Rao, H.M.; Zielinski, D.J.; Lu, Y.; Mainsah, B.O.; Potter, N.D.; Sommer, M.A.; Kopper, R.; et al. Psychophysiological Markers of Performance and Learning during Simulated Marksmanship in Immersive Virtual Reality. *J. Cogn. Neurosci.* **2021**, *33*, 1253–1270. [CrossRef] [PubMed]
8. Smets, E.; Garssen, B.; Bonke, B.d.; De Haes, J. The Multidimensional Fatigue Inventory (MFI) psychometric qualities of an instrument to assess fatigue. *J. Psychosom. Res.* **1995**, *39*, 315–325. [CrossRef]
9. Hart, S.G. NASA-task load index (NASA-TLX); 20 years later. In Proceedings of the Human Factors and Ergonomics Society Annual Meeting, Philadelphia, PA, USA, 1–5 October 2018; Sage Publications Sage: Los Angeles, CA, USA, 2006; Volume 50, pp. 904–908.
10. Tseng, B.Y.; Gajewski, B.J.; Kluding, P.M. Reliability, responsiveness, and validity of the visual analog fatigue scale to measure exertion fatigue in people with chronic stroke: A preliminary study. *Stroke Res. Treat.* **2010**, *2010*, 412964. [CrossRef]
11. Oostenveld, R.; Fries, P.; Maris, E.; Schoffelen, J.M. FieldTrip: Open source software for advanced analysis of MEG, EEG, and invasive electrophysiological data. *Comput. Intell. Neurosci.* **2011**, *2011*, 156869. [CrossRef] [PubMed]
12. Aldroubi, A.; Unser, M. *Wavelets in Medicine and Biology*; Routledge: Abingdon, UK, 2017.
13. Torrence, C.; Compo, G.P. A practical guide to wavelet analysis. *Bull. Am. Meteorol. Soc.* **1998**, *79*, 61–78. [CrossRef]
14. Makowski, D.; Pham, T.; Lau, Z.J.; Brammer, J.C.; Lespinasse, F.; Pham, H.; Schölzel, C.; Chen, S. NeuroKit2: A Python toolbox for neurophysiological signal processing. *Behav. Res. Methods* **2021**, *53*, 1689–1696. [CrossRef]
15. Khodadad, D.; Nordebo, S.; Müller, B.; Waldmann, A.; Yerworth, R.; Becher, T.; Frerichs, I.; Sophocleous, L.; Van Kaam, A.; Miedema, M.; et al. Optimized breath detection algorithm in electrical impedance tomography. *Physiol. Meas.* **2018**, *39*, 094001. [CrossRef]
16. Jas, M.; Larson, E.; Engemann, D.A.; Leppäkangas, J.; Taulu, S.; Hämäläinen, M.; Gramfort, A. A reproducible MEG/EEG group study with the MNE software: Recommendations, quality assessments, and good practices. *Front. Neurosci.* **2018**, *12*, 530. [CrossRef] [PubMed]
17. Guillot, A.; Collet, C.; Dittmar, A.; Delhomme, G.; Delemer, C.; Vernet-Maury, E. The physiological activation effect on performance in shooting. *J. Psychophysiol.* **2003**, *17*, 214–222. [CrossRef]
18. Marinescu, A.; Sharples, S.; Ritchie, A.; López, T.S.; McDowell, M.; Morvan, H. Exploring the relationship between mental workload, variation in performance and physiological parameters. *IFAC-PapersOnLine* **2016**, *49*, 591–596. [CrossRef]
19. Lee, K.; Liu, D.; Perroud, L.; Chavarriaga, R.; Millán, J.d.R. Endogenous control of powered lower-limb exoskeleton. In *Wearable Robotics: Challenges and Trends: Proceedings of the 2nd International Symposium on Wearable Robotics, WeRob2016, Segovia, Spain, 18–21 October 2016*; Springer: Berlin/Heidelberg, Germany, 2017; pp. 115–119.
20. Cheron, G.; Petit, G.; Cheron, J.; Leroy, A.; Cebolla, A.; Cevallos, C.; Petieau, M.; Hoellinger, T.; Zarka, D.; Clarinval, A.M.; et al. Brain oscillations in sport: Toward EEG biomarkers of performance. *Front. Psychol.* **2016**, *7*, 246. [CrossRef]
21. Paulus, M.P.; Potterat, E.G.; Taylor, M.K.; Van Orden, K.F.; Bauman, J.; Momen, N.; Padilla, G.A.; Swain, J.L. A neuroscience approach to optimizing brain resources for human performance in extreme environments. *Neurosci. Biobehav. Rev.* **2009**, *33*, 1080–1088. [CrossRef] [PubMed]

22. Hatfield, B.D.; Landers, D.M.; Ray, W.J. Cognitive processes during self-paced motor performance: An electroencephalographic profile of skilled marksmen. *J. Sport Exerc. Psychol.* **1984**, *6*, 42–59. [CrossRef]
23. Kerick, S.E.; McDowell, K.; Hung, T.M.; Santa Maria, D.L.; Spalding, T.W.; Hatfield, B.D. The role of the left temporal region under the cognitive motor demands of shooting in skilled marksmen. *Biol. Psychol.* **2001**, *58*, 263–277. [CrossRef]
24. Hillman, C.H.; Apparies, R.J.; Janelle, C.M.; Hatfield, B.D. An electrocortical comparison of executed and rejected shots in skilled marksmen. *Biol. Psychol.* **2000**, *52*, 71–83. [CrossRef]
25. Ishihara, T. Activation of abnormal EEG by mental work. *Rinsho Nohha (Clin. Electroencephalogr.)* **1966**, *8*, 26–34.
26. Sauseng, P.; Hoppe, J.; Klimesch, W.; Gerloff, C.; Hummel, F.C. Dissociation of sustained attention from central executive functions: Local activity and interregional connectivity in the theta range. *Eur. J. Neurosci.* **2007**, *25*, 587–593. [CrossRef]
27. Chuang, L.Y.; Huang, C.J.; Hung, T.M. The differences in frontal midline theta power between successful and unsuccessful basketball free throws of elite basketball players. *Int. J. Psychophysiol.* **2013**, *90*, 321–328. [CrossRef]
28. Aftanas, L.I.; Golocheikine, S.A. Human anterior and frontal midline theta and lower alpha reflect emotionally positive state and internalized attention: High-resolution EEG investigation of meditation. *Neurosci. Lett.* **2001**, *310*, 57–60. [CrossRef] [PubMed]
29. Park, D.C.; Lautenschlager, G.; Hedden, T.; Davidson, N.S.; Smith, A.D.; Smith, P.K. Models of visuospatial and verbal memory across the adult life span. *Psychol. Aging* **2002**, *17*, 299. [CrossRef] [PubMed]
30. Gevins, A.; Smith, M.E.; McEvoy, L.; Yu, D. High-resolution EEG mapping of cortical activation related to working memory: Effects of task difficulty, type of processing, and practice. *Cereb. Cortex* **1997**, *7*, 374–385. [CrossRef]
31. Jensen, O.; Tesche, C.D. Frontal theta activity in humans increases with memory load in a working memory task. *Eur. J. Neurosci.* **2002**, *15*, 1395–1399. [CrossRef]
32. Onton, J.; Delorme, A.; Makeig, S. Frontal midline EEG dynamics during working memory. *Neuroimage* **2005**, *27*, 341–356. [CrossRef] [PubMed]
33. Laukka, S.J.; Järvilehto, T.; Alexandrov, Y.I.; Lindqvist, J. Frontal midline theta related to learning in a simulated driving task. *Biol. Psychol.* **1995**, *40*, 313–320. [CrossRef] [PubMed]
34. Nakashima, K.; Sato, H. Relationship between frontal midline theta activity in EEG and concentration. *J. Hum. Ergol.* **1993**, *22*, 63–67.
35. Weber, E.; Doppelmayr, M. Kinesthetic motor imagery training modulates frontal midline theta during imagination of a dart throw. *Int. J. Psychophysiol.* **2016**, *110*, 137–145. [CrossRef] [PubMed]
36. Borghini, G.; Aricò, P.; Astolfi, L.; Toppi, J.; Cincotti, F.; Mattia, D.; Cherubino, P.; Vecchiato, G.; Maglione, A.G.; Graziani, I.; et al. Frontal EEG theta changes assess the training improvements of novices in flight simulation tasks. In Proceedings of the 2013 35th Annual International Conference of the IEEE Engineering in Medicine and Biology Society (EMBC), Osaka, Japan, 3–7 July 2013; pp. 6619–6622.
37. Long, Y.; Liu, Z.; Chan, C.K.Y.; Wu, G.; Xue, Z.; Pan, Y.; Chen, X.; Huang, X.; Li, D.; Pu, W. Altered temporal variability of local and large-scale resting-state brain functional connectivity patterns in schizophrenia and bipolar disorder. *Front. Psychiatry* **2020**, *11*, 422. [CrossRef] [PubMed]

sensors

Review

An Overview on Recent Advances in Biomimetic Sensors for the Detection of Perfluoroalkyl Substances

Fatemeh Ahmadi Tabar [1,2], Joseph W. Lowdon [2], Soroush Bakhshi Sichani [1], Mehran Khorshid [1], Thomas J. Cleij [2], Hanne Diliën [2], Kasper Eersels [2], Patrick Wagner [1,*] and Bart van Grinsven [2]

[1] Laboratory for Soft Matter and Biophysics ZMB, Department of Physics and Astronomy, KU Leuven, Celestijnenlaan 200 D, B-3001 Leuven, Belgium; fatemeh.ahmaditabar@kuleuven.be (F.A.T.); soroush.bakhshisichani@kuleuven.be (S.B.S.); mehran.khorshid@kuleuven.be (M.K.)

[2] Sensor Engineering Department, Faculty of Science and Engineering, Maastricht University, P.O. Box 616, 6200 MD Maastricht, The Netherlands; thomas.cleij@maastrichtuniversity.nl (T.J.C.); kasper.eersels@maastrichtuniversity.nl (K.E.); bart.vangrinsven@maastrichtuniversity.nl (B.v.G.)

* Correspondence: patrickhermann.wagner@kuleuven.be; Tel.: +32-16-32-21-79

Abstract: Per- and polyfluoroalkyl substances (PFAS) are a class of materials that have been widely used in the industrial production of a wide range of products. After decades of bioaccumulation in the environment, research has demonstrated that these compounds are toxic and potentially carcinogenic. Therefore, it is essential to map the extent of the problem to be able to remediate it properly in the next few decades. Current state-of-the-art detection platforms, however, are lab based and therefore too expensive and time-consuming for routine screening. Traditional biosensor tests based on, e.g., lateral flow assays may struggle with the low regulatory levels of PFAS (ng/mL), the complexity of environmental matrices and the presence of coexisting chemicals. Therefore, a lot of research effort has been directed towards the development of biomimetic receptors and their implementation into handheld, low-cost sensors. Numerous research groups have developed PFAS sensors based on molecularly imprinted polymers (MIPs), metal–organic frameworks (MOFs) or aptamers. In order to transform these research efforts into tangible devices and implement them into environmental applications, it is necessary to provide an overview of these research efforts. This review aims to provide this overview and critically compare several technologies to each other to provide a recommendation for the direction of future research efforts focused on the development of the next generation of biomimetic PFAS sensors.

Keywords: molecularly imprinted polymers; biomimetic sensors; polyfluoroalkyl substances; environmental pollution; aptamers

Citation: Ahmadi Tabar, F.; Lowdon, J.W.; Bakhshi Sichani, S.; Khorshid, M.; Cleij, T.J.; Diliën, H.; Eersels, K.; Wagner, P.; van Grinsven, B. An Overview on Recent Advances in Biomimetic Sensors for the Detection of Perfluoroalkyl Substances. *Sensors* **2024**, *24*, 130. https://doi.org/10.3390/s24010130

Academic Editor: Evgeny Katz

Received: 17 November 2023
Revised: 15 December 2023
Accepted: 23 December 2023
Published: 26 December 2023

1. Introduction

Polyfluoroalkyl substances (PFAS) are a category of organic molecules consisting of a fully fluorinated alkyl chain [1]. Perfluorooctanoic acid (PFOA), perfluorooctane sulfonate (PFOS), and hexafluoropropylene oxide dimer acid (HFPO-DA)—see Figure 1—are some of the most common and problematic PFAS with widespread use in different products such as semi-conductors, firefighting foams, lubricants, and non-stick coatings [2–4]. As they have been extensively used, they can be found in surface/drinking water and sediments [5,6]. Moreover, these compounds are capable of bioaccumulation in human and animal tissue, have high chemical and thermal stability, and are potentially carcinogenic and neurotoxic [7–9]. This has raised concerns and debates about the potential risks of PFAS species and forced legislative bodies to react [10–12]. For instance, in 2020, the European Commission (EC) decided to reframe the EU Drinking Water Directive and incorporated a maximum health advisory level of 0.1 μg/L for each individual type of PFAS molecules [13]. There are also short-chain PFAS molecules with only four carbon atoms such as perfluorobutanesulfonic acid (PFBS) and perfluorobutanoic acid (PFBA) that are considered as toxic; however, their serum half-life inside the human body ranges from a few days to less than a month [14]. This is considerably shorter than for PFOS, with an estimated half-life between 3.4 and 5 years; as a consequence, the bioaccumulation seems less critical.

Currently, PFOA detection typically relies on liquid chromatography paired with mass spectrometry [15–17]. Although this method is highly sensitive and selective, it usually requires elaborate sample preparation, expensive equipment, and well-trained personnel. As a result, it is not suitable for a fast and facile examination in routine environmental monitoring [18–20]. Consequently, it is highly desirable to have sensors available that can provide a facile, sensitive, selective, fast, quantitative, and cost-effective way of detecting PFOA directly in the field. User-friendly bio (mimetic) sensors can meet these demands.

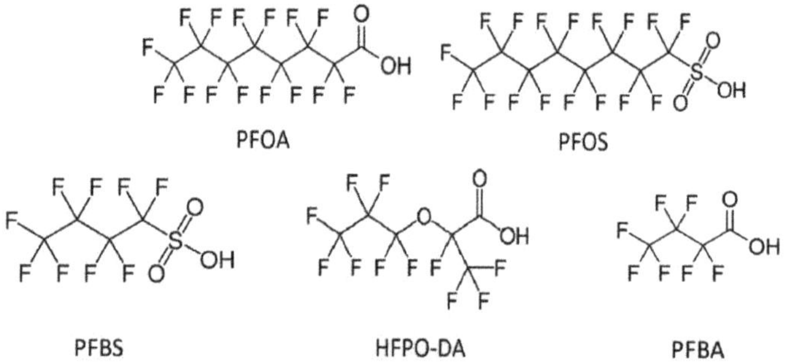

Figure 1. The chemical structures of perfluorooctanoic acid PFOA (molecular weight MW: 414 g/mol), perfluorooctane sulfonate PFOS (MW: 500 g/mol), perfluorobutanesulfonic acid PFBS (MW: 300 g/mol), hexafluoropropylene oxide dimer acid HFPO-DA (MW: 330 g/mol), and perfluorobutanoic acid PFBA (MW: 214 g/mol).

In general, bio- and chemosensors consist of two main components, the receptor layer and the transducer [21,22]. The receptor layer recognizes the target based on specific molecular interactions and is combined with a transducer, which converts the binding events between targets and receptors into interpretable data [23]. The receptor choice has a direct impact on the detection range and selectivity of the sensor. Therefore, the this review will focus on the different receptor types that have been developed for the detection of PFAS molecules. To date, biomimetic receptors such as molecularly imprinted polymers (MIPs), aptamers, and metal–organic frameworks (MOFs) have been synthesized for PFAS detection. The aim is to eventually integrate these receptors into portable devices for the onsite detection of these compounds [24]. Recently, P450-type enzymes were also identified that can biodegrade PFAS molecules, which may offer potential for enzymatic PFAS detection in future [25,26].

MIPs are polymeric matrices with predetermined recognition properties for a certain molecule or a set of similar molecules [27,28]. The applications of MIPs initially focused on separation and extraction processes [29]. In recent years, MIPs have also been applied in a wider context, including solid-phase extraction, drug delivery, catalysis, and environmental and chemical sensing [30–33]. A key benefit of MIP technology is that MIP-based sensors can be developed for a wide variety of targets [34–36]. This includes environmental contaminants, chemical and biological compounds, as well as industrial chemicals [35,37,38]. Regarding specifically fluorinated contaminants, we refer the reader to [39–43].

The specific binding interaction of MIPs with their targets leads to changes in physical characteristics including mass, electrochemical impedance, thermal resistance, and fluorescence, which can be used as transducer mechanisms in the sensor [44–48]. In comparison to biological receptors such as antibodies and enzymes, MIPs are stable over a wide pH and temperature range, while their sensitivity and selectivity towards their target are only marginally lower. They are relatively straightforward and low-cost to prepare with adjustable surface properties and applicability outside a laboratory environment [49–51].

While MIPs are chemically and physically more stable than natural receptors, there are still points of attention. First of all, incomplete extraction of template molecules can result in template leakage during the measurements. Secondly, the recognition properties of the binding sites can be heterogeneous, which may affect the reproducibility of analytical results [52]. These disadvantages can be overcome by a better control over the polymerization process for example by using nanoMIPs [53–55].

Aptamers are receptors based on nucleic acids (DNA or RNA) that bind a particular target analyte, or a group of target analytes, by folding into specific conformations, van der Waals and electrostatic interactions, and/or hydrogen bonds [56]. Aptamers can be obtained by a combinatorial selection process known as systematic evolution of ligands by exponential enrichment SELEX [57]. SELEX involves the progressive selection of oligonucleotide sequences from a large pool of randomly generated sequences towards high binding affinities between the desired target and oligonucleotide sequences. Aptamers typically contain 25–80 bases, which will fold into complex tertiary structures. Aptamers can be designed to recognize various analytes with high affinity and specificity, including organic dyes, toxins, and proteins [58]. Furthermore, it is possible to select aptamers in a way that they recognize only an individual target analyte, or to bind a set of analytes with a similar structure. Thanks to their small size, it is possible to incorporate different aptamers into the same sensor to detect multiple chemicals in parallel, with each aptamer selective towards a different analyte. Similar to MIPs, aptamers can also be combined with various transducer principles [59]. Unfortunately, aptamer development for a new target can be expensive and time consuming, and it is hard to scale up their synthesis to mass production. In comparison to MIPs, aptamers have less physical and chemical stability towards harsh environmental conditions, and they are prone to enzymatic degradation [60].

Metal–organic frameworks (MOFs) are porous materials consisting of rigid inorganic groups and flexible organic linker ligands. MOFs are potential candidates for selective chemical sensing with low detection limits owing to their extremely high surface area, variability of metal nodes, and modifiable organic linkers to provide adjustable binding sites [61,62]. By choosing different metal clusters for the organic linkers to coordinate around, surface characteristics and pore sizes can be adjusted [63]. Due to their stability, they can be used repeatedly to detect specific analytes [64]. MOFs with different metal centers can trap PFAS by strong affinity interactions, making them promising for the development of PFAS-detection platforms [65]. MOFs can furthermore be combined with MIPs to improve their sensing properties [66–68], while also the combination of MOFs with aptamers has been reported in the literature [69]. For strategies to combine molecular imprinting with aptamer technology, we refer to the recent review article by Zhou and coworkers [70].

The aim of this review is to present a better understanding of the advantages that each of these receptors (MIPs, aptamers, and MOFs) offers for PFAS determination and how to optimize the design of a sensor layer in view of the desired application. We therefore aim at providing an overview of the most recent innovations in biomimetic PFAS detection. Opportunities for each receptor type will be analyzed and compared to potential obstacles and challenges that sensors based on these type of receptors will face when implementing them into a real-life application. This way, this literature overview seeks to provide recommendations for future research towards the development of PFAS sensors for direct application in environmental screening.

2. PFAS Sensing with Receptors Made via Imprinting Technology

MIPs are synthetic polymer structures prepared by various polymerization methods between a crosslinker, template and one or more functional monomers in a porogenic solvent [71,72]. During the polymerization, specific interactions take place between the template and the monomer's functional groups [73]. Subsequently, when the template is removed with an appropriate solvent, molecular cavities are created whose shape, size, structure and functionality are complementary to the template analyte and are able to detect the target in another matrix through a "lock–key interaction"; see Figure 2 [74–76]. The sensitivity and selectivity of the resulting sensing tool strongly depend on the affinity of the imprinted polymer for the target [77].

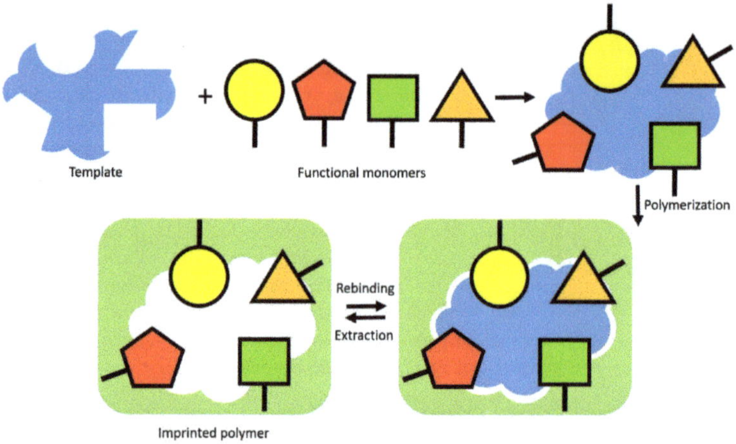

Figure 2. Schematic representation of the concept of molecular imprinting. The synthetic receptor formation starts by a stereochemical arrangement of functional monomers around a template of interest through self-assembly. The most adopted approach consist of adding an initiator and crosslinker to the pre-polymerization mixture and thermally or UV-induced polymerization. This creates a highly crosslinked polymeric network around the template that serves as a plastic mold to which the template can specifically rebind upon extraction.

Optimizing the stoichiometric ratios between the crosslinker, monomer, and template will improve this affinity and the binding capacity of the MIP [78]. The sensing capability of a MIP is usually compared with a non-imprinted polymer (NIP) that is synthesized in an identical manner, but without the presence of the template analyte to evaluate the effect of imprinting [79]. Clearly, the major challenge is to achieve MIPs with a high affinity and specificity for the target analyte [77].

Most approaches to synthesize MIPs are focused on free radical polymerization as this method offers a facile and low-cost route for creating large batches of MIPs. However, these particles are highly heterogeneous which makes it hard to create reproducible sensors. Therefore, more controllable methods of creating homogenous, high affinity MIPs have been developed by more controllable methods such as suspension, precipitation and emulsion polymerization [80]. However, all these methods will result in the creation of MIP particles that still need to be deposited on a planar sensing electrode. Therefore, methods to directly deposit MIPs onto a conductive sensing surface, such as electropolymerization, are becoming increasingly popular [81,82]. The polymerization method is of high importance as it affects the size, shape, homogeneity, thermal durability, and binding capacity of the resulting MIP particles or layers [83–86].

MIPs can also be synthesized directly on the sensor surface by depositing a thin polymer layer, which is imprinted with the target [55,87,88]. With this method, the binding cavities are mostly located at the outer layer of the substrates; see Figure 3 [89,90]. Surface imprinting can be accomplished directly on the electrode's surface, or the outer layer of a carrier such as nanoparticles and nanofibers. The MIP layer usually embeds only part of the template, which can be sufficient for the selective rebinding of the template after its removal. To date, there are several studies reporting on the use of surface-imprinted polymers for PFAS sensing [91–94].

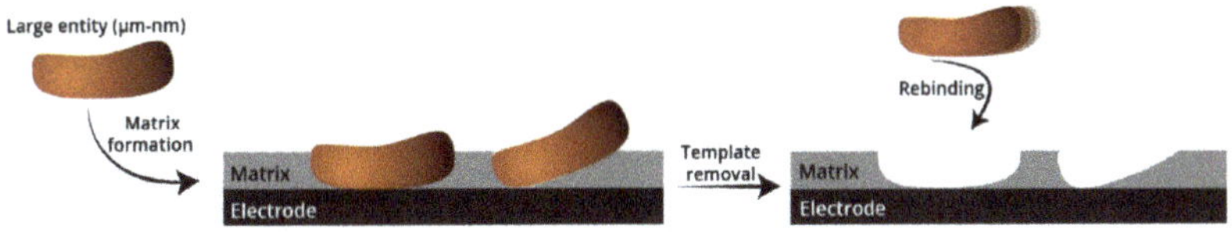

Figure 3. Schematic illustration of molecular imprinting directly on an electrode surface. As shown, the templates leave the imprinted cavities on the top of polymeric matrix and target analytes can rebind with the imprints. Figure reprinted from [95]. Copyright 2022, with permission from Elsevier.

The synthesis approach depends on the specific application that is targeted. In some cases, it is preferable to prepare polymers separately by for instance bulk or suspension polymerization to facilitate quality control, enable mass production, and control surface coverage. Nevertheless, these methods can have drawbacks including difficulty in controlling the layer thickness, embedding of the template, and cavity accessibility [96,97]. It is worth mentioning that one of the main challenges in conventional polymerization methods is the incorporation of polymerized MIPs into the sensors and imprinting of polymers on the electrode's surface is a feasible solution to this issue [98]. Molecular imprinting, directly on the electrode's surface, offers several benefits such as binding sites with better accessibility, faster binding kinetics, and faster mass transport [52]. In general, larger macromolecular entities (cells, proteins, bacteria, etc.) cannot be dissolved in a pre-polymerization mixture and therefore need to be imprinted on the surface of a solid substrate. In addition to the polymerization method, the selection of monomers will affect the performance of MIPs and depending on the monomers; the MIPs will function best in a specific pH range [99].

2.1. Imprinting with Conventional Polymerization Methods

Bulk free radical polymerization is the most common route to prepare MIPs for low-molecular-weight compounds such as PFAS, owing to its simplicity and low production costs for large amounts [100]. Once the polymerization is completed, the product is a solid polymeric structure that needs to be crushed, ground and then eluted with solvents. The synthesized product can be sieved with an appropriate mesh size to obtain particle sizes adapted to a particular application, which may vary from micrometer to submicrometer diameters [101,102]. A disadvantage is the grinding and sieving process, which may take long and result in significant product waste [43,103].

In 2023, Ahmadi Tabar et al. synthesized PFOA MIPs by bulk free radical polymerization and optimized the receptor by changing the molar ratio of the polymerization components (1/4/12 for template/monomer/crosslinker) to maximize the target affinity and selectivity [104]. Rebinding of PFOA to the MIPs was assessed by a thermal transducer known as heat transfer method HTM [105,106]. This method works by recording the thermal resistance between the chip and the sample with two temperature sensors while the chip is covered with a MIP layer. The temperature below the chip (T_1) is kept constant at, e.g., 37 °C using a temperature control unit, and the output temperature (T_2, above the MIPs layer) is measured; see Figure 4. Increasing the concentration of PFOA led to a concentration-dependent decrease in T_2 for the MIP chip, while temperature changes were negligible for its NIP counterpart; see Figure 5a. The effect size (%) was calculated by dividing the temperature changes by the initial temperature (°C), which is plotted in function of the PFOA concentration in Figure 5b. Increasing T_1 from 37 °C to 40 °C minimized the noise on the signal and lowered the detection limit LoD from 0.48 nM to 22 pM. This LoD is below the PFOA contamination level (0.1 µg/L: 0.24 nM) stated in the EU Drinking Water Directive [13].

The results also demonstrated that the sensor is selective with the cross selectivity below 30% for other PFAS molecules such as heptafluorobutyric acid (HFBA), and perfluorobutanesulfonic acid (PFBS). Furthermore, the sensor was able to detect PFOA in spiked environmental samples including river water and soil in the regulatorily relevant concentrations with LoD values of 91 and 154 pM, respectively. These results provided proof of the potential application of MIP-based sensors in routinely monitoring of environmental samples for PFAS contamination. The benefit of combining bulk MIPs with a thermal readout principle is that the synthesis approach is scalable and rebinding results in an easily interpretable increase in the thermal resistance, respectively a decrease in the temperature T_2.

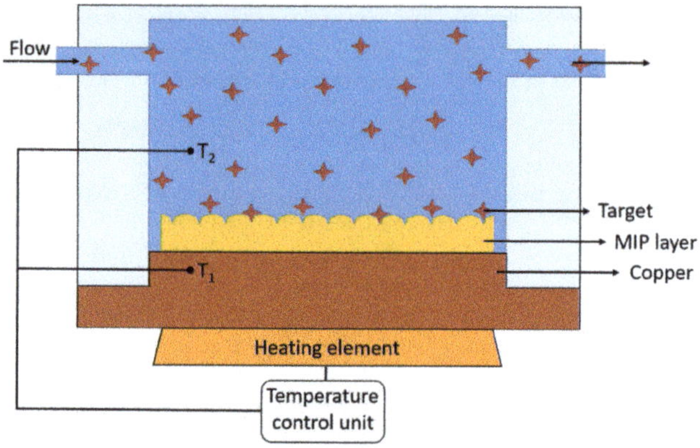

Figure 4. Schematic illustration of the heat transfer method setup. The temperature T_1 is constant by using a heating element and a temperature control unit. The temperature in the fluid (T_2) is varying by the changes on the MIPs layer.

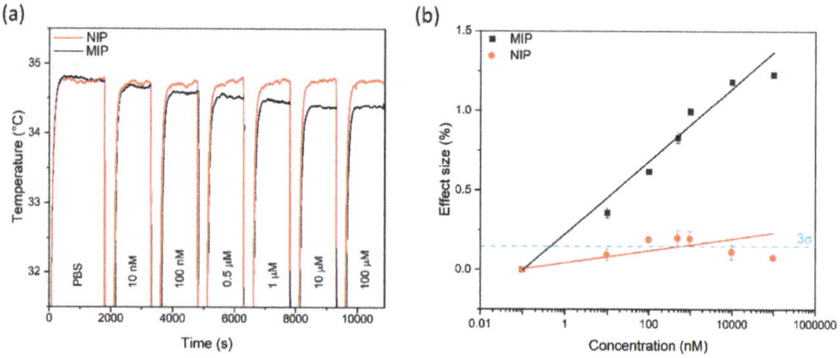

Figure 5. (**a**) Temperature response (T_2) for both MIP and NIP after exposure to different concentrations of PFOA in PBS (T_1 was kept constant at 37 °C). (**b**) Dose–response curve of MIP- and NIP-covered sensor chips obtained by HTM. The LoD for these measurements was calculated as 0.48 nM based on the intercept of the 3σ line with the MIP curve. Reproduced with permission from [104]. Copyright 2023, Elsevier (CC-BY).

Precipitation and emulsion polymerization are other techniques for synthesizing MIPs [107–109]. In these methodologies, the polymerization approach is the same as bulk polymerization, but the post-processing stages are not necessary, resulting in fewer steps and, more crucially, a lower risk of damaging the binding sites. The emulsion polymerization is rapid and has a mechanical dispersion system constantly working in the presence of a surfactant, and it can achieve continuous production [110]. There is no need for a surfactant for precipitation polymerization and, during this polymerization, the growing polymer segregates from the solution, finally forming MIP particles with micro or submicrometer dimensions [111,112].

Cao and coworkers prepared MIPs for the selective adsorption of PFOA in aqueous solutions by precipitation polymerization of acrylamide in the presence of PFOA as the template molecule [39]. The concentration of PFOA in Milli-Q water was measured by liquid chromatography with tandem mass spectrometry. The optimized MIPs showed a high affinity for PFOA, and the uptake percentage by the MIPs was 1.3–2.5-fold higher than that of the NIP when exposed to PFOA alone. The MIPs adsorbent showed a high selectivity for PFOA over other PFAS molecules such as PFOS and perfluorodecanoic acid. Furthermore, the reusability of the MIPs adsorbent was confirmed in five consecutive adsorption–desorption cycles without a notable decrease in the PFOA uptake. In summary, the results were promising in terms of selectivity, the sorption capacity of the resulting MIPs and the relatively low batch-to-batch variability, which also makes the MIPs promising candidates for PFOA detection. In this context, it is noteworthy that there is also literature on PFAS absorption using MIPs [43].

2.2. Imprinting by Electropolymerization

Surface imprinting of low-molecular-weight compounds can be achieved by depositing and imprinting a polymer layer directly on an electrode surface via electropolymerization, which is a particularly useful method in combination with electrochemical transducers [113,114]. By cycling the potential in a predetermined range with a given sweep rate, the electroactive monomer (such as aniline, o-phenylenediamine, and pyrrole) will be electropolymerized and the substrates will be coated by a very thin layer of polymer [81,92,115]. In this method, the potential range, the sweep rate and the composition of pre-polymerization mixture can control and optimize the adherence and morphology of the imprinted polymer layer on the electrode [116]. The main advantage of this method is that the polymer layer thickness is controllable in a reproducible manner with low batch-to-batch variability. In addition, it is possible to automate the process. Finally, it is feasible to obtain better binding capacity and sensitivity, which is the result of thinner and more homogenous layers. Additionally, because of the ultrathin MIPs film that is produced and the proximity of the imprinted cavities to the surface, this approach makes template removal facile. Therefore, electropolymerization is typically a straightforward and fast method that involves these phases: dissolution and interaction of an electroactive monomer with the given template in a solvent (the solvent can even be water with electrolytes), coating electrochemically, and finally the elution of the templates [30].

Clark et al. performed electropolymerization on a glassy carbon electrode surface by cyclic voltammetry in an aqueous solution containing o-phenylenediamine (o-PD) and PFOS [93]. Figure 6 shows a schematic illustration of fabrication process for the MIP-based sensing platform. After template removal by a water/methanol solution, they performed oxygen reduction (O_2 was dissolved in water) on the electrode, as illustrated in Figure 6b. In differential pulse voltammetry (DPV), the electrode revealed oxygen reduction peaks around −0.5 and −0.9 V and the first one was used as an electrochemical signal to plot the calibration curves. PFOS was able to associate with the MIPs (Figure 6c) and block the electrochemical signal of the oxygen redox reaction (Figure 6d). In this electrochemical spectroscopy technique, as the PFOS concentration increased, the effective electrode surface area decreased. This can be seen by the increase in the charge-transfer resistance (R_{ct}), which is the diameter of the semi-circle of the Nyquist plots in Figure 6e. The curve in Figure 6f is the change in R_{ct} with respect to the baseline against the logarithm of PFOS concentration. Furthermore, the sensor achieved a detection limit of 3.4 pM for PFOS using electrochemical impedance spectroscopy without a redox mediator such as ferrocene carboxylic acid. Clark et al. also revealed that two common environmental interferents (sodium chloride and humic acid) do not affect the sensor signal. Moreover, it was feasible to obtain reproducible results with matrices such as river water using impedance spectroscopy.

In a similar approach, Karimian and co-workers developed an electrochemical sensing platform for the determination of trace amounts of PFOS in water [94]. The sensor consisted of a gold electrode functionalized with a thin layer of MIPs, synthesized by electropolymerization of o-PD with PFOS as the template. The sensor was activated by template elution with adequate solvents. Ferrocene carboxylic acid (FcCOOH) was used as a redox probe, capable of generating analytically useful voltametric signals by competing for the recognition regions with PFOS, while PFOS itself is not electrochemically active. According to the observations, the voltametric signal at the MIP-coated electrode decreased gradually when the sensor was submerged in PFOS containing samples in deionized water, scaling inversely with the PFOS concentration. According to the selectivity results, other PFAS molecules including PFOA, HFBA, and PFBS caused maximally 20% change in the signal normalized to PFOS. The sensor also demonstrated a limit of detection of 0.04 nM, and an acceptable reproducibility and repeatability.

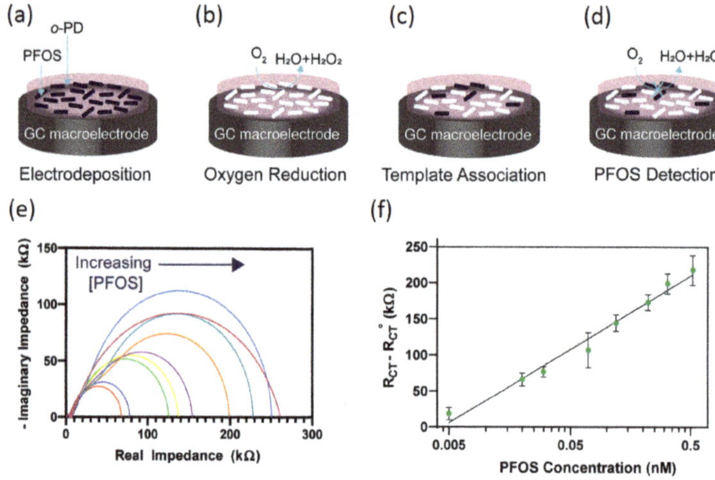

Figure 6. Schematic illustration of the PFOS detection procedure. (**a**) Electropolymerization of o-phenylenediamine (o-PD) on a glassy carbon macroelectrode. PFOS molecules are shown as black ovals and the white ovals are the biding cavities remained after PFOS removal. (**b**) Driving of oxygen reduction on the MIP-modified electrode. (**c**) Rebinding of the template molecule with the MIPs. (**d**) Blocking the electrochemical signal of the redox reaction by bound PFOS molecules. (**e**) The R_{ct} values increases with the increase in the PFOS concentration. (**f**) The normalized R_{ct} against the logarithm of PFOS concentration. Reproduced with permission from [93]. Copyright 2020, American Chemical Society (CC-BY).

2.3. Imprinting on Nanoparticles

In order for MIPs to work optimally, the number of binding cavities is essential. One strategy to increase the number of binding sites can be achieved by increasing the thickness of the imprinted polymer layer. However, this will reduce diffusion and therefore mass transport and interaction with the transducer [117,118]. One practical solution is to synthesize MIPs on the surface or the external layer of a particular carrier with a large surface area [96]. This improves elution and rebinding of template- and target molecules, it decreases the amount of non-accessible binding sites. In addition, it enhances both the availability of the target to the binding regions as well as the corresponding kinetics [118]. Polystyrene microspheres, silica nanoparticles, carbonaceous nanomaterials, and magnetic nanoparticles are examples of frequently employed carriers [119–122]. The surfaces of imprinted substances are controllable, and the recognition regions with high density are easily accessible by the targets, improving the adsorption capacity and effectiveness [118,123,124].

In 2022, Lu et al. modified pristine glassy carbon electrodes (GCE) with a thin layer of gold nanostars (AuNS) by drop-casting to increase the sensitivity of the electrode to the electrochemical probe (ferrocene carboxylic acid) [125]. These AuNs-coated GCEs were then coated with a layer of PFOS-imprinted o-PD using cyclic voltammetry for electropolymerization to enhance the sensitivity towards PFOS; see Figure 7. The interaction between the MIP layer and PFOS was analyzed by the oxidation peak of FcCOOH (Fe^{2+}/Fe^{3+}) using DPV. Figure 7 (left) indicates that the oxidation peak has entirely disappeared for the MIP/AuNS/GCE before PFOS removal. This means that the MIP layer is able to completely block the charge transfer between the working electrode and the solution. This voltametric sensor was able to detect PFOS with a limit of detection (LoD) of 0.015 nM calculated by using the 3σ method. The suggested sensing platform was also capable of detecting trace levels of PFOS in tap water. However, during the measurements significant interferences with perfluorobutanoic acid (PFBA) or PFBS were observed. This observation can be explained, by the smaller sizes of PFBA and PFBS molecules, enabling them to pass across the MIPs layer and occupy the PFOS-shaped cavities by non-specific binding. Therefore, it is required to first screen an unknown sample for the presence of small PFAS molecules such as PFBA and PFBS.

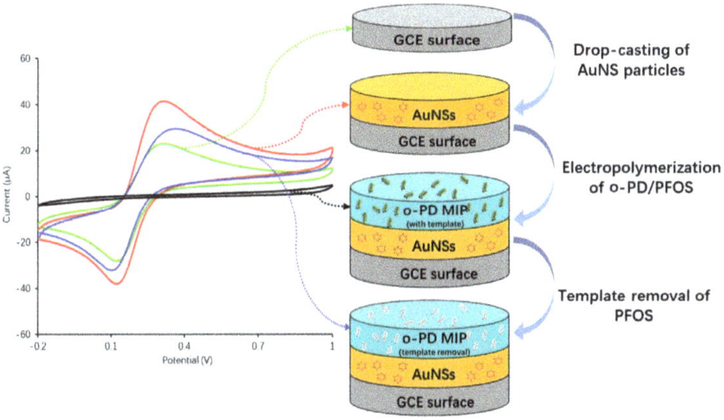

Figure 7. Schematic illustration of the voltametric sensor consisting of MIP and gold nanostars (AuNS) coatings for PFOS determination. Right: The GCE surface is first modified with AuNS and then electropolymerized with o-PD using cyclic voltammetry (CV). Left: The CV curve and the probes' oxidation peak for pristine GCE, AuNS/GCE, and MIP/AuNS/GCE before and after PFOS removal. Figure adapted from [125]. Copyright 2021, with permission from Elsevier.

Gao et al. prepared an electrochemical sensor for the detection of PFOS in real water samples [126]. The sensor (PFOS-MIPPDA/AuNPs/GCE) was made from a GCE modified with gold nanoparticles (AuNPs) and an electropolymerized molecularly imprinted polydopamine (DA) coating with PFOS as the template. PFOS detection was achieved by using DPV and $K_3[Fe(CN)_6]$ as the detection probe. The results revealed that the developed PFOS-MIPPDA/AuNPs/GCE sensor was able to determine PFOS with a nanomolar detection limit. The sensor also showed promising results for analyzing real water samples including tap, lake, and canal water.

Zheng et al. developed a photoluminescence sensor (PL) for selective detection and quantification of PFOA based on MIP-coated CdTe@CdS quantum dots (QDs) [127]. This optical sensor provided fast and sensitive detection of PFOA in the presence of common interferents by the PL quenching via target rebinding into the recognition cavities in the polymeric layer. Furthermore, the fabricated sensor demonstrated a good linearity in the range from 0.25 to 15.00 μmol/L with a PFOA detection limit of 25 nM.

2.4. Imprinting on Nanofibers

As mentioned above, one solution to improve the recognition performance of the sensor is to synthesize MIPs on the exterior layer of a particular carrier with a high surface-to-volume ratio. Electrospun fibers can be considered as promising carriers because of their large surface area. Wang et al. successfully prepared a MIPs MOFs (Co/Fe)-driven carbon nanofiber (Co/Fe@CNF) electrode for electrochemical determination of PFOA, more information on MoFs is provided in Section 3.2 [67]. MIPs were formed by electropolymerization of pyrrole with PFOA as template. Owing to the strong adsorption force between the imprinting sites of MIPs and PFOA, PFOA molecules could reach the surface of electrode. In DPV measurements, the peak at 0.2 V (corresponding to PFOA) was used to plot the calibration curves (Figure 8a). The response current of MIPs Co/Fe@CNF electrode increased with the increase in PFOA concentration (Figure 8b). Under optimum conditions, the resultant MIPs Co/Fe@CNF was able to determine PFOA with a linear response with respect to the logarithm of the PFOA concentration and the limit of detection was 1.07 nM. The as-developed sensor also worked properly for measuring PFOA in real wastewater samples and it was a promising candidate for the determination of PFOA in environmental water samples.

In recent literature, there are many more PFAS sensors based on MIPs and Table 1 provides a comparative overview on the polymer composition, the readout principle, the limit of detection, and the sample type, which has been used for the measurements.

Table 1. Comparison of the different MIP-based receptors for PFAS detection.

Target	Receptor Material	Receptor Type	Readout Principle	Limit of Detection	Sample Type	Ref.
PFOA	poly acrylamide	MIPs	HTM	22 pM	river water and soil extract	[104]
	poly VBT and PFDA		SPR sensor	2 pM	seawater	[11]
	CdTe@CdS/poly APTES		photoluminescence sensor	25 nM	river water and tap water	[127]
	AgI–BiOINFs/poly acrylamide		photoelectrochemical sensor	24 pM	river water and tap water	[8]
	poly pyrrole/ graphitic carbon nitride nanosheets		Electrochemi-luminescence sensor	24 pM	river water, tap water, and lake water	[41]
	poly pyrrole/Co/Fe@CNF	MIPs and MOFs	DPV	1.07 nM	wastewater	[67]
HFPO-DA	poly o-PD/gold electrode	MIPs	DPV	250 fM	river water	[92]
PFOS	poly APTES/SiO_2 NPs nanoparticles	MIPs	fluorescence quantification	11 nM	river water and tap water	[119]
	TiO_2 nanotube arrays/poly APTES		photoelectrochemical sensor	172 nM	river water, tap water, and mountain water	[128]
	polyaniline on paper		DC resistance measurements	2.4 pM	DI water	[46]
	phenolic resin		LC–MS/MS	12 pM	milk	[97]
	poly o-PD/GCE		DPV	0.05 nM	DI water	[2]
	G-UCNPs-SiO_2/poly APTES		fluorescence quantification	1 pM	human serum, egg, lake water	[89]
	Au/poly o-PD		DPV	0.04 nM	tap water	[94]
	poly o-PD/ AuNS/GCE		DPV	0.015 nM	tap water	[125]
	poly o-PD/GCE		EIS	3.4 pM	river water	[93]
	poly DA/AuNPs/GCE		DPV	4.2 nM	lake water, canal water, tap water	[126]
	CNW/poly o-PD		DPV and EIS	2.4 nM	tap and wastewater	[91]
	poly chitosan/carbon quantum dots		fluorescence spectrophotometry	0.8 fM	serum and urine	[31]

Abbreviations: HTM: heat transfer method; VBT: (Vinylbenzyl) trimethylammonium chloride; PFDA: 1H,1H,2H,2H-perfluorodecyl acrylate; SPR: surface plasmon resonance; APTES: 3-aminopropyltriethoxysilane; AgI–BiOINFs: AgI nanoparticles–BiOI nanoflake arrays; DC: direct current; DPV: differential pulse voltammetry; HFPO-DA: hexafluoropropylene oxide dimer acid; EIS: electrochemical impedance spectroscopy; CNW: B,N-codoped carbon nanowalls; LC–MS/MS: liquid chromatography–tandem mass spectrometry; Co-N-C: cobalt-embedded Nitrogen-doped Carbon.

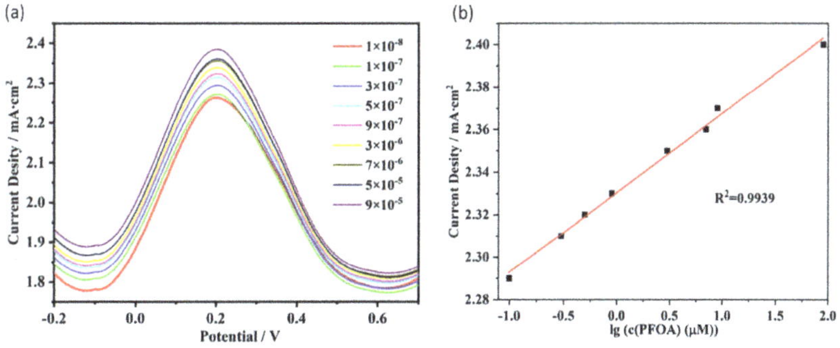

Figure 8. (**a**) Differential pulse voltammetry (DPV) of MIPs Co/Fe@CNF at different PFOA concentrations in molar units and (**b**) linear relationship between the current density and the logarithm of the PFOA concentration. Figure adapted from [67]. Copyright 2023, with permission from Elsevier.

3. PFAS Sensing with Other Synthetic Receptors

3.1. PFAS Sensing with Aptamers

As mentioned before, aptamer molecules undergo conformation changes in the presence of various target analytes and bind to them with high selectivity and affinity [129,130]. They are gaining increasing attention from researchers as a substitute to antibodies as specific elements for target molecule recognition owing to their flexibility, relatively small size, and easy chemical modification. Figure 9 indicates how the aptamer goes through conformational changes in the presence of a target molecule. Different kinds of reactions and physical factors can participate in the formation of aptamer-target complexes, namely hydrogen bonding, polar groups, shape complementarity, and van der Waals forces [58]. Aptasensors are a class of biosensors that combine a synthetic, biomimetic recognition element (aptamer) with chemical/physical transduction for precise detection of various target molecules. These sensors can potentially be used in environmental monitoring due to their high sensitivity and selectivity, high efficiency, and the ability to miniaturize these platforms [131]. Therefore, aptasensors can be developed for screening of PFAS and other existing pollutants in water.

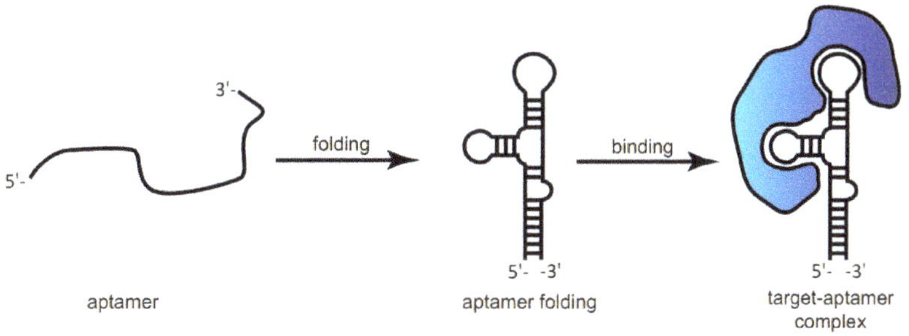

Figure 9. Schematically representing the formation of a target–aptamer complex. The aptamer folds into a 3D structure, upon which it binds to the target molecule. Reproduced with permission from [132]. Copyright 2017, Society for Neuroscience (CC-BY).

Park et al. demonstrated the potential use of DNA aptamers for detecting PFAS molecules and other fluorinated alternatives for the first time in a study published in 2022 [59]. The designed aptamer was capable of specifically binding PFOA and was integrated into a fluorescence-based aptasensor, able to detect PFOA with a LoD of 0.17 μM in water. The detection mechanism was based on quenching of the fluorescence of fluorescein by dabcyl and, by binding of PFOA, the aptamer changed its conformation so that the quenching stopped.

Figure 10a shows the predicted structure of the aptamer (with 30 bases) after binding to PFOA. The aptamer was mixed with PFOA solutions with different concentrations (0.5–50 μM) and, after 40 min, the fluorescence intensity was recorded. The fluorescence intensity was increasing with the increase in PFOA concentration (Figure 10b). The existence of interferents negligibly affected the aptamer performance, and the first proof of application was provided by testing the sensor in wastewater effluents. The fluorescence-based aptasensor was sufficiently sensitive for screening PFOA levels in water near accidental spills and industrial sites, where high concentrations of PFAS were anticipated. This work demonstrated the potential application of aptasensors for effective monitoring of the trace levels of different PFAS molecules and other fluorinated substances in water environments. The LoD is not yet low enough to measure concentrations below the regulatory limit (0.1 μg/L for each PFAS molecule and 0.24 nM for PFOA), but fluorescence-based sensors have the advantage that it is not necessary to immobilize the receptor on a solid support, everything can be performed in solution. To date, and to the best of our knowledge, there is no additional literature on PFAS-sensitive aptamers.

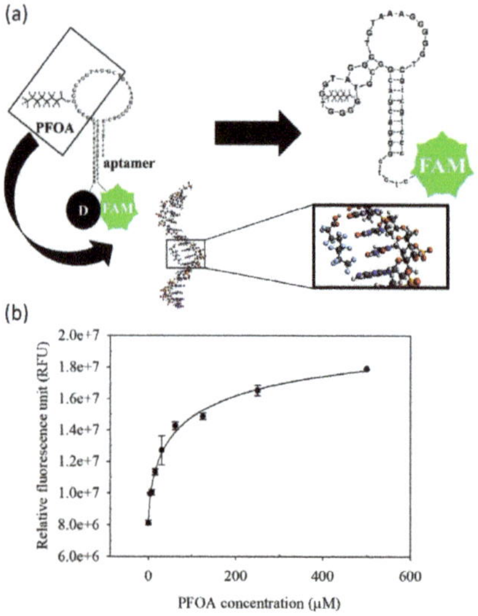

Figure 10. (**a**) The predicted 2D structures of aptamer after exposure to PFOA. (**b**) Fluorescence responses for binding of PFOA with different concentrations to the aptamer. The aptamer was modified with fluorescein (FAM) at 5′-end and dabcyl (D) at 3′-end, which was used as the quencher strand. Reproduced with permission from [60]. Copyright 2021, Elsevier (CC-BY-NC-ND).

3.2. PFAS Sensing with Metal–Organic Frameworks

Metal–organic frameworks (MOFs) are a type of crystalline porous nanomaterials made of metal ions and organic ligands. Because of their large specific areas, tunable pore size, straightforward synthesis routes, abundant functional groups, and chemical stability, MOFs are extensively used in diverse fields including separation, gas storage, drug delivery, electrochemical applications, catalysis, and importantly the detection of chemicals [67]. They have been applied in affinity-based determination of various analytes such as alcohols, ammonia, biomolecules, and recently fluorocarbon [133–135]. Firstly, the enormous surface area (ranging from 10^3 to 10^4 m^2 per gram of MOFs material) and porous structure of MOFs provide more interfaces and active sites for interaction with target molecules. Different types of interactions between MOFs and PFAS species exist, including redox, electrostatic, H-bonding, hydrophobic, and attractive intermolecular fluorine–fluorine (F–F) interactions as shown in Figure 11. Which interactions are at work depends on the precise type of the PFAS molecule and the design of the MOF structure. It is noteworthy that the negatively charged fluorine functionalities in two different molecules are responsible for an attractive force, which is known from experiments and quantum-chemistry calculations; see [Varadwaj, ChemPhysChem 2018]. Secondly, the organic ligands with versatile functional groups provide easy functionalization of MOFs with a broad range of molecules such as nucleic acids, enzymes, and nanoparticles. Finally, the diverse compositions of MOFs between metal and organic ligands offer a lot of functionality, such as catalytic activity, electrochemical activity, and optical activity. As a result, MOFs can be utilized as signal probes for different detection methods [136].

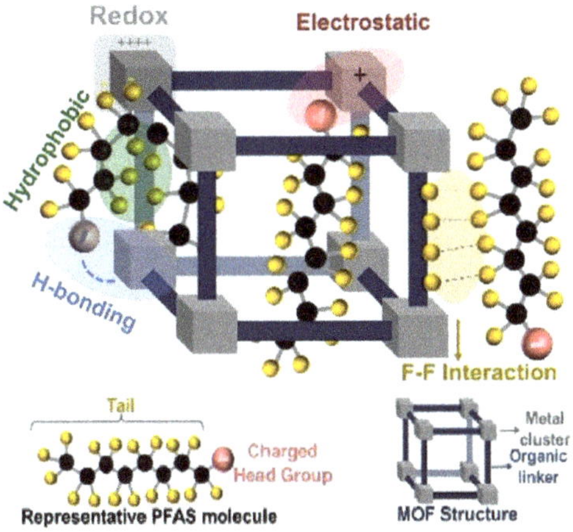

Figure 11. Schematic illustration of the different interactions between fluorinated MOFs and PFAS molecules including redox, electrostatic, hydrophobic, hydrogen bonding and F–F interactions. Reproduced with permission [66]. Copyright 2022, John Wiley and Sons (CC-BY-NC).

In 2020, Cheng and coworkers prepared a MOFs-based impedimetric sensor using a microfluidic platform for ultrasensitive in situ determination of PFOS [23]. The mesoporous MOFs Cr-MIL-101 (a chromium-based metal–organic framework) with high surface area and pore volumes was employed as the probe for capturing PFOS, which was based on the affinity of the chromium center toward both the fluorine and sulfonate functionalities. The MOFs capture probes were sandwiched between interdigitated microelectrodes in a microfluidic channel, forming an impedance sensor in a portable microfluidic device. This sensor directly measured PFOS concentrations by a proportional change in the electrical current as seen from the increase in the impedance signal. This microfluidic platform integrated with a MOFs-based sensor demonstrated ultra-sensitivity for the rapid in situ detection of PFOS with a LoD of 0.5 ng/L, corresponding to 1 pM at the molar scale. However, the selectivity of sensor towards other PFAS molecules was not yet studied.

Chen et al. designed a fluorescent MOFs array for optical sensing of multiple PFAS molecules in water samples [137]. The sensor array comprised three zirconium based porphyrinic coordination networks (PCNs) to determine PFAS molecules. The MOFs sensing array was also utilized to discriminate between six different PFAS by making a distinctive fluorescent response pattern for each molecule, according to their adsorptive affinity with the MOFs. The principal sensing mechanism was the quenching of the fluorescence emission of PCNs caused by the adsorption of PFAS. As an example, with increase in the PFOA concentration, the fluorescence emission of PCNs was quenched proportionally; see Figure 12. The calculated LoD for PFOA was 111 nM and it was in the same range for other PFAS molecules including PFOS. Importantly, the PCNs sensors showed a very fast response toward PFAS within only 10 s, owing to the ordered pore structure enabling rapid PFAS diffusion.

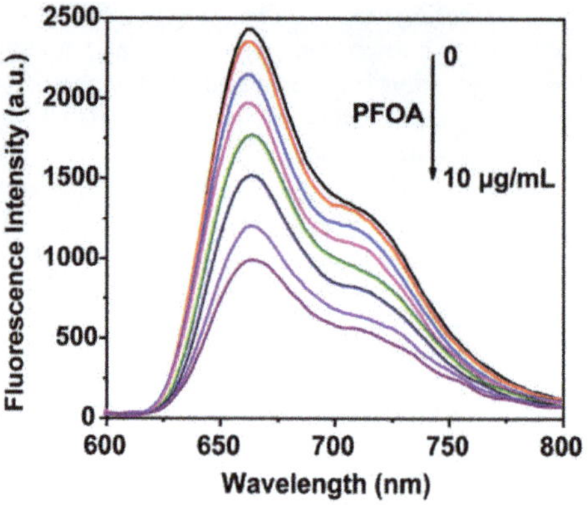

Figure 12. Fluorescence emission of the PCNs suspension at excitation wavelength λ_{ex} = 430 nm upon exposure to different concentrations of PFOA in water (0–10 μg/mL). Figure adopted from [137]. Copyright 2021, with permission from American Chemical Society.

Several other PFAS sensors based on MOFs exist, which are summarized in Table 2; a few of these platforms are able to detect PFAS in complex samples in the relevant concentration ranges. Despite these promising results, there are still some challenges associated with the use of MOFs in the sensing of PFAS species. The synthesis process generally requires harsh solvents and high temperatures, and thus, a "greener" synthesis approach should be applied. Furthermore, most MOFs are not stable in aqueous media, which will limit their applicability in sensing platforms. If these challenges can be overcome, MOFs may prove to be exceptionally advantageous towards solving the difficulties associated with PFAS pollution.

Table 2. Comparison of the different receptors based on MOFs and aptamer for PFAS detection.

Target	Receptor Material	Receptor Type	Readout Principle	Limit of Detection	Sample Type	Ref.
PFOA	DNA aptamer	aptamer	fluorescent quantification	0.17 μM	wastewater	[60]
	MOFs-coated probes	MOFs	mass spectrometry	26 pM	tap water, rainwater, and seawater	[138]
PFOS	MOFs Cr-MIL-10	MOFs	EIS	1 pM	groundwater	[23]
	zinc based MOFs		mass spectroscopy	1.28 nM	tap water and river water	[139]
	MOFs-derived Co-N-C nanosheets		colorimetric measurements	20 nM	river water, tap water, and lake water	[140]
PFAS	MIL-101(Cr)	MOFs	UHPLC–MS/MS	0.004–0.12 ng/L	tap water, river water, wastewater	[141]
	zirconium based porphyrinic coordination networks		fluorescent quantification	111 nM for PFOA	surface water and groundwater	[137]

Abbreviations: UHPLC–MS/MS: ultra-high-performance liquid chromatography–tandem mass spectrometry; Co-N-C: cobalt-embedded nitrogen-doped carbon.

4. Comparison between the Different Receptor Types

In the previous sections, a summary of recent studies on PFAS detection was provided. Different targets, receptor material, receptor type, readout principle, limit of detection, and sample type were discussed. MIP-based sensors seem to have the lowest LoDs when comparing them to other biomimetic PFAS sensing platforms. However, several challenges still lie ahead when it comes to incorporating these receptors into commercial devices. Real-world samples such as lake and river water contain very low concentrations of PFAS. Most of the user-friendly, handheld sensors can simply not reach the desired detection limits yet. Some of the more sensitive sensors on the other hand, are mostly focusing on the detection of PFOS and PFOA specifically, while there are more than 5000 different PFAS compounds identified [142]. Therefore, it is crucial to try to re-engineer these sensors towards the detection of a broader range of PFAS depending on the application. The next research phase should experiment with selectivity and intelligently design MIPs based on the envisioned application by integrating, e.g., computational studies into the design cycle [143].

The main challenge for industrialization, however, lays in upscaling the synthesis procedure towards mass production. In this regard, MIP-based receptors are more suitable for upscaling to industrial production due to their relatively low-cost and straightforward synthesis process. For all the receptors discussed in this review, it is essential to create large batches of sensors that are re-usable and provide accurate results in a reproducible manner. In a final step, the current lab-based prototypes should then be turned into handheld sensor solutions for on-site screening, combining for instance a dipstick-like sampler with a portable, smartphone-based transducer. The detection limits of the resulting sensors can be further optimized by combining biomimetic receptors with the most recent advances in the field of electrochemical and optical (quantum dots, fluorescence, etc.) MIP-based sensing [144–146].

5. Conclusions

PFAS molecules have attracted considerable attention worldwide as emerging pollutants because of their adverse effect on humans, aquatic life, and the environment. This review was written to provide readers with an overview of PFAS sensors based on synthetic receptors as these have the same benefits but overcome some of the drawbacks of natural bioreceptors. It summarizes different receptor layers used for selective determination of PFAS in the past few years.

MIP-based receptors are promising candidates, owing to their distinct ability to bind special targets with high selectivity. The sensors employing these receptors offer advantages with regard to their facile preparation, portability, user-friendliness, and cost-effectiveness. Synthesis of MIP-based receptors is possible via different approaches ranging from conventional polymerization methods such as bulk polymerization, to more advanced methods like electropolymerization. In many studies, nanomaterials have been used as substrate for electropolymerization improve the detection performance and response time of the sensor. In general, the MIP-based sensors demonstrate adequate sensing performance in terms of very low detection limit. Furthermore, many of the MIP-based sensing platforms introduced in this review show promising results for PFAS determination in real-world samples such as river water and tap water.

On the other hand, aptamers can also be used as recognition elements for PFAS detection but their detection limit needs to be lowered. There is only one publication on aptamers so far, dating from 2022, but more results in this context maybe expected in near future. Finally, the diverse compositions of MOFs between metal and organic ligands offer a lot of functionalities, making them promising candidates as signal probes for PFAS detection. Although using MOFs for PFAS determination is new, several recent studies show their capability at achieving highly sensitive and selective sensors for different PFAS species.

Considering all elements, both MIPs and MOFs are promising candidates to serve as receptors in on-site biomimetic PFAS sensors: both receptor types enable quantifying even subnanomolar concentrations, i.e., below the legally allowed limits. We see a slight advantage for MIPs because there are synthesis routes that enable continuous operation (nanoMIPs), or to fabricate large batches simultaneously and directly on transducer elements (electropolymerization). Moreover, it is facile to imprint MIP materials with a mixture of different PFAS molecules, so that the sensor will respond to a broad spectrum of these compounds. In the case of a positive sensor response, it will still be possible to perform a more selective analysis by chromatography and mass spectrometry. Such reduced selectivity can probably also be achieved by MOFs thanks to the fluorine–fluorine interaction. It is noteworthy that MOFs and MIPs can already be purchased from commercial suppliers and the step is small to adapt these materials towards PFAS detection. For on-site analysis with the analytical result ready within a few minutes, it is of course mandatory to make compact and low cost, but still accurate, readout techniques available. Here, we see a role for miniature photospectrometers and impedance analyzers. The costs have dropped tremendously in recent years and both transducers can be readout with a smartphone to arrive at a truly mobile application. A final element to bring sensor-based PFAS detection and quantification to the market would be accreditation of the instrument according to the norms of the International Organization for Standardization ISO. This still appears as a hurdle, but if sensor developers can show that sensor-derived data comply with the results of ISO-certified methods, this should become feasible.

Author Contributions: Conceptualization, F.A.T., K.E., P.W. and B.v.G.; investigation, F.A.T., K.E., P.W. and B.v.G.; writing—original draft preparation, F.A.T.; writing—review and editing, F.A.T., J.W.L., S.B.S., M.K., K.E., P.W. and B.v.G.; supervision, H.D., K.E., P.W. and B.v.G.; project administration, T.J.C., H.D., K.E., P.W. and B.v.G.; funding acquisition, H.D., K.E., P.W. and B.v.G. All authors have read and agreed to the published version of the manuscript.

Funding: The presented review is part of the project "High-tech sensors for monitoring the release of per- and poly-fluoroalkyl substances (PFAS) from recycled plastics in a circular economy (GPMU/20/063)", funded by KU Leuven and Maastricht University through the Global PhD Partnership between both universities.

Institutional Review Board Statement: Not applicable.

Informed Consent Statement: Not applicable, there are no studies on patients in this work.

Conflicts of Interest: The authors declare that they have no known competing financial interests or personal relationships that could have appeared to influence the work reported in this paper.

References

1. Chen, C.; Wang, J.; Yang, S.; Yan, Z.; Cai, Q.; Yao, S. Analysis of Perfluorooctane Sulfonate and Perfluorooctanoic Acid with a Mixed-Mode Coating-Based Solid-Phase Microextraction Fiber. *Talanta* **2013**, *114*, 11–16. [CrossRef] [PubMed]
2. Kazemi, R.; Potts, E.I.; Dick, J.E. Quantifying Interferent Effects on Molecularly Imprinted Polymer Sensors for Per- and Polyfluoroalkyl Substances (PFAS). *Anal. Chem.* **2020**, *92*, 10597–10605. [CrossRef] [PubMed]
3. Smaili, H.; Ng, C. Adsorption as a Remediation Technology for Short-Chain per- and Polyfluoroalkyl Substances (PFAS) from Water—A Critical Review. *Environ. Sci.* **2022**, *9*, 344–362. [CrossRef]
4. Yu, H.; Chen, Y.F.; Guo, H.Q.; Ma, W.T.; Li, J.; Zhou, S.G.; Lin, S.; Yan, L.S.; Li, K.X. Preparation of Molecularly Imprinted Carbon Microspheres by One-Pot Hydrothermal Method and Their Adsorption Properties to Perfluorooctane Sulfonate. *Chin. J. Anal. Chem.* **2019**, *47*, 1776–1784. [CrossRef]
5. Hassan, M.H.; Khan, R.; Andreescu, S. Advances in Electrochemical Detection Methods for Measuring Contaminants of Emerging Concerns. *Electrochem. Sci. Adv.* **2022**, *2*, e2100184. [CrossRef]
6. Jones, J.L.; Burket, S.R.; Hanley, A.; Shoemaker, J.A. Development of a Standardized Adsorbable Organofluorine Screening Method for Wastewaters with Detection by Combustion Ion Chromatography. *Anal. Methods* **2022**, *14*, 3501–3511. [CrossRef]
7. Bell, E.M.; De Guise, S.; McCutcheon, J.R.; Lei, Y.; Levin, M.; Li, B.; Rusling, J.F.; Lawrence, D.A.; Cavallari, J.M.; O'Connell, C.; et al. Exposure, Health Effects, Sensing, and Remediation of the Emerging PFAS Contaminants—Scientific Challenges and Potential Research Directions. *Sci. Total Environ.* **2021**, *780*, 146399. [CrossRef]
8. Gong, J.; Fang, T.; Peng, D.; Li, A.; Zhang, L. A Highly Sensitive Photoelectrochemical Detection of Perfluorooctanic Acid with Molecularly Imprined Polymer-Functionalized Nanoarchitectured Hybrid of AgI-BiOI Composite. *Biosens. Bioelectron.* **2015**, *73*, 256–263. [CrossRef]
9. Ranaweera, R.; An, S.; Cao, Y.; Luo, L. Highly Efficient Preconcentration Using Anodically Generated Shrinking Gas Bubbles for Per- and Polyfluoroalkyl Substances (PFAS) Detection. *Anal. Bioanal. Chem.* **2022**, *415*, 4153–4162. [CrossRef]

10. Cao, F.; Wang, L.; Ren, X.; Wu, F.; Sun, H.; Lu, S. The Application of Molecularly Imprinted Polymers in Passive Sampling for Selective Sampling Perfluorooctanesulfonic Acid and Perfluorooctanoic Acid in Water Environment. *Environ. Sci. Pollut. Res.* **2018**, *25*, 33309–33321. [CrossRef]
11. Pitruzzella, R.; Arcadio, F.; Perri, C.; Del Prete, D.; Porto, G.; Zeni, L.; Cennamo, N. Ultra-Low Detection of Perfluorooctanoic Acid Using a Novel Plasmonic Sensing Approach Combined with Molecularly Imprinted Polymers. *Chemosensors* **2023**, *11*, 211. [CrossRef]
12. Hill, N.I.; Becanova, J.; Lohmann, R. A Sensitive Method for the Detection of Legacy and Emerging Per- and Polyfluorinated Alkyl Substances (PFAS) in Dairy Milk. *Anal. Bioanal. Chem.* **2022**, *414*, 1235–1243. [CrossRef] [PubMed]
13. Directive (EU) 2020/2184 of the European Parliament and of the Council of 16 December 2020 on the Quality of Water Intended for Human Consumption (Recast). Available online: https://Eur-Lex.Europa.Eu/Eli/Dir/2020/2184/Oj (accessed on 14 November 2023).
14. Fenton, S.E.; Ducatman, A.; Boobis, A.; DeWitt, J.C.; Lau, C.; Ng, C.; Smith, J.S.; Roberts, S.M. Per- and Polyfluoroalkyl Substance Toxicity and Human Health Review: Current State of Knowledge and Strategies for Informing Future Research. *Environ. Toxicol. Chem.* **2021**, *40*, 606–630. [CrossRef] [PubMed]
15. Skaggs, C.S.; Logue, B.A. Ultratrace Analysis of Per- and Polyfluoroalkyl Substances in Drinking Water Using Ice Concentration Linked with Extractive Stirrer and High Performance Liquid Chromatography—Tandem Mass Spectrometry. *J. Chromatogr. A* **2021**, *1659*, 462493. [CrossRef]
16. Casey, J.S.; Jackson, S.R.; Ryan, J.; Newton, S.R. The Use of Gas Chromatography—High Resolution Mass Spectrometry for Suspect Screening and Non-Targeted Analysis of per- and Polyfluoroalkyl Substances. *J. Chromatogr. A* **2023**, *1693*, 463884. [CrossRef]
17. Gogoi, P.; Yao, Y.; Li, Y.C. Understanding PFOS Adsorption on a Pt Electrode for Electrochemical Sensing Applications. *ChemElectroChem* **2023**, *10*, 202201006. [CrossRef]
18. Gonzalez de Vega, R.; Cameron, A.; Clases, D.; Dodgen, T.M.; Doble, P.A.; Bishop, D.P. Simultaneous Targeted and Non-Targeted Analysis of per- and Polyfluoroalkyl Substances in Environmental Samples by Liquid Chromatography-Ion Mobility-Quadrupole Time of Flight-Mass Spectrometry and Mass Defect Analysis. *J. Chromatogr. A* **2021**, *1653*, 462423. [CrossRef]
19. Farooq, S.; Nie, J.; Cheng, Y.; Yan, Z.; Li, J.; Bacha, S.A.S.; Mushtaq, A.; Zhang, H. Molecularly Imprinted Polymers' Application in Pesticide Residue Detection. *Analyst* **2018**, *143*, 3971–3989. [CrossRef]
20. Tarannum, N.; Khatoon, S.; Dzantiev, B.B. Perspective and Application of Molecular Imprinting Approach for Antibiotic Detection in Food and Environmental Samples: A Critical Review. *Food Control* **2020**, *118*, 107381. [CrossRef]
21. Oprea, A.; Weimar, U. Gas Sensors Based on Mass-Sensitive Transducers Part 1: Transducers and Receptors—Basic Understanding. *Anal. Bioanal. Chem.* **2019**, *411*, 1761–1787. [CrossRef]
22. Lian, X.; Zhou, Y.J.; Zhang, H.F.; Li, M.; Huang, X.C. Luminescence Turn-on Detection by an Entanglement-Protected MOF Operating: Via a Divided Receptor-Transducer Protocol. *J. Mater. Chem. C* **2020**, *8*, 3622–3625. [CrossRef]
23. Cheng, Y.H.; Barpaga, D.; Soltis, J.A.; Shutthanandan, V.; Kargupta, R.; Han, K.S.; McGrail, B.P.; Motkuri, R.K.; Basuray, S.; Chatterjee, S. Metal-Organic Framework-Based Microfluidic Impedance Sensor Platform for Ultrasensitive Detection of Perfluorooctanesulfonate. *ACS Appl. Mater. Interfaces* **2020**, *12*, 10503–10514. [CrossRef] [PubMed]
24. Naseri, M.; Mohammadniaei, M.; Sun, Y.; Ashley, J. The Use of Aptamers and Molecularly Imprinted Polymers in Biosensors for Environmental Monitoring: A Tale of Two Receptors. *Chemosensors* **2020**, *8*, 32. [CrossRef]
25. Berhanu, A.; Mutanda, I.; Taolin, J.; Qaria, M.A.; Yang, B.; Zhu, D. A Review of Microbial Degradation of Per- and Polyfluoroalkyl Substances (PFAS): Biotransformation Routes and Enzymes. *Sci. Total Environ.* **2023**, *859*, 160010. [CrossRef] [PubMed]
26. Shahsavari, E.; Rouch, D.; Khudur, L.S.; Thomas, D.; Aburto-Medina, A.; Ball, A.S. Challenges and Current Status of the Biological Treatment of PFAS-Contaminated Soils. *Front. Bioeng. Biotechnol.* **2021**, *8*, 602040. [CrossRef]
27. Ali, G.K.; Omer, K.M. Molecular Imprinted Polymer Combined with Aptamer (MIP-Aptamer) as a Hybrid Dual Recognition Element for Bio(Chemical) Sensing Applications. Review. *Talanta* **2022**, *236*, 122878. [CrossRef]
28. Kamyab, H.; Chelliapan, S.; Tavakkoli, O.; Mesbah, M.; Bhutto, J.K.; Khademi, T.; Kirpichnikova, I.; Ahmad, A.; ALJohani, A.A. A Review on Carbon-Based Molecularly-Imprinted Polymers (CBMIP) for Detection of Hazardous Pollutants in Aqueous Solutions. *Chemosphere* **2022**, *308*, 136471. [CrossRef]
29. Ashley, J.; Shahbazi, M.A.; Kant, K.; Chidambara, V.A.; Wolff, A.; Bang, D.D.; Sun, Y. Molecularly Imprinted Polymers for Sample Preparation and Biosensing in Food Analysis: Progress and Perspectives. *Biosens. Bioelectron.* **2017**, *91*, 606–615. [CrossRef]
30. Rebelo, P.; Costa-Rama, E.; Seguro, I.; Pacheco, J.G.; Nouws, H.P.A.; Cordeiro, M.N.D.S.; Delerue-Matos, C. Molecularly Imprinted Polymer-Based Electrochemical Sensors for Environmental Analysis. *Biosens. Bioelectron.* **2021**, *172*, 112719. [CrossRef]
31. Jiao, Z.; Li, J.; Mo, L.; Liang, J.; Fan, H. A Molecularly Imprinted Chitosan Doped with Carbon Quantum Dots for Fluorometric Determination of Perfluorooctane Sulfonate. *Microchim. Acta* **2018**, *185*, 473. [CrossRef]
32. Yu, Q.; Deng, S.; Yu, G. Selective Removal of Perfluorooctane Sulfonate from Aqueous Solution Using Chitosan-Based Molecularly Imprinted Polymer Adsorbents. *Water Res.* **2008**, *42*, 3089–3097. [CrossRef] [PubMed]
33. Malik, A.A.; Nantasenamat, C.; Piacham, T. Molecularly Imprinted Polymer for Human Viral Pathogen Detection. *Mater. Sci. Eng. C* **2017**, *77*, 1341–1348. [CrossRef] [PubMed]
34. Cennamo, N.; D'Agostino, G.; Sequeira, F.; Mattiello, F.; Porto, G.; Biasiolo, A.; Nogueira, R.; Bilro, L.; Zeni, L. A Simple and Low-Cost Optical Fiber Intensity-Based Configuration for Perfluorinated Compounds in Water Solution. *Sensors* **2018**, *18*, 3009. [CrossRef] [PubMed]

35. Selvolini, G.; Marrazza, G. MIP-Based Sensors: Promising New Tools for Cancer Biomarker Determination. *Sensors* **2017**, *17*, 718. [CrossRef] [PubMed]
36. Peeters, M.; Troost, F.J.; van Grinsven, B.; Horemans, F.; Alenus, J.; Murib, M.S.; Keszthelyi, D.; Ethirajan, A.; Thoelen, R.; Cleij, T.J.; et al. MIP-Based Biomimetic Sensor for the Electronic Detection of Serotonin in Human Blood Plasma. *Sens. Actuators B Chem.* **2012**, *171–172*, 602–610. [CrossRef]
37. Cao, Y.; Feng, T.; Xu, J.; Xue, C. Recent Advances of Molecularly Imprinted Polymer-Based Sensors in the Detection of Food Safety Hazard Factors. *Biosens. Bioelectron.* **2019**, *141*, 111447. [CrossRef] [PubMed]
38. Whitcombe, M.J.; Chianella, I.; Larcombe, L.; Piletsky, S.A.; Noble, J.; Porter, R.; Horgan, A. The Rational Development of Molecularly Imprinted Polymer-Based Sensors for Protein Detection. *Chem. Soc. Rev.* **2011**, *40*, 1547–1571. [CrossRef]
39. Cao, F.; Wang, L.; Ren, X.; Sun, H. Synthesis of a Perfluorooctanoic Acid Molecularly Imprinted Polymer for the Selective Removal of Perfluorooctanoic Acid in an Aqueous Environment. *J. Appl. Polym. Sci.* **2016**, *133*, 43192. [CrossRef]
40. Fang, C.; Chen, Z.; Megharaj, M.; Naidu, R. Potentiometric Detection of AFFFs Based on MIP. *Environ. Technol. Innov.* **2016**, *5*, 52–59. [CrossRef]
41. Chen, S.; Li, A.; Zhang, L.; Gong, J. Molecularly Imprinted Ultrathin Graphitic Carbon Nitride Nanosheets-Based Electrochemiluminescence Sensing Probe for Sensitive Detection of Perfluorooctanoic Acid. *Anal. Chim. Acta* **2015**, *896*, 68–77. [CrossRef]
42. Dickman, R.A.; Aga, D.S. A Review of Recent Studies on Toxicity, Sequestration, and Degradation of per- and Polyfluoroalkyl Substances (PFAS). *J. Hazard. Mater.* **2022**, *436*, 129120. [CrossRef] [PubMed]
43. Tasfaout, A.; Ibrahim, F.; Morrin, A.; Brisset, H.; Sorrentino, I.; Nanteuil, C.; Laffite, G.; Nicholls, I.A.; Regan, F.; Branger, C. Molecularly Imprinted Polymers for Per- and Polyfluoroalkyl Substances Enrichment and Detection. *Talanta* **2023**, *258*, 124434. [CrossRef] [PubMed]
44. Ganesan, S.; Chawengkijwanich, C.; Gopalakrishnan, M.; Janjaroen, D. Detection Methods for Sub-Nanogram Level of Emerging Pollutants—Per and Polyfluoroalkyl Substances. *Food Chem. Toxicol.* **2022**, *168*, 113377. [CrossRef]
45. Pardeshi, S.; Dhodapkar, R. Advances in Fabrication of Molecularly Imprinted Electrochemical Sensors for Detection of Contaminants and Toxicants. *Environ. Res.* **2022**, *212*, 113359. [CrossRef] [PubMed]
46. Chi, T.Y.; Chen, Z.; Kameoka, J. Perfluorooctanesulfonic Acid Detection Using Molecularly Imprinted Polyaniline on a Paper Substrate. *Sensors* **2020**, *20*, 7301. [CrossRef] [PubMed]
47. Vu, O.T.; Nguyen, Q.H.; Nguy Phan, T.; Luong, T.T.; Eersels, K.; Wagner, P.; Truong, L.T.N. Highly Sensitive Molecularly Imprinted Polymer-Based Electrochemical Sensors Enhanced by Gold Nanoparticles for Norfloxacin Detection in Aquaculture Water. *ACS Omega* **2023**, *8*, 2887–2896. [CrossRef] [PubMed]
48. Lowdon, J.W.; Eersels, K.; Arreguin-Campos, R.; Caldara, M.; Heidt, B.; Rogosic, R.; Jimenez-Monroy, K.L.; Cleij, T.J.; Diliën, H.; van Grinsven, B. A Molecularly Imprinted Polymer-Based Dye Displacement Assay for the Rapid Visual Detection of Amphetamine in Urine. *Molecules* **2020**, *25*, 5222. [CrossRef]
49. Cennamo, N.; D'Agostino, G.; Porto, G.; Biasiolo, A.; Perri, C.; Arcadio, F.; Zeni, L. A Molecularly Imprinted Polymer on a Plasmonic Plastic Optical Fiber to Detect Perfluorinated Compounds in Water. *Sensors* **2018**, *18*, 1836. [CrossRef]
50. Ayerdurai, V.; Cieplak, M.; Kutner, W. Molecularly Imprinted Polymer-Based Electrochemical Sensors for Food Contaminants Determination. *Trends Anal. Chem.* **2023**, *158*, 116830. [CrossRef]
51. Jahanban-Esfahlan, A.; Roufegarinejad, L.; Jahanban-Esfahlan, R.; Tabibiazar, M.; Amarowicz, R. Latest Developments in the Detection and Separation of Bovine Serum Albumin Using Molecularly Imprinted Polymers. *Talanta* **2020**, *207*, 120317. [CrossRef]
52. Jamalipour Soufi, G.; Iravani, S.; Varma, R.S. Molecularly Imprinted Polymers for the Detection of Viruses: Challenges and Opportunities. *Analyst* **2021**, *146*, 3087–3100. [CrossRef] [PubMed]
53. McClements, J.; Bar, L.; Singla, P.; Canfarotta, F.; Thomson, A.; Czulak, J.; Johnson, R.E.; Crapnell, R.D.; Banks, C.E.; Payne, B.; et al. Molecularly Imprinted Polymer Nanoparticles Enable Rapid, Reliable, and Robust Point-of-Care Thermal Detection of SARS-CoV-2. *ACS Sens.* **2022**, *7*, 1122–1131. [CrossRef] [PubMed]
54. Stilman, W.; Campolim Lenzi, M.; Wackers, G.; Deschaume, O.; Yongabi, D.; Mathijssen, G.; Bartic, C.; Gruber, J.; Wübbenhorst, M.; Heyndrickx, M.; et al. Low Cost, Sensitive Impedance Detection of *E. coli* Bacteria in Food-Matrix Samples Using Surface-Imprinted Polymers as Whole-Cell Receptors. *Phys. Status Solidi A* **2022**, *219*, 2100405. [CrossRef]
55. Eersels, K.; Lieberzeit, P.; Wagner, P. A Review on Synthetic Receptors for Bioparticle Detection Created by Surface-Imprinting Techniques—From Principles to Applications. *ACS Sens.* **2016**, *1*, 1171–1187. [CrossRef]
56. Seo, H.B.; Gu, M.B. Aptamer-Based Sandwich-Type Biosensors. *J. Biol. Eng.* **2017**, *11*, 11. [CrossRef] [PubMed]
57. Zhou, W.; Jimmy Huang, P.J.; Ding, J.; Liu, J. Aptamer-Based Biosensors for Biomedical Diagnostics. *Analyst* **2014**, *139*, 2627–2640. [CrossRef] [PubMed]
58. Kudłak, B.; Wieczerzak, M. Aptamer Based Tools for Environmental and Therapeutic Monitoring: A Review of Developments, Applications, Future Perspectives. *Crit. Rev. Environ. Sci. Technol.* **2020**, *50*, 816–867. [CrossRef]
59. Park, J.; Yang, K.A.; Choi, Y.; Choe, J.K. Novel ssDNA Aptamer-Based Fluorescence Sensor for Perfluorooctanoic Acid Detection in Water. *Environ. Int.* **2022**, *158*, 107000. [CrossRef]
60. Mukunzi, D.; Habimana, J.d.D.; Li, Z.; Zou, X. Mycotoxins Detection: View in the Lens of Molecularly Imprinted Polymer and Nanoparticles. *Crit. Rev. Food Sci. Nutr.* **2022**, *63*, 6034–6068. [CrossRef]

61. Vanoursouw, T.M.; Rottiger, T.; Wadzinski, K.A.; Vanderwaal, B.E.; Snyder, M.J.; Bittner, R.T.; Farha, O.K.; Riha, S.C.; Mondloch, J.E. Adsorption of a PFAS Utilizing MOF-808: Development of an Undergraduate Laboratory Experiment in a Capstone Course. *J. Chem. Educ.* **2023**, *100*, 861–868. [CrossRef]
62. Li, R.; Alomari, S.; Stanton, R.; Wasson, M.C.; Islamoglu, T.; Farha, O.K.; Holsen, T.M.; Thagard, S.M.; Trivedi, D.J.; Wriedt, M. Efficient Removal of Per- And Polyfluoroalkyl Substances from Water with Zirconium-Based Metal-Organic Frameworks. *Chem. Mater.* **2021**, *33*, 3276–3285. [CrossRef]
63. FitzGerald, L.I.; Olorunyomi, J.F.; Singh, R.; Doherty, C.M. Towards Solving the PFAS Problem: The Potential Role of Metal-Organic Frameworks. *ChemSusChem* **2022**, *15*, e202201136. [CrossRef] [PubMed]
64. Hu, M.L.; Razavi, S.A.A.; Piroozzadeh, M.; Morsali, A. Sensing Organic Analytes by Metal-Organic Frameworks: A New Way of Considering the Topic. *Inorg. Chem. Front.* **2020**, *7*, 1598–1632. [CrossRef]
65. Menger, R.F.; Funk, E.; Henry, C.S.; Borch, T. Sensors for Detecting Per- and Polyfluoroalkyl Substances (PFAS): A Critical Review of Development Challenges, Current Sensors, and Commercialization Obstacles. *Chem. Eng. J.* **2021**, *417*, 129133. [CrossRef] [PubMed]
66. Karbassiyazdi, E.; Kasula, M.; Modak, S.; Pala, J.; Kalantari, M.; Altaee, A.; Esfahani, M.R.; Razmjou, A. A Juxtaposed Review on Adsorptive Removal of PFAS by Metal-Organic Frameworks (MOFs) with Carbon-Based Materials, Ion Exchange Resins, and Polymer Adsorbents. *Chemosphere* **2023**, *311*, 136933. [CrossRef] [PubMed]
67. Wang, Y.; Ren, R.; Chen, F.; Jing, L.; Tian, Z.; Li, Z.; Wang, J.; Hou, C. Molecularly Imprinted MOFs-Driven Carbon Nanofiber for Sensitive Electrochemical Detection and Targeted Electro-Fenton Degradation of Perfluorooctanoic Acid. *Sep. Purif. Technol.* **2023**, *310*, 123257. [CrossRef]
68. Pirot, S.M.; Omer, K.M.; Alshatteri, A.H.; Ali, G.K.; Shatery, O.B.A. Dual-Template Molecularly Surface Imprinted Polymer on Fluorescent Metal-Organic Frameworks Functionalized with Carbon Dots for Ascorbic Acid and Uric Acid Detection. *Spectrochim. Acta Part A Mol. Biomol. Spectrosc.* **2023**, *291*, 122340. [CrossRef]
69. Lv, M.; Zhou, W.; Tavakoli, H.; Bautista, C.; Xia, J.; Wang, Z.; Li, X.J. Aptamer-Functionalized Metal-Organic Frameworks (MOFs) for Biosensing. *Biosens. Bioelectron.* **2021**, *176*, 112947. [CrossRef]
70. Zhou, Q.; Xu, Z.; Liu, Z. Molecularly Imprinting-Aptamer Techniques and Their Applications in Molecular Recognition. *Biosensors* **2022**, *12*, 576. [CrossRef]
71. Wu, Y.; Li, Y.; Tian, A.; Mao, K.; Liu, J. Selective Removal of Perfluorooctanoic Acid Using Molecularly Imprinted Polymer-Modified TiO_2 Nanotube Arrays. *Int. J. Photoenergy* **2016**, *2016*, 7368795. [CrossRef]
72. Abbasian Chaleshtari, Z.; Foudazi, R. A Review on Per- and Polyfluoroalkyl Substances (PFAS) Remediation: Separation Mechanisms and Molecular Interactions. *ACS ES T Water* **2022**, *2*, 2258–2272. [CrossRef]
73. Karadurmus, L.; Bilge, S.; Sınağ, A.; Ozkan, S.A. Molecularly Imprinted Polymer (MIP)-Based Sensing for Detection of Explosives: Current Perspectives and Future Applications. *Trends Anal. Chem.* **2022**, *155*, 116694. [CrossRef]
74. Caldara, M.; van Wissen, G.; Cleij, T.J.; Diliën, H.; van Grinsven, B.; Eersels, K.; Lowdon, J.W. Deposition Methods for the Integration of Molecularly Imprinted Polymers (MIPs) in Sensor Applications. *Adv. Sens. Res.* **2023**, *2*, 2200059. [CrossRef]
75. Cennamo, N.; D'Agostino, G.; Arcadio, F.; Perri, C.; Porto, G.; Biasiolo, A.; Zeni, L. Measurement of MIPs Responses Deposited on Two SPR-POF Sensors Realized by Different Photoresist Buffer Layers. *IEEE Trans. Instrum. Meas.* **2020**, *69*, 1464–1473. [CrossRef]
76. Hasseb, A.A.; Abdel Ghani, N.d.T.; Shehab, O.R.; El Nashar, R.M. Application of Molecularly Imprinted Polymers for Electrochemical Detection of Some Important Biomedical Markers and Pathogens. *Curr. Opin. Electrochem.* **2022**, *31*, 100848. [CrossRef]
77. Metwally, M.G.; Benhawy, A.H.; Khalifa, R.M.; El Nashar, R.M.; Trojanowicz, M. Application of Molecularly Imprinted Polymers in the Analysis of Waters and Wastewaters. *Molecules* **2021**, *26*, 6515. [CrossRef] [PubMed]
78. Jamieson, O.; Mecozzi, F.; Crapnell, R.D.; Battell, W.; Hudson, A.; Novakovic, K.; Sachdeva, A.; Canfarotta, F.; Herdes, C.; Banks, C.E.; et al. Approaches to the Rational Design of Molecularly Imprinted Polymers Developed for the Selective Extraction or Detection of Antibiotics in Environmental and Food Samples. *Phys. Status Solidi A* **2021**, *218*, 2100021. [CrossRef]
79. Irshad, M.; Iqbal, N.; Mujahid, A.; Afzal, A.; Hussain, T.; Sharif, A.; Ahmad, E.; Athar, M.M. Molecularly Imprinted Nanomaterials for Sensor Applications. *Nanomaterials* **2013**, *3*, 615–637. [CrossRef]
80. Yu, H.; Chen, H.; Fang, B.; Sun, H. Sorptive Removal of Per- and Polyfluoroalkyl Substances from Aqueous Solution: Enhanced Sorption, Challenges and Perspectives. *Sci. Total Environ.* **2023**, *861*, 160647. [CrossRef]
81. Crapnell, R.D.; Hudson, A.; Foster, C.W.; Eersels, K.; van Grinsven, B.; Cleij, T.J.; Banks, C.E.; Peeters, M. Recent Advances in Electrosynthesized Molecularly Imprinted Polymer Sensing Platforms for Bioanalyte Detection. *Sensors* **2019**, *19*, 1204. [CrossRef]
82. Wackers, G.; Cornelis, P.; Putzeys, T.; Peeters, M.; Tack, J.; Troost, F.; Doll, T.; Verhaert, N.; Wagner, P. Electropolymerized Receptor Coatings for the Quantitative Detection of Histamine with a Catheter-Based, Diagnostic Sensor. *ACS Sens.* **2021**, *6*, 100–110. [CrossRef] [PubMed]
83. Yang, J.; Li, Y.; Wang, J.; Sun, X.; Cao, R.; Sun, H.; Huang, C.; Chen, J. Molecularly Imprinted Polymer Microspheres Prepared by Pickering Emulsion Polymerization for Selective Solid-Phase Extraction of Eight Bisphenols from Human Urine Samples. *Anal. Chim. Acta* **2015**, *872*, 35–45. [CrossRef] [PubMed]
84. Yang, Y.; Shen, X. Preparation and Application of Molecularly Imprinted Polymers for Flavonoids: Review and Perspective. *Molecules* **2022**, *27*, 7355. [CrossRef] [PubMed]

85. Cui, F.; Zhou, Z.; Zhou, H.S. Molecularly Imprinted Polymers and Surface Imprinted Polymers Based Electrochemical Biosensor for Infectious Diseases. *Sensors* **2020**, *20*, 996. [CrossRef] [PubMed]
86. Caldara, M.; Lowdon, J.W.; Rogosic, R.; Arreguin-Campos, R.; Jimenez-Monroy, K.L.; Heidt, B.; Tschulik, K.; Cleij, T.J.; Diliën, H.; Eersels, K.; et al. Thermal Detection of Glucose in Urine Using a Molecularly Imprinted Polymer as a Recognition Element. *ACS Sens.* **2021**, *6*, 4515–4525. [CrossRef]
87. Tretjakov, A.; Syritski, V.; Reut, J.; Boroznjak, R.; Volobujeva, O.; Öpik, A. Surface Molecularly Imprinted Polydopamine Films for Recognition of Immunoglobulin G. *Microchim. Acta* **2013**, *180*, 1433–1442. [CrossRef]
88. Stilman, W.; Yongabi, D.; Bakhshi Sichani, S.; Thesseling, F.; Deschaume, O.; Putzeys, T.; Pinto, T.C.; Verstrepen, K.; Bartic, C.; Wübbenhorst, M.; et al. Detection of Yeast Strains by Combining Surface-Imprinted Polymers with Impedance-Based Readout. *Sens. Actuators B Chem.* **2021**, *340*, 129917. [CrossRef]
89. Tian, L.; Guo, H.; Li, J.; Yan, L.; Zhu, E.; Liu, X.; Li, K. Fabrication of a Near-Infrared Excitation Surface Molecular Imprinting Ratiometric Fluorescent Probe for Sensitive and Rapid Detecting Perfluorooctane Sulfonate in Complex Matrix. *J. Hazard. Mater.* **2021**, *413*, 125353. [CrossRef]
90. Dong, C.; Shi, H.; Han, Y.; Yang, Y.; Wang, R.; Men, J. Molecularly Imprinted Polymers by the Surface Imprinting Technique. *Eur. Polym. J.* **2021**, *145*, 110231. [CrossRef]
91. Pierpaoli, M.; Szopińska, M.; Olejnik, A.; Ryl, J.; Fudala-Ksiażek, S.; Łuczkiewicz, A.; Bogdanowicz, R. Engineering Boron and Nitrogen Codoped Carbon Nanoarchitectures to Tailor Molecularly Imprinted Polymers for PFOS Determination. *J. Hazard. Mater.* **2023**, *458*, 131873. [CrossRef]
92. Glasscott, M.W.; Vannoy, K.J.; Kazemi, R.; Verber, M.D.; Dick, J.E. µ-MIP: Molecularly Imprinted Polymer-Modified Microelectrodes for the Ultrasensitive Quantification of GenX (HFPO-DA) in River Water. *Environ. Sci. Technol. Lett.* **2020**, *7*, 489–495. [CrossRef]
93. Clark, R.B.; Dick, J.E. Electrochemical Sensing of Perfluorooctanesulfonate (PFOS) Using Ambient Oxygen in River Water. *ACS Sens.* **2020**, *5*, 3591–3598. [CrossRef] [PubMed]
94. Karimian, N.; Stortini, A.M.; Moretto, L.M.; Costantino, C.; Bogialli, S.; Ugo, P. Electrochemosensor for Trace Analysis of Perfluorooctanesulfonate in Water Based on a Molecularly Imprinted Poly(o-Phenylenediamine) Polymer. *ACS Sens.* **2018**, *3*, 1291–1298. [CrossRef] [PubMed]
95. Dery, L.; Zelikovich, D.; Mandler, D. Electrochemistry of Molecular Imprinting of Large Entities. *Curr. Opin. Electrochem.* **2022**, *34*, 100967. [CrossRef]
96. Mahmoudpour, M.; Torbati, M.; Mousavi, M.M.; de la Guardia, M.; Ezzati Nazhad Dolatabadi, J. Nanomaterial-Based Molecularly Imprinted Polymers for Pesticides Detection: Recent Trends and Future Prospects. *Trends Anal. Chem.* **2020**, *129*, 115943. [CrossRef]
97. Ren, J.; Lu, Y.; Han, Y.; Qiao, F.; Yan, H. Novel Molecularly Imprinted Phenolic Resin–Dispersive Filter Extraction for Rapid Determination of Perfluorooctanoic Acid and Perfluorooctane Sulfonate in Milk. *Food Chem.* **2023**, *400*, 134062. [CrossRef]
98. Mostafiz, B.; Bigdeli, S.A.; Banan, K.; Afsharara, H.; Hatamabadi, D.; Mousavi, P.; Hussain, C.M.; Keçili, R.; Ghorbani-Bidkorbeh, F. Molecularly Imprinted Polymer-Carbon Paste Electrode (MIP-CPE)-Based Sensors for the Sensitive Detection of Organic and Inorganic Environmental Pollutants: A Review. *Trends Environ. Anal. Chem.* **2021**, *32*, 00144. [CrossRef]
99. Peeters, M.; Troost, F.J.; Mingels, R.H.G.; Welsch, T.; van Grinsven, B.; Vranken, T.; Ingebrandt, S.; Thoelen, R.; Cleij, T.J.; Wagner, P. Impedimetric Detection of Histamine in Bowel Fluids Using Synthetic Receptors with pH-Optimized Binding Characteristics. *Anal. Chem.* **2013**, *85*, 1475–1483. [CrossRef]
100. ul Gani Mir, T.; Malik, A.Q.; Singh, J.; Shukla, S.; Kumar, D. An Overview of Molecularly Imprinted Polymers Embedded with Quantum Dots and Their Implementation as an Alternative Approach for Extraction and Detection of Crocin. *ChemistrySelect* **2022**, *7*, 202200829. [CrossRef]
101. Akgönüllü, S.; Kılıç, S.; Esen, C.; Denizli, A. Molecularly Imprinted Polymer-Based Sensors for Protein Detection. *Polymers* **2023**, *15*, 629. [CrossRef]
102. Wang, L.; Zhi, K.; Zhang, Y.; Liu, Y.; Zhang, L.; Yasin, A.; Lin, Q. Molecularly Imprinted Polymers for Gossypol via Sol-Gel, Bulk, and Surface Layer Imprinting-A Comparative Study. *Polymers* **2019**, *11*, 602. [CrossRef] [PubMed]
103. Li, T.; Li, X.; Liu, H.; Deng, Z.; Zhang, Y.; Zhang, Z.; He, Y.; Yang, Y.; Zhong, S. Preparation and Characterization of Molecularly Imprinted Polymers Based on β-Cyclodextrin-Stabilized Pickering Emulsion Polymerization for Selective Recognition of Erythromycin from River Water and Milk. *J. Sep. Sci.* **2020**, *43*, 3683–3690. [CrossRef] [PubMed]
104. Tabar, F.A.; Lowdon, J.W.; Caldara, M.; Cleij, T.J.; Wagner, P.; Diliën, H.; Eersels, K.; van Grinsven, B. Thermal Determination of Perfluoroalkyl Substances in Environmental Samples Employing a Molecularly Imprinted Polyacrylamide as a Receptor Layer. *Environ. Technol. Innov.* **2023**, *29*, 103021. [CrossRef]
105. van Grinsven, B.; Eersels, K.; Peeters, M.; Losada-Pérez, P.; Vandenryt, T.; Cleij, T.J.; Wagner, P. The Heat-Transfer Method: A Versatile Low-Cost, Label-Free, Fast, and User-Friendly Readout Platform for Biosensor Applications. *ACS Appl. Mater. Interfaces* **2014**, *6*, 13309–13318. [CrossRef] [PubMed]
106. Wagner, P.; Bakhshi Sichani, S.; Khorshid, M.; Lieberzeit, P.; Losada-Pérez, P.; Yongabi, D. Bioanalytical Sensors Using the Heat-Transfer Method HTM and Related Techniques. *tm-Tech. Mess.* **2023**, *90*, 761–785. [CrossRef]
107. Lowdon, J.W.; Diliën, H.; van Grinsven, B.; Eersels, K.; Cleij, T.J. Colorimetric Sensing of Amoxicillin Facilitated by Molecularly Imprinted Polymers. *Polymers* **2021**, *13*, 2221. [CrossRef]

108. Wang, Z.; Zhang, Z.; Yan, R.; Fu, X.; Wang, G.; Wang, Y.; Li, Z.; Zhang, X.; Hou, J. Facile Fabrication of Snowman-like Magnetic Molecularly Imprinted Polymer Microspheres for Bisphenol A via One-Step Pickering Emulsion Polymerization. *React. Funct. Polym.* **2021**, *164*, 104911. [CrossRef]
109. Chen, H.; Son, S.; Zhang, F.; Yan, J.; Li, Y.; Ding, H.; Ding, L. Rapid Preparation of Molecularly Imprinted Polymers by Microwave-Assisted Emulsion Polymerization for the Extraction of Florfenicol in Milk. *J. Chromatogr. B* **2015**, *983–984*, 32–38. [CrossRef]
110. Zhao, G.; Liu, J.; Liu, M.; Han, X.; Peng, Y.; Tian, X.; Liu, J.; Zhang, S. Synthesis of Molecularly Imprinted Polymer via Emulsion Polymerization for Application in Solanesol Separation. *Appl. Sci.* **2020**, *10*, 2868. [CrossRef]
111. Pardeshi, S.; Singh, S.K. Precipitation Polymerization: A Versatile Tool for Preparing Molecularly Imprinted Polymer Beads for Chromatography Applications. *RSC Adv.* **2016**, *6*, 23525–23536. [CrossRef]
112. Alizadeh, T.; Memarbashi, N. Evaluation of the Facilitated Transport Capabilities of Nano- and Micro-Sized Molecularly Imprinted Polymers (MIPs) in a Bulk Liquid Membrane System. *Sep. Purif. Technol.* **2012**, *90*, 83–91. [CrossRef]
113. Rehman, A.U.; Crimi, M.; Andreescu, S. Current and Emerging Analytical Techniques for the Determination of PFAS in Environmental Samples. *Trends Environ. Anal. Chem.* **2023**, *37*, 00198. [CrossRef]
114. Islam, G.J.; Arrigan, D.W.M. Voltammetric Selectivity in Detection of Ionized Perfluoroalkyl Substances at Micro-Interfaces between Immiscible Electrolyte Solutions. *ACS Sens.* **2022**, *7*, 2960–2967. [CrossRef] [PubMed]
115. Clark, R.B.; Dick, J.E. Towards Deployable Electrochemical Sensors for Per- And Polyfluoroalkyl Substances (PFAS). *ChemComm* **2021**, *57*, 8121–8130. [CrossRef] [PubMed]
116. Moro, G.; Cristofori, D.; Bottari, F.; Cattaruzza, E.; De Wael, K.; Moretto, L.M. Redesigning an Electrochemical MIP Sensor for PFOS: Practicalities and Pitfalls. *Sensors* **2019**, *19*, 4433. [CrossRef]
117. Chi, H.; Liu, G. Carbon Nanomaterial-Based Molecularly Imprinted Polymer Sensors for Detection of Hazardous Substances in Food: Recent Progress and Future Trends. *Food Chem.* **2023**, *420*, 136100. [CrossRef]
118. Gao, M.; Gao, Y.; Chen, G.; Huang, X.; Xu, X.; Lv, J.; Wang, J.; Xu, D.; Liu, G. Recent Advances and Future Trends in the Detection of Contaminants by Molecularly Imprinted Polymers in Food Samples. *Front. Chem.* **2020**, *8*, 616326. [CrossRef]
119. Feng, H.; Wang, N.; Trant, T.; Yuan, L.; Li, J.; Cai, Q. Surface Molecular Imprinting on Dye-(NH_2)-SiO_2 NPs for Specific Recognition and Direct Fluorescent Quantification of Perfluorooctane Sulfonate. *Sens. Actuators B Chem.* **2014**, *195*, 266–273. [CrossRef]
120. Steigerwald, J.M.; Peng, S.; Ray, J.R. Novel Perfluorooctanesulfonate-Imprinted Polymer Immobilized on Spent Coffee Grounds Biochar for Selective Removal of Perfluoroalkyl Acids in Synthetic Wastewater. *ACS EST Eng.* **2022**, *3*, 520–532. [CrossRef]
121. Du, L.; Wu, Y.; Zhang, X.; Zhang, F.; Chen, X.; Cheng, Z.; Wu, F.; Tan, K. Preparation of Magnetic Molecularly Imprinted Polymers for the Rapid and Selective Separation and Enrichment of Perfluorooctane Sulfonate. *J. Sep. Sci.* **2017**, *40*, 2819–2826. [CrossRef]
122. Lin, L.; Guo, H.; Lin, S.; Chen, Y.; Yan, L.; Zhu, E.; Li, K. Selective Extraction of Perfluorooctane Sulfonate in Real Samples by Superparamagnetic Nanospheres Coated with a Polydopamine-Based Molecularly Imprinted Polymer. *J. Sep. Sci.* **2021**, *44*, 1015–1025. [CrossRef] [PubMed]
123. Du, L.; Cheng, Z.; Zhu, P.; Chen, Q.; Wu, Y.; Tan, K. Preparation of Mesoporous Silica Nanoparticles Molecularly Imprinted Polymer for Efficient Separation and Enrichment of Perfluorooctane Sulfonate. *J. Sep. Sci.* **2018**, *41*, 4363–4369. [CrossRef] [PubMed]
124. Guo, H.; Liu, Y.; Ma, W.; Yan, L.; Li, K.; Lin, S. Surface Molecular Imprinting on Carbon Microspheres for Fast and Selective Adsorption of Perfluorooctane Sulfonate. *J. Hazard. Mater.* **2018**, *348*, 29–38. [CrossRef] [PubMed]
125. Lu, D.; Zhu, D.Z.; Gan, H.; Yao, Z.; Luo, J.; Yu, S.; Kurup, P. An Ultra-Sensitive Molecularly Imprinted Polymer (MIP) and Gold Nanostars (AuNS) Modified Voltammetric Sensor for Facile Detection of Perfluorooctance Sulfonate (PFOS) in Drinking Water. *Sens. Actuators B Chem.* **2022**, *352*, 131005. [CrossRef]
126. Gao, Y.; Gou, W.; Zeng, W.; Chen, W.; Jiang, J.; Lu, J. Determination of Perfluorooctanesulfonic Acid in Water by Polydopamine Molecularly Imprinted/Gold Nanoparticles Sensor. *Microchem. J.* **2023**, *187*, 108378. [CrossRef]
127. Zheng, L.; Zheng, Y.; Liu, Y.; Long, S.; Du, L.; Liang, J.; Huang, C.; Swihart, M.T.; Tan, K. Core-Shell Quantum Dots Coated with Molecularly Imprinted Polymer for Selective Photoluminescence Sensing of Perfluorooctanoic Acid. *Talanta* **2019**, *194*, 1–6. [CrossRef] [PubMed]
128. Tran, T.T.; Li, J.; Feng, H.; Cai, J.; Yuan, L.; Wang, N.; Cai, Q. Molecularly Imprinted Polymer Modified TiO_2 Nanotube Arrays for Photoelectrochemical Determination of Perfluorooctane Sulfonate (PFOS). *Sens. Actuators B Chem.* **2014**, *190*, 745–751. [CrossRef]
129. Yang, S.; Teng, Y.; Cao, Q.; Bai, C.; Fang, Z.; Xu, W. Electrochemical Sensor Based on Molecularly Imprinted Polymer-Aptamer Hybrid Receptor for Voltammetric Detection of Thrombin. *J. Electrochem. Soc.* **2019**, *166*, B23–B28. [CrossRef]
130. Hayat, A.; Marty, J.L. Aptamer Based Electrochemical Sensors for Emerging Environmental Pollutants. *Front. Chem.* **2014**, *2*, 41. [CrossRef]
131. Karimzadeh, Z.; Mahmoudpour, M.; Guardia, M.d.l.; Ezzati Nazhad Dolatabadi, J.; Jouyban, A. Aptamer-Functionalized Metal Organic Frameworks as an Emerging Nanoprobe in the Food Safety Field: Promising Development Opportunities and Translational Challenges. *Trends Anal. Chem.* **2022**, *152*, 116622. [CrossRef]
132. Wolter, O.; Mayer, G. Aptamers as Valuable Molecular Tools in Neurosciences. *J. Neurosci.* **2017**, *37*, 2517–2523. [CrossRef] [PubMed]
133. Assen, A.H.; Yassine, O.; Shekhah, O.; Eddaoudi, M.; Salama, K.N. MOFs for the Sensitive Detection of Ammonia: Deployment of Fcu-MOF Thin Films as Effective Chemical Capacitive Sensors. *ACS Sens.* **2017**, *2*, 1294–1301. [CrossRef] [PubMed]

134. Daniel, M.; Mathew, G.; Anpo, M.; Neppolian, B. MOF Based Electrochemical Sensors for the Detection of Physiologically Relevant Biomolecules: An Overview. *Coord. Chem. Rev.* **2022**, *468*, 214627. [CrossRef]
135. Wang, G.D.; Li, Y.Z.; Shi, W.J.; Zhang, B.; Hou, L.; Wang, Y.Y. A Robust Cluster-Based Eu-MOF as Multi-Functional Fluorescence Sensor for Detection of Antibiotics and Pesticides in Water. *Sens. Actuators B Chem.* **2021**, *331*, 129377. [CrossRef]
136. Varadwaj, A.; Varadwaj, P.R.; Marques, H.M.; Yamashita, K. Revealing Factors Influencing the Fluorine-Centered Non-Covalent Interactions in Some Fluorine-Substituted Molecular Complexes: Insights from First-Principles Studies. *ChemPhysChem* **2018**, *19*, 1486–1499. [CrossRef] [PubMed]
137. Chen, B.; Yang, Z.; Qu, X.; Zheng, S.; Yin, D.; Fu, H. Screening and Discrimination of Perfluoroalkyl Substances in Aqueous Solution Using a Luminescent Metal-Organic Framework Sensor Array. *ACS Appl. Mater. Interfaces* **2021**, *13*, 47706–47716. [CrossRef]
138. Suwannakot, P.; Lisi, F.; Ahmed, E.; Liang, K.; Babarao, R.; Gooding, J.J.; Donald, W.A. Metal-Organic Framework-Enhanced Solid-Phase Microextraction Mass Spectrometry for the Direct and Rapid Detection of Perfluorooctanoic Acid in Environmental Water Samples. *Anal. Chem.* **2020**, *92*, 6900–6908. [CrossRef]
139. Wang, S.; Niu, H.; Zeng, T.; Zhang, X.; Cao, D.; Cai, Y. Rapid Determination of Small Molecule Pollutants Using Metal-Organic Frameworks as Adsorbent and Matrix of MALDI-TOF-MS. *Microporous Mesoporous Mater.* **2017**, *239*, 390–395. [CrossRef]
140. Li, Y.; Lu, Y.; Zhang, X.; Cao, H.; Huang, Y. Cobalt-Embedded Nitrogen-Doped Carbon Nanosheets with Enhanced Oxidase-like Activity for Detecting Perfluorooctane Sulfonate. *Microchem. J.* **2022**, *181*, 107814. [CrossRef]
141. Jia, Y.; Qian, J.; Pan, B. Dual-Functionalized MIL-101(Cr) for the Selective Enrichment and Ultrasensitive Analysis of Trace Per-And Poly-Fluoroalkyl Substances. *Anal. Chem.* **2021**, *93*, 11116–11122. [CrossRef]
142. Tian, Q.; Sun, M. Analysis of GenX and Other Per- and Polyfluoroalkyl Substances in Environmental Water Samples. In *Separation Science and Technology (New York)*; Elsevier Inc.: Amsterdam, The Netherlands, 2019; Volume 11, pp. 355–370.
143. Wang, Y.; Darling, S.B.; Chen, J. Selectivity of Per- and Polyfluoroalkyl Substance Sensors and Sorbents in Water. *ACS Appl. Mater. Interfaces* **2021**, *13*, 60789–60814. [CrossRef] [PubMed]
144. Wang, L.; Pagett, M.; Zhang, W. Molecularly imprinted polymer (MIP) based electrochemical sensors and their recent advances in health applications. *Sens. Actuators Rep.* **2023**, *5*, 100153. [CrossRef]
145. Díaz-Álvarez, M.; Martin-Esteban, A. Molecularly Imprinted Polymer-Quantum Dot Materials in Optical Sensors: An Overview of Their Synthesis and Applications. *Biosensors* **2021**, *11*, 79. [CrossRef] [PubMed]
146. Bowei, L.; Qi, J.; Liu, F.; Zhao, R.; Arabi, M.; Ostovan, A.; Song, J.; Wang, X.; Zhang, Z.; Chen, L. Molecular imprinting-based indirect fluorescence detection strategy implemented on paper chip for non-fluorescent microcystin. *Nature* **2023**, *14*, 6553.

sensors

Review

Photoplethysmography for the Assessment of Arterial Stiffness

Parmis Karimpour, James M. May and Panicos A. Kyriacou *

Research Centre for Biomedical Engineering, City, University of London, London EC1V 0HB, UK;
parmis.karimpour@city.ac.uk (P.K.); james.may.1@city.ac.uk (J.M.M.)
* Correspondence: p.kyriacou@city.ac.uk

Abstract: This review outlines the latest methods and innovations for assessing arterial stiffness, along with their respective advantages and disadvantages. Furthermore, we present compelling evidence indicating a recent growth in research focused on assessing arterial stiffness using photoplethysmography (PPG) and propose PPG as a potential tool for assessing vascular ageing in the future. Blood vessels deteriorate with age, losing elasticity and forming deposits. This raises the likelihood of developing cardiovascular disease (CVD), widely reported as the global leading cause of death. The ageing process induces structural modifications in the vascular system, such as increased arterial stiffness, which can cause various volumetric, mechanical, and haemodynamic alterations. Numerous techniques have been investigated to assess arterial stiffness, some of which are currently used in commercial medical devices and some, such as PPG, of which still remain in the research space.

Keywords: vascular ageing; arterial stiffness; photoplethysmography; PPG; cardiovascular disease; CVD

Citation: Karimpour, P.; May, J.M.; Kyriacou, P.A. Photoplethysmography for the Assessment of Arterial Stiffness. *Sensors* **2023**, *23*, 9882. https://doi.org/10.3390/s23249882

Academic Editor: Tibor Hianik

Received: 24 October 2023
Revised: 8 December 2023
Accepted: 13 December 2023
Published: 17 December 2023

1. Introduction

The primary aim of this review is to cover the current state-of-the-art technology and approaches in detecting arterial stiffness and monitoring vascular ageing. The review aims to evaluate current trends, especially Photoplethysmography (PPG)-based approaches, and to offer suggestions for future research. The concept of using non-invasive measuring techniques to evaluate vascular stiffness is presented along with the benefits and drawbacks of the available modalities.

Globally, cardiovascular disease (CVD) is the leading cause of mortality, accounting for approximately 17.9 million fatalities in 2019 [1]. As vessels age, the heart and blood vessels undergo changes, leading to degeneration and an increased risk of CVD. The arterial system's ability to function may be harmed or even rendered ineffective by diseases that might develop as a result of vascular ageing. The diseases can impact upper and lower vasculature. Around 20% of people in the United Kingdom (UK) aged between 55 and 75 suffer from peripheral arterial disease (PAD), which is a narrowing of the artery vessels due to factors such as wall thickening or the accumulation of fatty deposits [2,3]. PAD patients typically have no outward symptoms, making early detection difficult. In extreme cases, this may necessitate even an amputation because in the absence of treatment, cardiovascular death may occur. Cardiovascular death can occur due to a lack of blood reaching the organs/tissues when the arteries become narrow or obstructed, a state known as gangrene. Affected limbs may have decreased perfusion and hence suffer from oxygen starvation, immobility, and increased discomfort [4]. Early detection can prevent cardiovascular death, and is crucial for asymptomatic patients. Advancements in technology and techniques to monitor vascular health can provide important information to avoid limb loss and other severe pathologies.

An artery wall is a muscular tube lined with smooth tissue consisting of three main layers: the intima, media, and adventitia [5]. Vascular ageing refers to the alteration of the mechanical and structural properties of the arterial wall with ageing. Collagen and elastin are the two fibres that make up the artery walls. Collagen replaces the elastin fibres as the vessels age, and the resulting collagen bridges cause the vessels to lose some of their elasticity. High blood pressure (BP), hypertension, can be brought on by arterial wall damage that restricts blood flow. Heart attacks, coronary artery disease (CAD), and arrhythmias are just a few of the many disorders that can develop as vessels age. As mentioned before, PAD, hypertension, and atherosclerosis (a build-up of plaque that obstructs an artery and impairs peripheral circulation) resulting in tissue damage and rest pain can develop. With so many diseases linked to the ageing of the vessels, it is crucial to evaluate vascular ageing effectively and at the early stages if possible. Since changes to the arteries' elastic properties lead to an increase in arterial stiffness [6], the evaluation of vascular ageing relies heavily on the measurement of arterial stiffness. Figure 1 illustrates a comparison between a healthy and unhealthy vessel in terms of elasticity.

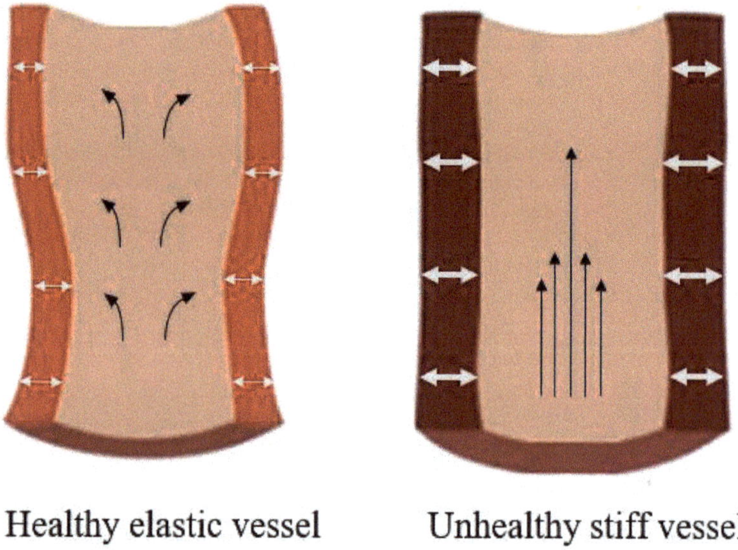

Figure 1. Comparison of healthy elastic vessel (**left image**) with an aged, unhealthy stiff vessel (**right image**). The unhealthy vessel comprises thicker walls and a uniform flow due to the stiffer walls. Meanwhile, the elasticity of the healthy vessel allows for an increased pulsatile flow. Modified from LaRocca et al. [7].

Non-invasive imaging techniques such as ultrasound and Magnetic Resonance Imaging (MRI) are common in clinical practice. Angiography, a more invasive technique, in which a contrast agent is injected into the vessel [8], is also utilised. These techniques, while routinely used, are perhaps not ideal when rapid and early screening might be the desirable way forward. Such techniques can be cumbersome, expensive, operator-dependent, and sometimes invasive. In addition, patients must be referred by a general practitioner (GP) for hospital-based screening, which might delay detection and increase patient concern. The use of non-invasive sensing modalities for the assessment of vascular disease has attracted a lot of attention from researchers throughout the years. Despite many efforts, a quick and relatively simple screening method that is clinically acceptable is not yet available.

Photoplethysmography (PPG) is an optical technology that has risen in popularity. It is suggested that PPG can be a solution when developing novel techniques for assessing vascular ageing. Arterial stiffness can cause haemodynamic, mechanical, and volumetric changes. In aged vessels, the unhealthy vessels become stiff and thicken, becoming less elastic and making it more difficult for blood to pass through. Blood flow is altered in ways that can damage downstream organs. A higher pulse pressure results from higher systolic arterial pressure and lower diastolic arterial pressure. Unhealthy vessels experience volumetric changes as a result of having less pulsatile expansion than healthy vessels. These changes can influence the PPG signal. Therefore, PPG-based technology has the potential to evaluate viscoelastic properties of arteries, including monitoring haemodynamic and volumetric blood changes, and hence evaluate vascular ageing.

The current methods utilised in clinical settings to evaluate vascular health have limitations. This literature review aims to explore state-of-the-art non-invasive techniques for assessing vascular ageing through measurements of arterial stiffness. By examining prior research conducted in in vitro and in in vivo settings, while emphasising key discoveries and outcomes along with the strengths and weaknesses of each technique, this review proposes recommendations for assessing vascular health in the future. While existing methods are inconvenient for GP-level screening, PPG has the potential to provide a non-invasive optical solution for assessing vascular ageing. PPG has gathered attention for its ability to assess viscoelastic properties, monitor haemodynamic changes, and detect volumetric alterations in blood flow. As a result, it offers promise in assessing vascular health by addressing the shortcomings of existing techniques. Nonetheless, further research is crucial for understanding the full scope of the capabilities and limitations of PPG before its widespread adoption in clinical practice.

2. Methods

Literature was identified through searches conducted in PubMed, Embase, Scopus, Web of Science, and Cochrane Library. Keywords such as "vascular ageing" and "non-invasive", and "arterial stiffness", were used in assessing the titles and abstracts. Several permutations of "photoplethysmography" were also employed. Only publications from 1990 to November 2023 were included to assess the shift from traditional to recent technologies. Papers published in the English language were only included in the search. This yielded a total of 543 papers from all the databases. Studies involving in vivo and in vitro experiments met the inclusion criteria. A total of 266 duplicate papers were eliminated. Furthermore, 213 papers were omitted because their titles and/or abstracts were unrelated to the study. Therefore, 64 papers were included for analysis by meeting the inclusion criteria (Figure 2). The searches were performed in November 2023.

Out of the 64 studies reviewed, 32 evaluated arterial stiffness using PPG-based methods. Another 32 studies concentrated on alternative measurement techniques. Two PPG-based studies found in the search overlapped with another method; they have been included in the overall count for PPG-based procedures. The split for the mean number of publications found per decade can be seen in Figure 3. Between 2010 and 2019, demand rose for arterial stiffness measurement techniques, particularly those using PPG-based methods, compared to the period from 2000 to 2009. The search and inclusion criteria revealed a lack of interest in arterial stiffness assessment techniques from 1990 to 1999.

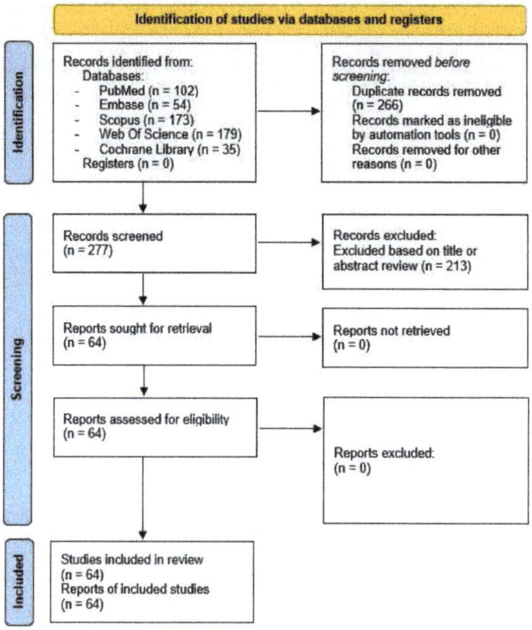

Figure 2. Flow diagram from PRISMA [9] illustrating the search strategy for the literature review procedure. The PRISMA flow diagram facilitated study selection. It was used to present the review clearly, mapping the number of papers and eliminating duplicates in databases. Records deemed irrelevant based on title or abstract were excluded from the screened total. Subsequently, 64 papers underwent eligibility assessment, all of which have been included in the review.

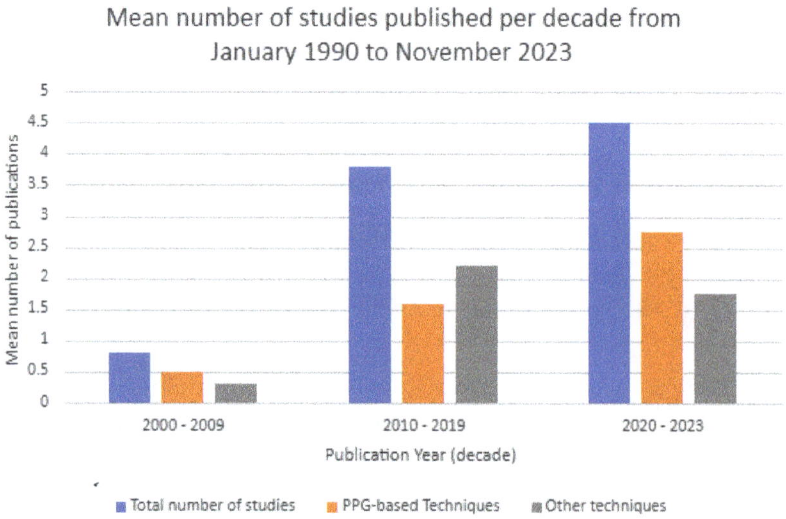

Figure 3. Bar chart showing the mean number of publications released per decade for the years between 2000 and 2023 that met the inclusion criteria. The total number of publications has been split between PPG-derived methodologies and other techniques used. No relevant studies were found for the years between 1990 and 1999.

3. Current State-of-the-Art Techniques and Technologies in Vascular Ageing

There are currently non-invasive methods and tools available to measure arterial stiffness. The calculation of indices, such as the augmentation index (AIx), ambulatory arterial stiffness index, and cardio-ankle vascular index (CAVI), and the measurement of Pulse Wave Velocity (PWV) are examples of current methodologies. Additionally, imaging methods such as ultrasound and MRI have been investigated over the years. Research has also been conducted utilising Laser Doppler Flowmetry (LDF) and near-infrared spectroscopy (NIRS). Researchers have examined the use of already-available instruments such as the Arteriograph device (TensioMed Ltd., Budapest, Hungary), the Mobil-O-Graph 24 h Pulse Wave Analysis (PWA) Monitor Device (I.E.M. GmbH, Stolberg, Germany), and more, including oscillometric devices, as well as computational approaches using algorithms and software programmes. More information on these subjects is covered later in the review, including their advantages and limitations.

3.1. The Recent Use of Measurements of Pulse Wave Velocity to Assess Arterial Stiffness

PWV is considered the gold standard for assessing arterial stiffness. PWV assesses the speed at which arterial pressure waves move through the aorta and large arteries. Recently, different techniques have been employed to analyse PWV in the assessment of arterial stiffness. Research has been conducted in areas linking arterial stiffness to body fat percentage (BFP), aortic valve replacement, children and adolescents with inflammatory bowel disease (IBD), and coronary stenosis. The use of impedance cardiography in assessing arterial stiffness using PWV has also been explored.

Gong et al. [10] conducted a study to examine the relationship between BFP and arterial stiffness. Eligible patients were categorised into four groups based on gender and arterial stiffness, and carotid femoral PWV (cfPWV) was measured. Each patient's BP, height, waist circumference, and body weight were also recorded. Given the relatively limited sample size, it was challenging to confirm a cause-and-effect relationship between BFP and cfPWV. PWV Tonometry-based devices were also utilised to measure the cfPWV prior to this investigation. Ji et al. [11] used cfPWV measurements to assess arterial stiffness. The study focused on the operator's placement, the distance being measured, and the tonometer position. Manual measurements were made of the distances between the suprasternal notch (SNN) and the remote detection site (femoral artery), the distance between the sternal notch and the proximal (carotid artery) detection point, and the distance between the femoral artery and the carotid artery detection point (cf-distance). Western nations make extensive use of cfPWV, and the study's tonometry-based apparatus is frequently used to assess aortic PWV. The device allowed for the measurement of central BP; however, questions over the accuracy of doing so with non-invasive methods remain.

Cantürk et al. [12] investigated the severity of aortic stenosis after an aortic valve replacement. The patients were between the ages of 43 and 75. PWV was measured at baseline and six months after having an aortic valve replaced. However, a limited sample size was employed, and the study's methodology was different from that of earlier research, making comparison challenging. A larger investigation is needed for more conclusive results.

The arterial stiffness of children and adolescents with IBD has also been examined. Based on the carotid artery and the femoral artery, the Vicorder device (Skidmore Medical Limited, Bristol, UK) was used to take cfPWV measurements. A causal interpretation of the findings was constrained because it was a cross-sectional study of a relatively small and heterogeneous sample. Additionally, no comparison groups were available, meaning the outcomes had to be interpreted in relation to pre-determined reference values. Future research on the long-term effects of CVDs in patients with IBD should make use of bigger sample numbers [13].

Liu et al. [14] used multidetector computed tomography angiography to determine coronary stenosis using brachial-ankle PWV (baPWV). Although only elastic arterial stiffness should be recorded, one limitation of baPWV is that the parameter captures both muscle and elastic arterial stiffness. In another study, PWV was examined to determine whether it reflects central or peripheral arterial stiffness. However, one of the study's shortcomings was that a longitudinal follow-up was not conducted to examine subsequent CVD occurrences. The results' applicability to other populations is also uncertain despite the sample size being rather large [15]. In another study, Yufu et al. [16] measured the aortic PWA and percent mean pulse amplitude (%MPA). These tests were performed on 33 patients at the right brachial artery on both sides of the ankle. However, the study omitted the consideration of additional factors impacting %MPA beyond aortic stiffness, such as femoral stiffness.

Scudder et al. [17] recorded dual impedance signals using a spot electrode arrangement on 78 adults aged 19 to 78. The pulse transit time (PTT), which was obtained by measuring the aortic flow onset and arrival times of peripheral pulse waveforms, was used to calculate PWV. PTT has an inverse relationship to PWV and represents the amount of time it takes for a pulse wave to travel a known distance. The study had drawbacks despite the authors' claim that d-ICG is a reliable approach to measure arterial stiffness, reasonably inexpensive, and easy to use due to minimal expertise being required. Firstly, the sample size was small and comprised, mostly, young healthy adults. Secondly, because of the longer pulse trajectory through distal arteries, such as the tibial artery, d-ICG was unable to provide direct assessments of central aortic stiffness (as with most non-invasive measures of PWV). Furthermore, it was advised that specific software configurations needed to be changed in order to collect the whole signal range and use the first derivative to determine timing characteristics.

3.2. The Recent Use of Arterial Stiffness Index Calculations to Assess Arterial Stiffness

The use of indices for measuring arterial stiffness has been studied by numerous researchers. This section summarises the recent utilisation of different indices, including the CAVI, ambulatory arterial stiffness index, AIx, arterial velocity–pulse index (AVI), arterial pressure–volume index (API), compliance index (CI), arterial stiffness index (ASI), and ankle-brachial index (ABI).

In order to determine arterial stiffness, Miyoshi and Ito [18] used the CAVI, which measured the distance from the origin at the aorta to the ankle and calculated the time it took for the pulse pressure wave to travel from the aortic valve to the ankle in order to obtain the PWV from the heart to the ankle. Additionally, the brachial artery in the upper arm was used to gauge BP. Following that, the Bramwell–Hill formula was applied to express the link between the PWV and change in volume. The fact that the CAVI can analyse arterial characteristics by splitting them into BP and arterial stiffness is one of its advantages over other methods for measuring arterial stiffness. Additionally, it measures the ascending aorta, suggesting that the arterial stiffness measurement may be more closely related to cardiac function. The CAVI has the potential to be helpful in clinical settings because of its relative simplicity and flexibility, though more research is needed.

Souza-Neto et al. [19] evaluated the ambulatory arterial stiffness index for the measurement of arterial stiffness in heart transplant patients. It was determined that the ambulatory arterial stiffness index presented a non-invasive way to indicate hypertension. However, the accuracy was questioned due to the study's small sample size. Moreover, the index was not compared with standard methods such as PWV or AIx. In rheumatoid arthritis (RA) patients, Klocke et al. [20] used radial PWA to determine AIx. The subjects were between the ages of 18 and 50, and the results were contrasted with that of a healthy control group. Despite this, validation is still needed from larger research that might examine further elements of the illness. Investigations into known physiological factors that can affect the AIx are also necessary. As an illustration, it was stated that AIx was lower in men and that it negatively correlated with height and heart rate and positively correlated with age and peripheral BP; however, these results must be investigated further.

Zhang et al. [21] measured AVI and API to evaluate early atherosclerosis. As stated in Table 1, baPWV was unable to predict the presence of early atherosclerosis. However, this could be explained as, according to Vlachopoulos et al. [22], baPWV is the average arterial stiffness between the brachial and ankle arteries, which reduces sensitivity in detecting changes in central arterial stiffness. Despite this, the study had some drawbacks, including a sample size with a high proportion of patients who had taken cardiovascular medications such as anti-hypertensive medications. The effects of such medications remain unknown [21]. Similarly, the study conducted by Wang et al. [23] comparing clinical features in patients with chronic kidney disease (CKD) using measurements of arterial stiffness in various arteries did not consider the effects of medications. To assess local vascular stiffness, CI was utilised to analyse the correlation between volume and pressure changes at the fingertip. There were 186 CKD patients and 46 healthy subjects. The study had limitations in terms of the types of cardiovascular risk factors and, as mentioned earlier, the medications examined; not all risk factors were considered. Furthermore, because it was a cross-sectional study, causality could not be established.

Table 1. Studies that used state-of-the-art techniques and technologies to assess vascular ageing.

Study	Year	Device(s)/Techniques	Measurement(s)	Number of Subjects	Major Findings
Gong et al. [10]	2023	Complior	Carotid femoral pulse wave velocity (cfPWV)	2063	In women over 60 years, body fat percentage (BFP) and cfPWV were correlated. There was no correlation detected for men or women under the age of 60.
Ji et al. [11]	2018	Not stated	cfPWV	Not stated	The method was able to successfully assess aortic pulse wave velocity (PWV) and measure central blood pressure (BP).
Cantürk et al. [12]	2017	Not stated	PWV	38	The mean PWV improved in 20 individuals, as indicated by an absolute decrease in PWV detected 6 months after having aortic valve replacement compared to a baseline, and deteriorated in the other 18 subjects. Generally found that aortic valve replacement had no effect on PWV mean and that aortic stenosis was related to baseline PWV.
Lurz et al. [13]	2017	Vicorder device	cfPWV	25	All subjects were found to have a normal cfPWV, suggesting that cfPWV does not change in subjects under 18 years of age.
Liu et al. [14]	2011	Computed tomography angiography	Brachial-ankle PWV (baPWV) Coronary Artery Calcium (CAC) score	654	In 127 patients, there was at least one coronary artery that was stenotic. In comparison to the normal control group, the stenotic group's mean baPWV and mean CAC were considerably greater. The study found a link between coronary atherosclerosis and baPWV.
Tsuchikura et al. [15]	2010	Automated device	PWV	2806	In comparison to heart-carotid PWV (hcPWV), heart-brachial PWV (hbPWV), and femoral-ankle PWV (faPWV), it was discovered that baPWV had the strongest association with heart-femoral PWV (hfPWV). According to the findings, baPWV exhibited central stiffness as opposed to peripheral arterial stiffness.
Yufu et al. [16]	2004	Not stated	Aortic pulse wave analysis (PWA) Percent mean pulse amplitude (%MPA)	33	Higher arterial stiffness was indicated by a low %MPA. The %MPA was suggested as a novel atherosclerosis marker by the authors. It was discovered that the %MPA was lower in subjects with coronary artery stenosis than in those without it. The findings showed that brachial %MPA offered prognostic values for coronary atherosclerosis in people at risk for cardiovascular disease (CVD).

Table 1. *Cont.*

Study	Year	Device(s)/Techniques	Measurement(s)	Number of Subjects	Major Findings
Scudder et al. [17]	2021	Dual impedance cardiography	PWV Pulse transit time (PTT)	78	Measurements of PWV showed a considerable positive connection with advancing age.
Miyoshi and Ito [18]	2016	Not stated	Cardio-ankle vascular index (CAVI)	Not stated	The study found that CAVI allows for the quantitative evaluation of disease progression, with higher CAVI values indicating a worse prognosis compared to lower CAVI values. CAVI was independent of BP at the time of measurement (unlike PWV) and indicated the overall stiffness of the artery from the aortic origin to the ankle. The study concluded that CAVI was a better identifier of disease severity than baPWV.
Souza-Neto et al. [19]	2016	Not stated	Ambulatory arterial stiffness index	85	The study found that those with a risk factor such as hypertension would have arterial stiffness. It was shown that arterial stiffness was substantially correlated with diabetes, hypertension, peripheral arterial disease (PAD), and coronary artery disease (CAD). Gender did not appear to impact arterial stiffness. Contrary to popular belief, age did not seem to be associated with arterial stiffness.
Klocke et al. [20]	2003	Not stated	Augmentation index (AIx)	14	The findings suggested that rheumatoid arthritis (RA) and increased arterial stiffness were related.
Zhang et al. [21]	2017	Oscillometric device	Arterial velocity–pulse index (AVI) Arterial pressure–volume index (API) BaPWV	183	It was discovered that baPWV, which was measured separately, was unable to foretell the existence of early atherosclerosis. BaPWV was found to be comparable to AVI in predicting CAD.
Wang et al. [23]	2011	Not stated	Compliance index (CI)	232	Measurements were made for both CI derived from digital volume pulse (DVP) and PWV-DVP; chronic kidney disease (CKD) patients had lower CI-DVP and greater PWV-DVP values than the healthy control group. Additionally, it was discovered that patients with late-stage CKD had lower CI-DVP levels than those with early-stage CKD. The more cardiovascular risk factors there were, the lower the CI-DVP was, according to the data.

Table 1. *Cont.*

Study	Year	Device(s)/Techniques	Measurement(s)	Number of Subjects	Major Findings
Choy et al. [24]	2010	Oscillometric automated digital BP instrument	Arterial stiffness index (ASI) Ankle-brachial index (ABI) Arterial wave pattern	895	The risk of stroke was six times higher when the ASI was abnormal. Both ABI and arterial wave pattern showed strong correlations with stroke. However, there was a synergistic impact when assessing the risk of stroke when all three parameters were considered.
Naessen et al. [25]	2023	Ultrasound	Common carotid artery Intima thickness Intima/media (I/M) thickness ratio Intima–media thickness	63	It was discovered that compared to healthy subjects, patients with pulmonary arterial hypertension had an intima that was 56% thicker and an I/M ratio that was 128% higher. Patients with pulmonary arterial hypertension showed a thicker intima and greater I/M ratios than those with left ventricular heart failure with a reduced ejection fraction.
Li et al. [26]	2016	Real-time shear wave elastography	Longitudinal elasticity modulus PWV	179	PWV and systolic and diastolic BP were observed to be higher in AIS patients than in the control group. It was demonstrated that shear wave elastography could accurately and non-invasively measure the arterial wall's longitudinal elastic modulus and assess arterial stiffness.
Bjällmark et al. [27]	2010	Conventional and Ultrasonographic strain measures	Common carotid artery elasticity	20	It was concluded that 2D strain imaging through ultrasonography was proven superior to traditional vascular stiffness measurements for determining the elastic characteristics of the common carotid artery.
Kang et al. [28]	2011	Cardiac magnetic resonance imaging (MRI)	Pulmonary artery distensibility index	35	The study aimed to determine whether the pulmonary artery stiffness estimated based on right heart catheterisation and Cardiac MRI-derived pulmonary artery distensibility corresponded.
Ha et al. [29]	2018	4D flow MRI	Aorta's turbulent kinetic energy (TKE)	42	The study examined the extent and degree of turbulent blood flow in a healthy aorta and determined whether age has an impact on the turbulence level. TKE was a measurement of turbulence intensity. All the healthy subjects experienced turbulent flow in the aorta, and both groups' aortas were similar overall. However, when compared to the younger participants, the older subjects had 73% greater total TKE in the ascending aorta. This was associated with age-related dilation of the ascending aorta, which increases the volume available for the

Table 1. *Cont.*

Study	Year	Device(s)/Techniques	Measurement(s)	Number of Subjects	Major Findings
					generation of turbulence. It was determined that age-related geometric changes influenced the development of turbulent blood flow in the aortas of healthy subjects.
Sorelli et al. [30]	2018	Periflux 5000 Laser doppler flowmetry (LDF) system	Peripheral pulse Microvascular perfusion	54	A multi-Gaussian decomposition approach was applied to the LDF signals, and the algorithm proved effective at reconstructing the shape of the LDF pulses.
Rogers et al. [31]	2023	Near-infrared spectroscopy (NIRS) vascular occlusion testing	Age-related metabolic and microvascular function changes	34	It was concluded that the microvascular hyperaemic response and skeletal muscle metabolism decline with age for women.
Silva et al. [32]	2021	Mobil-O-Graph 24 h PWA Monitor Device Dual energy X-ray absorptiometry	Body composition PWV AIx Pulse Pressure Amplification Index Central Pulse Pressure	124	It was determined that arterial stiffness in the elderly is directly correlated with BFP. The findings could be related to an increased risk of cardiovascular disease.
Perrault et al. [33]	2019	SphygmoCor-Px System VP-1000 system EndoPAT 2000 system	PWV AIx Reactive hyperemia index (RHI)	40	A comparison was made between the three devices. A high level of PWV reliability was attained for both the VP-100 and SphygmoCor, as shown by a low coefficient of variation. AIx had a larger coefficient of variation when using the SphygmoCor or EndoPAT than PWV. The lack of association between RHI and AIx suggests that endothelial and artery parameters have different functional properties.
Markakis et al. [34]	2021	SphygmoCor-Px System Mobil-O-Graph 24 h PWA Monitor Device	Peripheral BP Central BP PWV Artificial Intelligence (AI)	57	Both devices were compared with patients experiencing hemodynamic shock in intensive care units (ICUs). The results showed that a lack of extra-vascular diseases made invasive procedures more reliable. However, the authors concluded that non-invasive techniques are practical and can be employed as extra monitoring techniques for shock patients. With the Mobil-O-Graph 24 h PWA Monitor Device and the SphygmoCor, full haemodynamic evaluations were successful in 48 patients and 29 patients, respectively. However, across the two devices, variations in the PWA were found.

Table 1. *Cont.*

Study	Year	Device(s)/Techniques	Measurement(s)	Number of Subjects	Major Findings
Costa et al. [35]	2019	SphygmoCor-Px System	PWV	151	The PWV mean was observed to be higher in hypertensive patients. This finding suggests a clear relationship between arterial stiffness and increased PWV in the presence of hypertension. It was also discovered that arterial stiffness was more common in males who were older and had more risk factors than females.
Sridhar et al. [36]	2007	PeriScope	cfPWV baPWV	3969	Patients with RA had the highest heart rates whereas those with end-stage renal disease (ESRD) had the highest systolic BP. The PWV was discovered to be higher among people at higher risk of developing CVD, including those with CAD, diabetes mellitus, ESRD, and RA, compared to the healthy control group.
Komine et al. [37]	2012	Oscillometric BP device Form PWV/ABI vascular diagnostic device SonoSite 180 Plus	BP baPWV cfPWV Carotid arterial compliance	173	The study demonstrated that arterial stiffness can be assessed solely through cuff pressure oscillometric BP measurement. Similar to cfPWV and carotid arterial compliance, the estimated API had repeatability.
Hoffmann et al. [38]	2019	Oscillometric BP monitor	Heart rate Peripheral BP Central BP PWV	8	Long-term space flight's vascular ageing biomarkers were assessed at baseline, 4 days, and 8 days post a 6-month International Space Station mission. Heart rate rose significantly 4 days after the return, but not on day 8, in comparison to the baseline. Additionally, central systolic BP also increased 4 days post-return versus the baseline measurement. PWV had an insignificant increase from baseline 4 days post-return and remained elevated on day 8. Overall, no clinically significant changes in early vascular ageing biomarkers were found in the evaluated cosmonauts.
Juganaru et al. [39]	2019	Anteriograph instrument	Waist-to-hip ratio	313	The device was able to identify patients at cardiovascular risk before any clinical indicators found arterial stiffness.

Table 1. *Cont.*

Study	Year	Device(s)/Techniques	Measurement(s)	Number of Subjects	Major Findings
Osman et al. [40]	2017	Anteriograph instrument NICOM	Arterial stiffness PWV AIx Cardiac output Stroke volume Total peripheral resistance	33	Ultrasound scans were taken at five gestational windows between 11 and 40 weeks of pregnancy. It was discovered that normal pregnancy is associated with significant alterations in the maternal cardiovascular system. Arterial stiffness changes were observed in all measurements during healthy pregnancy, and the aortic PWV showed a significant variation during pregnancy.
Kostis et al. [41]	2021	Computational algorithm	Arterial stiffness 1 (AS1) Arterial stiffness 2 (AS2)	14097	Both indices were able to predict the occurrence of strokes. The study found that the indices derived from pulse pressure were more accurate at predicting the occurrence of stroke than pulse pressure or chronological age alone.
Negoita et al. [42]	2018	Computational algorithm: semi-automatic vendor-independent software Vivid E95 ultrasound	PWV	12	The study successfully developed software to trace the luminal diameter and blood velocity in the human ascending aorta by drawing from ultrasound images. The technique could calculate the PWV in the ascending aortas of adults.

The ASI, which was used to quantify arterial distensibility, ABI, and the arterial wave pattern using an oscillometric automated digital BP instrument, was studied by Choy et al. [24]. The subjects were split into two groups: 266 newly diagnosed stroke patients, ranging in age from 26 to 98, and a control group of 629 volunteers, all of whom were older than 30. A greater ASI indicated a higher likelihood of stroke incidence. However, ASI was determined based on BP readings, which can change depending on the participant's health at the time of the examination. Subjects who had previously experienced a transient ischemic attack or a mild stroke may not have been detected despite efforts to rule out past stroke experience based on a questionnaire. Additionally, the effects of medications on ASI were not investigated.

3.3. The Recent Use of Imaging Techniques to Assess Arterial Stiffness

3.3.1. Ultrasonography

One of the key techniques for assessing vascular age is ultrasound, an acoustical imaging modality, which uses high-frequency sound waves. Researchers have employed ultrasound to assess arterial stiffness, yielding promising results. However, ultrasound methods tend to be expensive and require expertise to operate.

Naessen et al. [25] conducted a study involving 30 healthy individuals (median age, 62; range, 27 to 82), 19 patients with pulmonary arterial hypertension (median age, 53; range, 27 to 84), and 14 patients with left ventricular heart failure with a reduced ejection fraction (median age, 67; range, 48 to 82). The healthy subjects were non-smokers and had no prior history of heart or arterial disorders. Additionally, the subjects were not taking any medications that would have impacted the arterial wall. The study disproved the common belief that vascular changes in pulmonary arterial hypertensions are only related to lung vasculature.

Li et al. [26] utilised ultrasound imaging techniques to assess arterial stiffness in patients with acute ischemic stroke (AIS) using real-time shear wave elastography to measure longitudinal elasticity modulus. With this technology, 50,000 images could be captured every second. Furthermore, radio frequency ultrasonography technology was used to calculate the PWV of the bilateral carotid arteries. The results demonstrated that shear wave elastography can be used in vascular applications. Future research is necessary to evaluate risk factors and assign various weights to arterial stiffness in longitudinal and circumferential directions because the study was unable to analyse all deviations.

Age-dependent elasticity variations in the common carotid artery elasticity served as the foundation for the comparison between conventional and ultrasonographic strain measures in a study conducted by Bjällmark et al. [27]. The evaluation involved 10 younger subjects, between the ages of 25 and 28, and 10 older subjects, between the ages of 50 and 59. The conclusion drawn was that two-dimensional (2D) strain imaging exhibited greater accuracy compared to traditional measurements. However, concerns arose regarding the stiffness indices, as these were derived from BP and lumen diameter measurements obtained at different sites, casting doubt on the accuracy of the variables.

3.3.2. Magnetic Resonance Imaging

MRI, which uses magnetic fields to create images, has also been investigated by researchers for the assessment of arterial stiffness.

The study conducted by Kang et al. [28] used cardiac MRI to calculate the pulmonary artery distensibility index to assess pulmonary artery stiffness. However, the study had certain limitations. Firstly, despite the disease being rare, the subject group was small. Secondly, among the thirty-five patients under observation, only three were noted to have severe pulmonary regurgitation (PR), which can suggest that the pulmonary artery distensibility index would have been overestimated.

The impact of age on turbulent blood flow was assessed by Ha et al. [29] using four-dimensional (4D) flow MRI. As stated in Table 1, all subjects experienced turbulent flow in the aorta. However, this was based on a small sample size of twenty healthy males aged between 67 and 74 and twenty-two healthy males aged between 20 and 26. Although MRI does not expose patients to radiation, some may experience claustrophobia, rendering it inappropriate for a general evaluation of vascular ageing.

3.4. Laser Doppler Flowmetry and Near-Infrared Spectroscopy

The medical technique of LDF uses the concept of Doppler shift, which is the change in the frequency of light (in this case, laser). Sorelli et al. [30] assessed the vascular ageing using the Periflux 5000 LDF system (Perimed, Järfälla, Sweden). On the right hallux pulp of the individuals, microvascular perfusion was recorded. A supervised classifier was trained and validated using over 20,935 models of pulse waves. Although using LDF is less expensive than using other imaging techniques, it can have signal processing and motion artefact issues [43].

NIRS is an analytical technique that uses a broad spectrum of near-infrared light to illuminate the region of interest and measures the light that is absorbed, transmitted, reflected, or scattered. Age-related metabolic and microvascular function changes were assessed by Rogers et al. [31] using NIRS vascular occlusion testing. NIRS signals were recorded from 17 younger and 17 middle-aged and older women. The Framingham risk calculator was used to determine a 10-year risk. Due to the cross-sectional nature of the study, it was not possible to evaluate the causal temporal relationship between age and the results. Additionally, the thickness of the adipose tissue was not assessed, which might have reduced the absolute NIRS signals. Finally, the melanin levels were not considered, which might have affected the NIRS signal.

3.5. The Use of Existing Devices to Assess Arterial Stiffness

Commercial devices which measure arterial stiffness do exist. These devices include the Mobil-O-Graph 24 h PWA Monitor Device, SphygmoCor-Px System (AtCor Medina, Sydney, Australia), VP-1000 system (Omron Healthcare, Hoffman Estates, IL, USA), EndoPAT 2000 system (Itamar Medical, Franklin, MA, USA), PeriScope (M/S Genesis Medical Systems, Hyderabad, India), Form PWV/ABI vascular diagnostic device (Omron Healthcare, Kyoto, Japan), SonoSite 180 Plus (SonoSite Inc., Washington, DC, USA), Anteriograph, and NICOM (Cheetah Medical, Portland, OR, USA).

The Mobil-O-Graph 24 h PWA Monitor Device was used by Silva et al. [32] to evaluate the link between body composition and arterial stiffness. Dual-energy X-ray absorptiometry on a Hologic bone densitometry machine (Model Discovery A, Waltham, MA, USA) was used to obtain the body compositions of the participants. However, the study was constrained because it was carried out in a single location with community elders, lacking applicability to a broad and diverse population. Additionally, the study ignored the individuals' usage of medications such as anti-hypertensives, which would have affected the central circulation parameters.

Although there are instruments that can distinguish between healthy and unhealthy vessels, no device is ideal. Perrault et al. [33] compared the measurement capabilities of the SphygmoCor-Px System, the VP-1000 system, and the EndoPAT 2000 system in healthy subjects ranging in age from 23 to 71. The outputs of the instruments differed numerically, making it difficult to compare the outcomes. Furthermore, because there is a lack of knowledge regarding changes to the parameters in relation to illness progression, it makes it difficult to track disease progression or the efficacy of a particular intervention. Markakis et al. [34] evaluated the SphygmoCor-Px System and the Mobil-O-Graph 24 h PWA Monitor Device in terms of feasibility. Invasive and non-invasive measurements were performed within 24 h of admission and again 48 hours later for comparison purposes on patients experiencing hemodynamic shock in intensive care units (ICUs). It was concluded, as illustrated in Table 1, that non-invasive procedures can be used as part of an additional monitoring method while invasive techniques are more trustworthy. However, the results of the study are questioned because of the small sample size. The SphygmoCor-Px has also been used in studies to assess vascular ageing by acquiring the PWV. Costa et al. [35]

used the SphygmoCor-Px in a study conducted on in arterial hypertension individuals. Chi-square analysis was performed as the primary statistical analysis. Despite the small size representing a community and the requirement for a larger size, the study was able to demonstrate its ability to measure arterial stiffness non-invasively using a gold-standard instrument.

Sridhar et al. [36] used the PeriScope to measure arterial stiffness by obtaining the PWV of 988 healthy controls and 2988 who had a high risk of developing CVD. Using an oscillometric approach, the PeriScope simultaneously assessed the cfPWV and baPWV. It was concluded that those with a higher risk of developing CVD had a higher PWV than the healthy control group. In another study, Komine et al. [37] used an oscillometric BP device to evaluate arterial stiffness. An inflatable cuff was used to assess the BP of the individuals and the Form PWV/ABI vascular diagnostic device was used to obtain the baPWV and cfPWV. The SonoSite 180 Plus was used in the study to obtain images using ultrasound. It should be noted that the study did have restrictions. For instance, the oscillometric cuff pressure was used to take an indirect measurement of arterial volume. Between the brachial blood vessel and cuff, the size of the muscle and fat has unknown effects on the arterial volume. To determine whether the technique could be utilised to measure arterial stiffness when illnesses are present, more research is needed. In a more recent study, Hoffmann et al. [38] used a oscillometric BP monitor to assess early vascular ageing biomarkers in cosmonauts, aged between 41 and 51 (7 males and 1 woman), undergoing long-term space flights. To summarise, it was found that the cosmonauts had not undergone any clinically significant changes despite comparisons to the baseline measurements (represented by 65 to 90 days prior to the flight). Long-term flights in deep space should reportedly be looked upon in the future. Due to the study's limited sample size, it is recommended to conduct a follow-up examination several years later. The current follow-up duration in the study was deemed too short to adequately identify any delayed onset of vascular diseases.

Juganaru et al. [39] collected arterial stiffness parameters from 184 males and 129 females, ranging in age from 18 to 53, using the Anteriograph instrument. According to the study, a high waist-to-hip ratio may be a vascular stiffness risk factor. The early asymptomatic detection of vascular atherosclerosis could be aided by screening healthy people with high waist-to-hip ratios. Despite the device's non-invasiveness, quick deployment, and repeatability in assessing artery stiffness parameters, incorrect cuff positioning can produce false findings. Osman et al. [40] evaluated arterial stiffness changes in low-risk pregnant women using the Arteriograph. Non-invasive evaluations were performed on low-risk pregnant women. A cuff was placed on the right arm over the brachial artery to use the Arteriograph. The NICOM was also used in the study for measurements including those of cardiac output, stroke volume, and total peripheral resistance. Advantageously, the Arteriograph instrument has been extensively utilised in pregnancy research and has been validated in invasive and non-invasive measurements in non-pregnant populations. Additionally, to reduce bias in the results, only one expert who was trained on using both the Arteriograph and NICOM made the recordings. The authors were able to identify developments or changes in arterial stiffness over a longer length of time because the longitudinal investigation was based on five distinct occasions. This study was based on a small number of participants; however, a larger population is required in future studies.

3.6. The Recent Use of Computational Algorithms to Assess Arterial Stiffness

Over the years, researchers have created various computational algorithms to predict arterial stiffness. Computational algorithms navigate around the challenges associated with gathering extensive in vivo datasets along with the associated time and costs [44]. Kostis et al. [41] predicted arterial stiffness from pulse pressure using an algorithm. To test whether arterial age predicted stroke better than chronological age, two indices of arterial stiffness were created by the algorithm and adjusted for specified demographics. The study was constrained, nevertheless, because the algorithms had been demographically adjusted and hence could not be applied to other datasets. As a result, the approach could not be used with different datasets that included, for instance, different age and gender groupings. Having said that, and according to the authors, the approach may be used to design and carry out new randomised clinical trials.

To assess local arterial stiffness, a semi-automatic vendor-independent software was created by Negoita et al. [42] using a Vivid E95 ultrasound (GE Healthcare, Illinois, US) to collect images. The edges of the luminal arterial walls (M-mode) and blood velocity were determined by the software, and diameter and velocity waveforms were extracted from the ultrasound images. The study was assessed on healthy volunteers aged between 22 and 32, with four of them being females. The technique was vendor-independent; therefore, it could be used to analyse ultrasound images of diameter and velocity recorded on any ultrasound machine as long as they have been saved in the Digital Imaging and Communications in Medicine (DICOM) format.

4. Using Photoplethysmography for Arterial Stiffness Assessment

PPG is a widely used non-invasive optical technique. It aids in studying and monitoring pulsations associated with changes in blood volume in a peripheral vascular bed. Over the last thirty years, the number of published articles on PPG has significantly increased, covering both basic and applied research. Throughout these publications, PPG has been praised as a non-invasive, low-cost, and simple optical technique for measuring physiological parameters applied at the surface of the skin.

The popularity of this topic can be attributed to the realisation that PPG has important implications for a wide range of applications. Amongst many, it aids in blood oxygen detection, cardiovascular assessment, and vital sign monitoring. In addition, the significant contribution of PPG in wearable devices has exponentially elevated the popularity and usability of PPG.

Currently, there exists a large body of literature that contributes new knowledge on the relationship between PPG pulse morphology, PWA, and pulse feature extraction with the physiological status of peripheral blood vessels. This encompasses aspects such as ageing, stiffness, BP and compliance, and microvascular disease, amongst others. There are also significant efforts in the utilisation of the PPG for the detection of heart arrhythmias such as Atrial Fibrillation (AF). Researchers are continuing to strive to combine the PPG sensory capabilities of wearables, such as smartwatches, with Artificial Intelligence (AI) in delivering ubiquitous health monitoring solutions that go beyond the current available heart rate wearables [45].

PPG sensors comprise Light Emitting Diode(s) (LEDs) and photodector(s). The emitted light, which is made to transverse the skin, is reflected, absorbed, and scattered in the tissue and blood. The modulated light level, which emerges, is measured using a suitable photodetector. For example, it is possible for the hand to be directly transilluminated where the light source, usually in the broad region of 450 nm to 960 nm, is on one side of the skin and the detector is on the other side. This method, also called the transmission mode, is limited to areas such as the finger, the ear lobe, or the toe. However, when light is directed into the skin, a proportion is backscattered, emerging near the light source. The light source and the photodetector can be positioned side by side. This method, also called the reflection mode, allows measurements on virtually any skin area. The intensity of reflected and backscattered light reaching the photodetector in either reflection or transmission mode is measured. The variations in the photodetector current are assumed to correlate to blood volume changes beneath the probe. These variations are electronically amplified and recorded as a voltage signal called the photoplethysmogram (Figure 4) [45].

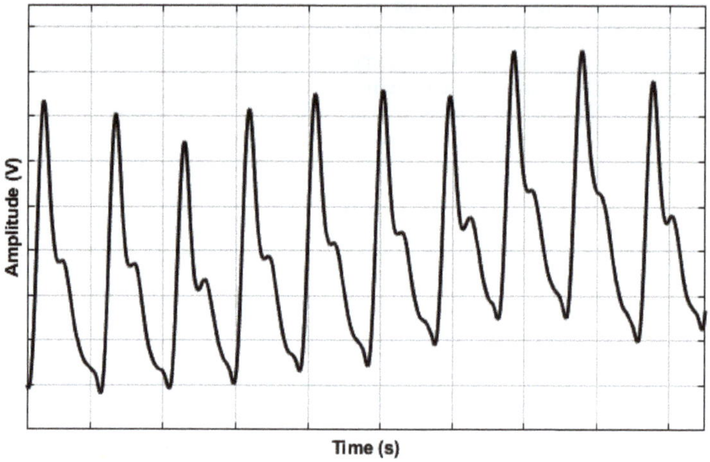

Figure 4. An example photoplethysmogram depicting the systolic and diastolic peaks with a dicrotic notch [45].

The PPG signal can be impacted by a number of variables such as temperature variations, the measurement site, perfusion status, and motion artefacts. As mentioned earlier, the transmittance or reflectance mode of the PPG sensor can differ depending on the anatomical measurement site, as shown in Figure 5. Thus, the wavelength of the light source(s) must be accounted for depending on the mode and distance that the light must penetrate. For example, red and infrared light reaches deeper than green light. Peripheral vasoconstriction can cause low-quality signals while good skin contact has demonstrated high-quality signals [46,47].

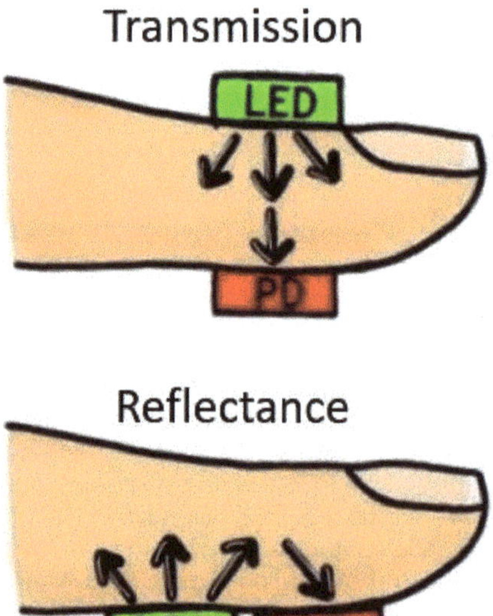

Figure 5. PPG sensor configuration modes. Transmittance mode configuration (**top image**) shows light source (LED) and photodetector (PD) placed on opposite sides of the body site. In reflectance mode (**bottom image**), the light source (LED) and photodetector (PD) are adjacent.

Understanding the constraints of PPG monitoring holds significance. Since PPG detects light, it faces drawbacks such as susceptibility to interference from ambient sources, impacting measurement accuracy. Motion artifacts also pose challenges [48], though utilising post-processing algorithms can mitigate the interferences. Additionally, variations in skin tones can impact the signal as PPG relies on light–tissue interactions. Considering these factors is crucial when developing PPG-based sensors for experiments, whether in vivo or in vitro.

4.1. The Recent Use of Photoplethysmography in Studies to Assess Arterial Stiffness

The evaluation of arterial stiffness using PPG has recently become popular. Many researchers have attempted to understand arterial stiffness using PPG in in vivo settings. Some in vivo research focuses on specific conditions including pregnancy, obesity, and diseases such as CAD. Studies have also compared PPG-based devices to one another and to other modalities to assess the viability of using PPG-based devices. The topics will be discussed in further detail below.

4.2. Existing Photoplethysmography-Based Devices to Assess Arterial Stiffness

PPG is a well-established optical technology; many researchers have adopted or developed new PPG-based devices, such as through research based at university laboratories. Some researchers have conducted in vivo studies to distinguish between healthy and unhealthy patients bilaterally (such as between healthy and PAD subjects) or by obtaining pulse waveforms from one measuring site, such as the left index finger. This section will introduce novel and existing PPG-based devices in further detail.

One of the earliest PPG research studies was conducted by Allen and Murray [49] on a group of healthy people. Firstly, PPGs were recorded from the right (R) and left (L) sides of six peripheral sites (that is, L and R ears, L and R thumbs, and L and R toes). To validate the electronic matching of right-to-left channels, a set of validation data was first gathered. Secondly, the healthy volunteers provided a set of physiological data derived from PPGs. These were the root mean square error (RMSE), which measured the differences between the right and left side, and cross-correlation analysis, which measured the degree of similarity. Allen and Murray [49] found that in healthy individuals, the right and left sides of the body were highly correlated, as perhaps expected. This work paved the way for further research that was conducted by Bentham et al. [50]. While it was already proven that in healthy individuals the PPGs from the right and left were highly correlated, Bentham et al. [50] obtained multi-site finger and toe PPG recordings from 43 healthy control patients and 31 PAD subjects to carry out another bilateral study. Beat-to-beat normalisation amplitude variability and pulse arrival time (PAT) were assessed in the frequency domain using magnitude-squared coherence (MSC) and in the time domain using two statistical techniques. When the results from the two subject groups were analysed, patients with PAD had a different signal on one side of the body compared to the other, unlike the healthy subjects. The work conducted by Bentham et al. [50] highlighted the possibility of distinguishing between healthy and PAD subjects. The clinical demographic dataset that was gathered for the study was nonetheless limited, and only a few fundamental variability variables were investigated in terms of PWA.

Brillante et al. [51] used PPG to non-invasively measure arterial stiffness from the left index finger in healthy people ranging in age from 18 to 67. The research focused on the impacts of categories such as age, gender, and race on different indices including stiffness index (SI) and reflection index (RI). Analyses based on simple correlation, Spearman's correlation, and multivariate regression were performed. Although the study concluded comparisons between the different categories and the indices, the study's sample of healthy adults over 65 was underrepresented. Similar to the study by Brillante et al. [51], where it was found that there were no differences between genders in terms of arterial stiffness measurements, the results obtained by Jannasz et al. [52] supported the notion that gender had no bearing on the likelihood of developing atherosclerosis. It should be emphasised, nevertheless, that the results were primarily focused on female participants, which may have influenced the findings. In another study conducted by Tąpolska et al. [53], the subjects were separated according to age, gender, and weight. The Pulse Trace PCA 2 device (Micro Medial, Rochester, UK) was used by placing a reader on the index finger to evaluate SI using PPG techniques. Although it was concluded that SI was more useful than RI, both can be used in clinical practice.

The work carried out by Tanaka et al. [54] using PPG signals taken from an occluded finger has made it possible to take measurements of the small artery and arteriole in the future. The study used Bland–Altman plots to evaluate the degree of agreement between the finger arterial stiffness index (FSI) and finger arterial elasticity index (FEI). Regression analysis, linear analysis, and bi-logarithmic analysis were used. The work has provided confidence in measuring arterial stiffness in smaller vessels.

Wowern et al. [55] conducted an in vivo trial on people of various ages and genders, including pregnant women, using the SphygmoCor-Px System and Meridian digital pulse wave analysis (DPA) (Salcor AB, Uppsala, Sweden). The experiment's goal was to understand how repeatable the arterial stiffness parameters determined by DPA were. Measurements were obtained from the left index finger. Second derivatives of the wave reflections were used to analyse the PPG signal (obtained from the Meridian DPA), and Bland–Altman plots were employed for statistical analysis. Despite our suggesting that DPA is a valuable tool to gauge vascular health, it still must be further examined because none of the DPA variables produced optimum repeatability, making it impossible to rely solely on such methods. In another in vivo study, carried out by Djurić et al. [56], a PPG sensor was used to measure blood flow using scalar coefficients. The encouraging results obtained from the study in differentiating between different age groups (above and below 50 years) have shown that PPG could be used in the future for vascular ageing measurements. It should be noted that the results were preliminary and based on a limited number of samples. Furthermore, the study was based on healthy volunteers without including any patients with vascular diseases. Future work should consider more categories for the age groups, as well as the impact of vascular diseases.

Huotari et al. [57] tested a transmission-probe-based PPG device created in a university lab. The sensor captured pulse waveforms, which were then mathematically decomposed to determine the arterial stiffness. However, because of the complexity of the hemodynamic features, it was challenging to calculate the arterial stiffness indices of arteries. Determining the link between PPG-derived indices and indices generated from pressure and flow pulses was a difficulty. The study demonstrated that arterial stiffness assessments using PPG had a promising future.

4.3. Use of Photoplethysmography on Specific Conditions

Researchers have implemented PPG-based techniques to assess arterial stiffness in patients with specific health conditions, such as CAD, heart transplantation, diabetes, hypertensive, and obesity, and high-risk patients. Pregnancy and cerebral pulsatility have also been investigated by researchers. The topics will be discussed in further detail.

Zekavat et al. [58] evaluated the association of ASI with BP and CAD. Multivariable COX proportional hazards and additive linear regression were among the analysis models used. The results led to a lack of confidence that PPG-derived ASI could predict CAD risk, leaving a gap for future research. Arterial stiffness changes in heart transplant patients have been compared by Sharkey et al. [59]. The study was conducted on 20 children with heart transplantation and on a healthy control group of 161 children. Data were collected bilaterally from the ear lobes, index fingers, and great toes. The PPG signal collected from the children with heart transplants was normalised and compared to the normalised PPG signal from the control group. For statistical analysis, multivariate (that is, binary logistic regression (BLR)) and univariate analyses (that is, the Mann–Whitney U test) were performed. This study suggested the possibility to measure arterial stiffness at different body sites. Research has also been conducted on diabetic patients whereby there is a possibility to distinguish between diabetic and non-diabetic individuals using an arterial stiffness monitoring system based on PPG technology [60]. Furthermore, the second derivative of PPG has been investigated as a potential indicator of arterial stiffness. In a study conducted on 260 patients, it was found that the arterial stiffness progression differed in the diabetic and non-diabetic stages [61].

The second derivate of a PPG signal and the PWV were compared in a study by Bortolotto et al. [62] that examined vascular ageing evaluation in hypertensive participants. The study involved 524 patients with hypertension and 140 with atherosclerosis alteration, which included coronary heart disease, peripheral vascular disease, and abdominal aortic aneurysm. The second derivative of the PPG was suggested as a potential tool for assessing vascular ageing in hypertensives. The length of the vascular segment may have been overstated by the PWV approach, which should be considered despite the fact that PWV was a better indicator of the presence of atherosclerosis alteration than the second derivative of PPG.

Korneeva and Drapkina [63] investigated the possibility of using PPG on obese patients with high BP by assessing arterial stiffness. The main objective was to provide statins, namely atorvastatin and rosuvastatin, to these patients and track the development of vascular stiffness. In an additional effort, Drapkina and Ivashkin [64] used a PPG device attached to a finger to conduct a pulse wave study on arterial stiffness in obese and high-BP patients. Prior to this research, Drapkina [65] used a finger PPG device to examine arterial stiffness in high-risk patients with high BP. The study was carried out similarly to that by Korneeva and Drapkina [63], albeit for high-risk patients. Both sets of results supported one another, noting that high-risk or obese patients had increased arterial stiffness.

Other circumstances, such as pregnancy-related circumstances, have also been analysed using PPG. The study conducted by Wowern et al. [66] involved PPG signals being collected from the left index fingers of healthy pregnant women. For analysis, linear and polynomial mixed effects were used to account for gestational age, and analysis of variance (ANOVA) and analysis of covariance (ANCOVA) were used to account for age influences. Yet, there was uncertainty and a lack of trust in the ability to identify pathological haemodynamic changes during pregnancy, thus postulating the necessity to investigate pathological changes that can occur during pregnancy.

In two independent acute interventions (a cold pressure test and one involving mild lower-body negative pressure), Lefferts et al. [67] investigated the effects of cerebrovascular pulsatility in terms of acute increases in arterial stiffness in middle-aged and young adults. A mean BP reading was acquired using PPG for continuous BP monitoring. Cerebrovascular hemodynamics at rest, during the cold pressure test, and during the lower-body negative pressure intervention were evaluated in 15 middle-aged people between the ages of 47 and 61 and in 15 young adults with genders matched to the middle-aged. The measurements, however, were not evaluated constantly, resulting in a little variation in the timings of the measurements. The middle-aged adult group also included a small number of participants who had CVD risk factors, such as obesity and anti-hypertensive medication usage, which could have improved validity but affected the findings. Future research should involve older subjects, as the study only focused on those under 61 years of age.

4.4. Comparing Current Photoplethysmography-Based Instruments and Other Measurement Methods

While some researchers have used custom-made PPG-based instruments for their work, others have compared several existing instruments. Complior (Alam Medical, Saint-Quentin-Fallavier, France), PulsePen (DiaTecne, Milan, Italy), and PulseTrace (GP Supplies, Borehamwood, UK) are the three devices that have been compared. PPG and measurement indices that gauge arterial stiffness, such the ABI, ASI, SI, and AIx, have also been examined. Additionally, imaging PPG (iPPG) and contact PPG (cPPG) have been compared.

Salvi et al. [68] compared current commercial PPG devices, namely Complior, PulsePen, and PulseTrace. In contrast to Complior and PulsePen, which used aortic PWV based on the interval between carotid and femoral pressure waves, PulseTrace was evaluated using SI measures. There remains a need to standardise PWV measurements and establish a reference value by contrasting various devices. Djeldjli et al. [69] compared iPPG with cPPG. The experiment was based on healthy participants. Two probes were used to record the signals for the cPPG, one on the right earlobe and the other on the index finger. The study's small sample size and primary focus on a certain age range and skin type limited the results to this application alone. Large-scale population testing is necessary to assess the impact of the measuring site on the measurements.

Kock et al. [70] compared the ABI to PPG measurements for arterial stiffness in elderly patients. Many analysis models were used, such as bivariate and multivariate linear regression, the Shapiro–Wilk test to analyse residuals, and winsorisation. Measures of central tendency and data dispersion were used to denote quantitative variables, whereas absolute frequencies and percentages were used to represent qualitative variables. It was concluded that the ABI did not relate to PPG indicators, and more research is needed to establish standardised procedures for vascular assessment. In another study, using a fingertip PPG device, Murakami et al. [71] compared the ASI to the well-known baPWV. Through statistical analyses such as the non-parametric Wilcoxon rank sum test, Receiver Operating Characteristic (ROC) curve, and the Area under the Curves (AUCs), the instrument was shown to be capable of measuring arterial stiffness in accordance with baPWV. To ascertain the possibility of distinguishing between subjects with high and low cardiovascular risk, Clarenbach et al. [72] compared the SI acquired through PPG to the AIx obtained through radial tonometry. The study involved 62 individuals who had either chronic obstructive pulmonary disease (COPD) or obstructive sleep apnea, and 21 healthy volunteers served as controls. The subjects were between the ages of 18 and 75. Whilst both devices were successful, the SI acquired from the PPG device could further distinguish between intermediate and high-risk people, rendering it more effective in clinical settings.

4.5. The Use of Photoplethysmography in Computational Models

Researchers have incorporated PPG into computational models to predict vascular health, as well as analysing incident and reflected waves from the PPG waveform. The advantages and disadvantages of using PPG in computational models are discussed.

Machine learning (ML) and deep learning (DL) were applied to the PPG signals of a database in a study conducted by Dall'Olio et al. [73]. The approach involved preprocessing data on the raw PPG signals, through steps such as detrending, demodulating, and denoising. The DL employed the entire signal to predict healthy vascular ageing (by bypassing the feature extraction stage), whereas the ML relied on known extracted features from the PPG signal. For DL, it was possible to bypass the feature extraction stage as several convolutional neural networks (CNNs) were applied to the entire PPG signal as an input. Although CNNs with 12 hidden layers or fewer showed good performance, more complicated structures cannot be trained on a common laptop; subsequent research should explore more complex structures using a feasible technique. Due to the black-box approach being used, the study was restricted in its ability to compare its findings to those of other ML methodologies. In a study conducted by Shin et al. [74], DL was applied to the PPG pulses of individuals ranging in age from 20 to 80. DL offers an advantage over manually recognising features during the assessment of vascular ageing since it has the potential to produce features from PPG waveforms automatically. It is anticipated that more databases will become accessible in the future, resulting in the DL approach performing better in the assessment of vascular ageing. Nonetheless, the employment of computational models in wearable technology raises additional issues such as the computing power that may affect the ability to obtain an immediate diagnosis. Resolving computation model issues, such as determining the optimal number of hidden layers or the trade-off between the number of parameters and the volume of training data, is crucial [73].

Park and Shin [75] evaluated vascular ageing using an artificial-neural-network-based regression model to analyse incident and reflected waves from a PPG waveform. Their study report claimed that a trustworthy single PPG-based technique for assessing arterial stiffness had not yet been developed. The Gaussian mixture model was used to deconstruct each waveform into incident and reflected waves after the recorded PPG signals were segmented for each beat. Since the measurements were based on nasal PPG rather than the more typical finger PPG measurements, it was challenging to generalise the findings. It is unclear how this has an impact on the findings. Future research is needed to understand the effects of various measurement sites. Assessing the model's performance in relation to risk factors that hasten vascular illnesses such as atherosclerosis, which were not considered in the study, is imperative.

In recent years, researchers have also shown interest in remote health monitoring, particularly through wearables [76]. This is driven by the ageing population, projected to increase by 10% in the next five years [77]. Remote PPG (rPPG), monitoring cardiovascular activity via facial video, has gained attention. Despite being labelled non-invasive and low-cost, rPPG faces challenges such as a low signal-to-noise (SNR) ratio. Lian et al. [78] proposed employing signal processing methods to enhance rPPG by reducing noise and improving accuracy. This involved data fusion, region of interest selection, and heart rate estimation. Furthermore, rPPG attracts attention for fatigue detection. Zhao et al. [79] utilised CNN for rPPG-based learning fatigue classification through multi-source feature fusion. However, the study used a self-collected dataset and lacked confirmation on enhancing fatigue detection accuracy with a larger sample size using DL. Previous research explored rPPG for estimating blood oxygen saturation (SpO_2). Casalino et al. [80] credited rPPG as being portable, enabling continuous SpO_2 monitoring. However, while rPPG is gaining popularity, it has not yet been applied rigorously in any large studies relating to vascular ageing.

4.6. Combining Photoplethysmography with Other Modalities and Using Photoplethysmography to Assess Novel Developments

Some researchers have attempted to fuse PPG with other modalities, such as an electrocardiogram (ECG), to overcome some of the current drawbacks. In the same respect, the novel device known as the single continuous passive leg movement (sPLM) has been assessed as a possible screening technique using PPG technology.

By combining multi-site PPG and ECG, Perpetuini et al. [81] performed an in vivo assessment of vascular stiffness. Ten ECG leads and eight PPG probes were used to collect signals. Signals could be simultaneously gathered from numerous places using various PPG probes. The ECG served as a reference for single-pulse PPG evaluation and averaging. Pressure cuffs were offered to ensure robust optode-to-skin connection. Given that the PPG–ECG system could record back-reflection signals at a significant inter-optode distance, it was hypothesised that PPG signals could be collected from large arteries. Additionally, numerous sites were monitored at once due to the large number of probes used. However, the results were not always interpretable when collecting multiple PPG signals, and significant computing resources may be required [82].

PPG was used in a study by Hydren et al. [83] in an effort to evaluate sPLM. All subjects were male and were split into two groups consisting of 12 younger and 12 older subjects. The instructions provided by Gifford and Richardson [84] for using the sPLM were followed. Although it is highlighted in Table 2 that a decline in vascular function brought on by ageing is sensitive to sPLM, further studies are required with larger samples prior to labelling the sPLM as a clinical tool for monitoring vascular ageing [83].

Table 2. PPG-based studies for the assessment of arterial stiffness.

Study	Year	Device(s)/Technique	Parameters Measured	Number of Subjects	Major Findings
Allen and Murray [49]	2000	Bilateral photoplethysmography (PPG) study	Waveform of the pulses from six peripheral sites (including ears, thumbs, and big toes).	40	The system validation data showed low levels of root mean square error (RMSE), indicating good right-to-left channel matching. According to the bilateral study, normal patients' pulses from their right and left sides were highly correlated at every segmental level (that is, the ears, thumbs, and toes).
Bentham et al. [50]	2018	Bilateral multi-site finger and toe PPG study	Multi-site finger and toe PPG	74	It was shown, using time-domain analysis, that those with PAD had lower normalised amplitude variability and significantly higher pulse arrival time (PAT) variability at the toe sites. In the frequency domain analysis, patients with PAD had noticeably decreased magnitude-squared coherence (MSC) values across a variety of frequency bands. It was discovered that the right toe had a different signal from the left toe while comparing the left and right sides of the body for the ear, finger, and toe. This resulted in the conclusion that PAD was at least present in one leg.
Brillante et al. [51]	2008	PPG	Peak to peak Stiffness index (SI) Reflection index (RI)	152	Age was found to significantly correlate with SI and RI, with race serving as an independent predictor of SI. At the same age, it was discovered that men had higher BP than women. There were no discernible differences between men and women in any of the arterial stiffness measures.
Jannasz et al. [52]	2023	SphygmoCor XCEL (AtCor Medina, Sydney, Australia)	Central and regional PWV cfPWV	118	cfPWV was discovered to be more reliable than regional PWV in determining arterial stiffness. There was insufficient evidence that gender affected the risk factors for atherosclerosis.
Tapolska et al. [53]	2019	Pulse Trace PCA 2 device	SI RI	295	Patients between the ages of 40 and 54 showed the greatest benefit from SI. The authors concluded that since SI and RI can be assessed non-invasively, they have the potential to be used in routine clinical practice to identify individuals who are at risk of developing future cardiovascular problems. Though RI can still be used as a supplementary measurement, it was found that it seemed to be less useful than SI.
Tanaka et al. [54]	2011	PPG	Finger arterial elasticity index (FEI) Finger arterial stiffness index (FSI)	199	The indices were gathered by occluding the finger, and future measurements of the stiffness of the small artery and arteriole can be made using the indices.

Table 2. *Cont.*

Study	Year	Device(s)/Technique	Parameters Measured	Number of Subjects	Major Findings
Wowern et al. [55]	2015	SphygmoCor-Px System Meridian digital pulse wave analysis (DPA)	PPG signals DPA Ejection Elasticity Index (EEI) Dicrotic Index (DI) Dicrotic Dilation Index (DDI)	112	It was discovered that the EEI should be used for large artery stiffness estimation while the DI and DDI should be used for small artery stiffness estimation.
Djurić et al. [56]	2023	PPG sensor	Blood flow Scalar coefficients	117	The study allowed for the analysis of amplitude changes in blood pulse waves. The scalar coefficient ratios declined with age, distinguishing between those above and below 50 years of age.
Huotari et al. [57]	2009	Transmission probe-based PPG device	Pulse waveforms from the left index finger and second toe	Not stated	It was discovered that PPG waveform analysis, as opposed to ultrasound analysis, offered greater details about artery structure and function.
Zekavat et al. [58]	2019	Finger PPG-derived ASI	ASI PPG BP	Approximately 500,000	It was determined that PPG-derived ASI was an inappropriate proxy for CAD risk but a genetically casual risk factor for BP.
Sharkey et al. [59]	2018	Bilateral PPG setup	Electrocardiogram (ECG) PPG PAT	181	A reduced PAT and thus higher arterial stiffness were discovered in heart transplant patients.
Wei et al. [60]	2013	PPG Radial pulse	cfPWV Spring constants	70	The PPG-based spring constant can distinguish between normal and pathological characteristics in both non-diabetic and diabetic individuals.
Park [61]	2023	SA-3000P (Medicore Co., Seoul, Republic of Korea) VS-1000 (Fukuda Denshi, Tokyo, Japan)	Second derivative of PPG signals CAVI	276	The second derivative of PPG, requiring a single transducer, proved simpler and yielded results in under 2 min. It was concluded that CAVI could not replace the second derivative of PPG.
Bortolotto et al. [62]	2000	Complior (Colson, Garges les Gonesses, France) Fukuda FCP-3166 (Fukuda, Tokyo, Japan)	cfPWV Second derivative of PPG signals	664	In patients over 60 with atherosclerosis, PWV and the second derivative of PPG remained higher, whereas in patients aged 60 with atherosclerosis, only PWV remained greater. While the index of the second derivative of PPG was correlated with age and other atherosclerosis risk factors, PWV was associated with age and arterial hypertension. The study found that aortic PWV more accurately captured changes in arterial compliance caused by ageing, high BP, and atherosclerosis than the second derivative of PPG.

Table 2. *Cont.*

Study	Year	Device(s)/Technique	Parameters Measured	Number of Subjects	Major Findings
Korneeva and Drapkina [63]	2015	PPG-based device	SI RI AIx Systolic BP	82	Obese patients with high BP showed vascular stiffness; based on pulse wave characteristics of the PPG signal, both treatments improved the arterial stiffness parameters, reducing SI and RI. Nonetheless, it was noted that compared to atorvastatin, rosuvastatin reduced the AIx.
Drapkina and Ivashkin [64]	2014	PPG-based device	SI RI AIx Systolic BP	82	The study found increased arterial stiffness in obese patients with high BP according to pulse wave analysis.
Drapkina [65]	2014	PPG-based device	SI RI AIx Systolic BP	82	The study found increased arterial stiffness in high-risk patients with high BP.
Wowern et al. [66]	2019	PPG	DPA	139	It was concluded that DPA reflects longitudinal changes in arterial compliance in normal pregnancy. In uncomplicated pregnancy, it was found that arterial stiffness changes in both large and small arteries significantly with gestational age.
Lefferts et al. [67]	2021	PPG Tonometry Ultrasound Doppler	BP cfPWV Carotid stiffness Blood velocity pulsatility index	30	Despite younger adults having larger carotid dilations, pulsatile damping decreased in both middle-aged and young adult groups. According to the study's findings, the carotid diameter and cerebrovascular pulsatility are altered differently in young and middle-aged adults. The study suggested that changes in intracranial cerebral pulsatility could be slowed down by cerebrovascular characteristics.
Salvi et al. [68]	2008	Complior PulsePen PulseTrace	Aortic PWV SI	50	Complior and PulsePen were found to be accurate at estimating PWV via Bland–Altman analysis whereas PulseTrace was found to be an unsuitable substitute for PWV.

Table 2. *Cont.*

Study	Year	Device(s)/Technique	Parameters Measured	Number of Subjects	Major Findings
Djeldjli et al. [69]	2021	Imaging PPG (iPPG) Contact PPG (cPPG)	16 features of the pulse wave relating to arterial stiffness and BP	12	A comparison was made between the iPPG and cPPG, whereby cPPG recorded signals using contact probes and iPPG used a quick camera to record signals remotely. High agreement between the results and the features captured by the reference sensors was observed. The contact and contactless PPG features (from two separate sites) were found to share strong correlations. The non-invasive iPPG approach provided quantitative data on the underlying mechanisms of waveform shape from various body sites and allowed for a remote means to evaluate waveform features.
Kock et al. [70]	2019	ABI PPG	Peak to peak time	93	The results showed that ABI was not related to PPG indicators.
Murakami et al. [71]	2019	Rossmax International LTD SB200 pulse oximeter (Taiwan, China)	ASI baPWV	18	The fingertip PPG device was found to be capable of measuring arterial stiffness in line with the baPWV.
Clarenbach et al. [72]	2012	PPG Radial tonometry	SI AIx	83	Both devices distinguished between people with high and low cardiovascular risk accurately. However, unlike AIx, SI could also distinguish between people at intermediate and high risk, making it potentially more effective in sizable clinical research. Low failure rates were present in both devices.
Dall'Olio et al. [73]	2020	Machine learning (ML) Deep learning (DL)	Raw PPG signal	4769	It was discovered that individuals do not age at the same rate. The study reported that women had better health than men in terms of vascular ageing.
Shin et al. [74]	2022	DL	PPG pulse signals	752	The model outperformed earlier models at the time without the requirement for an additional feature detection process.
Park and Shin [75]	2022	Artificial neural network-based regression model	Incident and reflected waves from PPG waveform	757	It was determined that the reflected wave's features may be used to assess vascular ageing, with the amplitude-related feature of the reflected wave being more favourable than the time-related feature in doing so.
Lian et al. [78]	2023	3 Lead ADS1292 ECG (Texas Instruments, Dallas, TX, USA) Remote PPG (rPPG)	ECG Heart rate	30	ECG signals were obtained as reference values. Video recordings from the rPPG were transferred for analysis; however, the results were limited to the Han ethnicity, lacking diversity.

Table 2. *Cont.*

Study	Year	Device(s)/Technique	Parameters Measured	Number of Subjects	Major Findings
Zhao et al. [79]	2023	700-MAX-ECG MONITOR (Maxim Integrated, San Jose, CA, USA) rPPG	ECG Heart rate	12	Unlike PPG and ECG, which may cause skin irritation and discomfort, rPPG offers a solution by eliminating adherence to the body.
Casalino et al. [80]	2023	rPPG	Blood oxygen saturation (SpO2)	10	Experimental results showed no significant differences with slight head movements. Further work is needed to explore potential light-condition impacts on real-time measurements.
Perpetuini et al. [81]	2019	Multi-site ECG and PPG	baPWV	78	It was discovered that the results were age-sensitive and had high signal quality since cross-talk effects were absent. The baPWV was found to be consistent when the results were compared to those from a commercial pulse sensor device, Enverdis Vascular Explorer (Düsseldorf, Germany).
Hydren et al. [83]	2019	Finger PPG-based device (Finapres Medical Systems, Amsterdam, The Netherlands) Doppler ultrasound	BP Single continuous pass leg movement (sPLM) Leg blood flow Leg vascular conductance	24	It was found that age-related decline in peripheral vascular function was found to be susceptible to sPLM. It was concluded that sPLM simplified the process and reduced the amount of equipment needed to just perform a Doppler ultrasound as compared to a standard passive leg movement.

5. Discussion and Conclusions

The measurement and detection of vascular ageing is important, whether it be in a hospital setting, a GP clinic, or even a home setting. Early detection and guidance to individuals can offer patients time to change their lifestyles and to postpone the onset and progression of diseases such as atherosclerosis or PAD. Unfortunately, no device has yet been proven ideal. It is important that a future device should be relatively inexpensive, non-invasive, accurate, and user-friendly. As such, an easy-to-use measuring tool can allow for measurements to take place at home without the need to be referred by the GP to a clinical setting for a trained specialist to take readings, saving time and resources.

Predominantly, imaging techniques used to assess vascular ageing are expensive, often impractical, and require a specialist. As stated before, procedures such as MRI and angiography must be carried out in a hospital setting where qualified professionals are on hand to take images and recordings. Despite the fact that these methods exist, patients still rely on GP referrals, which might delay the detection of vascular ageing and cause patients' concern, especially in more remote areas where GP access might be difficult in the first place. Unfortunately, there is not a single imaging technique that can accommodate all patients; although MRI is inappropriate for claustrophobic patients, other methods can expose patients to radiation, and certain imaging modalities, such as angiography, are more invasive. Home monitoring is gaining popularity as wearable devices expand. Therefore, to continuously monitor vascular ageing in real time, it is crucial to have a practical, expert-independent, relatively inexpensive diagnostic instrument that can be integrated into smart homes or wearable technology.

There are commercially available instruments to assess arterial stiffness, such as the SphygmoCor-Px System and the Mobil-O-Graph 24 h PWA Monitor Device; however, their use has certain limits. Nonetheless, potentially, there is a possibility of using commercially available tools at home to monitor progress on a frequent basis while, if necessary, maintaining the option to go to a clinical setting for more precise monitoring.

Researchers are increasingly turning to PPG-based technology, which might provide an answer when developing and inventing novel devices to evaluate vascular ageing and arterial stiffness. Due to PPG's simplicity of use, longitudinal studies could be performed to understand whether vascular ageing treatments can alter arterial stiffness and whether a PPG-based device can detect vascular ageing over time. PPG can easily be incorporated into devices and wearable technology, as has already been proven. PPG is suited for universal screening since it is easy to use, is relatively inexpensive, and does not need special training to operate. Another consideration, when creating novel technologies, should be the utilisation of a multi-sensor approach. Combining various sensors with PPG should be considered, as relying solely on PPG may not be the optimal solution. Employing a multimodal approach has the potential to address the challenges associated with PPG and result in a more reliable sensor technology for assessing vascular ageing.

To acquire more conclusive results from in vivo investigations, it is recommended that future studies use bigger sample numbers. Longitudinal studies should also be conducted to assess novel methodologies over the long term. Within study protocols, and as part of the recommended application of PPG devices, a strong contact pressure should be established because measurements can be affected by the quality of the contact [47]. Future research should focus on standardising measuring methods, expanding the databases accessible to computational models, and understanding the potential impacts of medications such as anti-hypertensive medications. Additionally, it is suggested that in the future, an in vitro system that can simulate the mechanical dynamics of a vascular system in a controlled environment be built. This way, pathologies can be introduced to imitate CVD disorders to verify the accuracy of novel devices.

In conclusion, this paper has reviewed and examined the current state-of-the-art techniques and technologies in the assessment of vascular ageing while highlighting the popularity of the well-established optical technique of PPG. Also, studies exploiting recently developed and commercially available devices have been discussed. In this review paper, the benefits and drawbacks of each study have been also discussed. Given that CVD is the main cause of mortality worldwide, it is only a matter of time before cost-effective, reliable screening based on PPG is needed.

Author Contributions: Writing—original draft, P.K.; writing—review and editing, J.M.M. and P.A.K.; supervision, P.A.K. and J.M.M. All authors have read and agreed to the published version of the manuscript.

Funding: This research received no external funding.

Institutional Review Board Statement: Not applicable.

Informed Consent Statement: Not applicable.

Data Availability Statement: Data sharing not applicable.

Conflicts of Interest: The authors declare no conflict of interest.

References

1. Cardiovascular Diseases. Available online: https://www.who.int/health-topics/cardiovascular-diseases (accessed on 2 February 2023).
2. Dhaliwal, G.; Mukherjee, D. Peripheral arterial disease: Epidemiology, natural history, diagnosis and treatment. *Int. J. Angiol.* **2007**, *16*, 36–44. [CrossRef] [PubMed]
3. Kyle, D.; Boylan, L.; Wilson, L.; Haining, S.; Oates, C.; Sims, A.; Guri, I.; Allen, J.; Wilkes, S.; Stansby, G. Accuracy of Peripheral Arterial Disease Registers in UK General Practice: Case-Control Study. *J. Prim. Care Community Health* **2020**, *11*, 2150132720946148. [CrossRef] [PubMed]
4. Buttolph, A.; Sapra, A. *Gangrene;* StatPearls Publishing: Treasure Island, FL, USA, 2023. Available online: http://www.ncbi.nlm.nih.gov/books/NBK560552/ (accessed on 8 August 2023).
5. Taki, A.; Kermani, A.; Ranjbarnavazi, S.M.; Pourmodheji, A. Chapter 4—Overview of Different Medical Imaging Techniques for the Identification of Coronary Atherosclerotic Plaques. In *Computing and Visualization for Intravascular Imaging and Computer-Assisted Stenting;* The Elsevier and MICCAI Society Book Series; Academic Press: Cambridge, MA, USA, 2017; pp. 79–106. [CrossRef]
6. Xu, X.; Wang, B.; Ren, C.; Hu, J.; Greenberg, D.A.; Chen, T.; Xie, L.; Jin, K. Age-related Impairment of Vascular Structure and Functions. *Aging Dis.* **2017**, *8*, 590–610. [CrossRef] [PubMed]
7. LaRocca, T.J.; Martens, C.R.; Seals, D.R. Nutrition and other lifestyle influences on arterial aging. *Ageing Res. Rev.* **2017**, *39*, 106–119. [CrossRef] [PubMed]
8. Omeh, D.J.; Shlofmitz, E. *Angiography;* StatPearls Publishing: Treasure Island, FL, USA, 2022. Available online: http://www.ncbi.nlm.nih.gov/books/NBK557477/ (accessed on 24 November 2022).
9. PRISMA. Available online: http://prisma-statement.org/prismastatement/flowdiagram.aspx (accessed on 27 November 2023).
10. Gong, J.; Han, Y.; Gao, G.; Chen, A.; Fang, Z.; Lin, D.; Liu, Y.; Luo, L.; Xie, L. Sex-specific difference in the relationship between body fat percentage and arterial stiffness: Results from Fuzhou study. *J. Clin. Hypertens.* **2023**, *25*, 286–294. [CrossRef] [PubMed]
11. Ji, H.; Xiong, J.; Yu, S.; Chi, C.; Bai, B.; Teliewubai, J.; Lu, Y.; Zhang, Y.; Xu, Y. Measuring the Carotid to Femoral Pulse Wave Velocity (Cf-PWV) to Evaluate Arterial Stiffness. *J. Vis. Exp. JoVE* **2018**, 57083. [CrossRef]
12. Cantürk, E.; Çakal, B.; Karaca, O.; Omaygenç, O.; Salihi, S.; Özyüksel, A.; Akçevin, A. Changes in Aortic Pulse Wave Velocity and the Predictors of Improvement in Arterial Stiffness Following Aortic Valve Replacement. *Ann. Thorac. Cardiovasc. Surg.* **2017**, *23*, 248–255. [CrossRef]
13. Lurz, E.; Aeschbacher, E.; Carman, N.; Schibli, S.; Sokollik, C.; Simonetti, G.D. Pulse wave velocity measurement as a marker of arterial stiffness in pediatric inflammatory bowel disease: A pilot study. *Eur. J. Pediatr.* **2017**, *176*, 983–987. [CrossRef]
14. Liu, C.-S.; Li, C.-I.; Shih, C.-M.; Lin, W.-Y.; Lin, C.-H.; Lai, S.-W.; Li, T.-C.; Lin, C.-C. Arterial stiffness measured as pulse wave velocity is highly correlated with coronary atherosclerosis in asymptomatic patients. *J. Atheroscler. Thromb.* **2011**, *18*, 652–658. [CrossRef]
15. Tsuchikura, S.; Shoji, T.; Kimoto, E.; Shinohara, K.; Hatsuda, S.; Koyama, H.; Emoto, M.; Nishizawa, Y. Brachial-ankle Pulse Wave Velocity as an Index of Central Arterial Stiffness. *J. Atheroscler. Thromb.* **2010**, *17*, 658–665. [CrossRef]
16. Yufu, K.; Takahashi, N.; Anan, F.; Hara, M.; Yoshimatsu, H.; Saikawa, T. Brachial arterial stiffness predicts coronary atherosclerosis in patients at risk for cardiovascular diseases. *Jpn. Heart J.* **2004**, *45*, 231–242. [CrossRef] [PubMed]
17. Scudder, M.R.; Jennings, J.R.; DuPont, C.M.; Lockwood, K.G.; Gadagkar, S.H.; Best, B.; Jasti, S.P.; Gianaros, P.J. Dual impedance cardiography: An inexpensive and reliable method to assess arterial stiffness. *Psychophysiology* **2021**, *58*, e13772. [CrossRef] [PubMed]
18. Miyoshi, T.; Ito, H. Assessment of Arterial Stiffness Using the Cardio-Ankle Vascular Index. *Pulse* **2016**, *4*, 11–23. [CrossRef] [PubMed]

19. de Souza-Neto, J.D.; de Oliveira, Í.M.; Lima-Rocha, H.A.; Oliveira-Lima, J.W.; Bacal, F. Hypertension and arterial stiffness in heart transplantation patients. *Clinics* **2016**, *71*, 494–499. [CrossRef]
20. Klocke, R.; Cockcroft, J.; Taylor, G.; Hall, I.; Blake, D. Arterial stiffness and central blood pressure, as determined by pulse wave analysis, in rheumatoid arthritis. *Ann. Rheum. Dis.* **2003**, *62*, 414–418. [CrossRef] [PubMed]
21. Zhang, Y.; Yin, P.; Xu, Z.; Xie, Y.; Wang, C.; Fan, Y.; Liang, F.; Yin, Z. Non-Invasive Assessment of Early Atherosclerosis Based on New Arterial Stiffness Indices Measured with an Upper-Arm Oscillometric Device. *Tohoku J. Exp. Med.* **2017**, *241*, 263–270. [CrossRef] [PubMed]
22. Vlachopoulos, C.; Xaplanteris, P.; Aboyans, V.; Brodmann, M.; Cífková, R.; Cosentino, F.; De Carlo, M.; Gallino, A.; Landmesser, U.; Laurent, S.; et al. The role of vascular biomarkers for primary and secondary prevention. A position paper from the European Society of Cardiology Working Group on peripheral circulation: Endorsed by the Association for Research into Arterial Structure and Physiology (ARTERY) Society. *Atherosclerosis* **2015**, *241*, 507–532. [CrossRef] [PubMed]
23. Wang, M.-C.; Wu, A.-B.; Cheng, M.-F.; Chen, J.-Y.; Ho, C.-S.; Tsai, W.-C. Association of Arterial Stiffness Indexes, Determined From Digital Volume Pulse Measurement and Cardiovascular Risk Factors in Chronic Kidney Disease. *Am. J. Hypertens.* **2011**, *24*, 544–549. [CrossRef]
24. Choy, C.-S.; Wang, D.Y.-J.; Chu, T.-B.; Choi, W.-M.; Chiou, H.-Y. Correlation Between Arterial Stiffness Index and Arterial Wave Pattern and Incidence of Stroke. *Int. J. Gerontol.* **2010**, *4*, 75–81. [CrossRef]
25. Naessen, T.; Einarsson, G.; Henrohn, D.; Wikström, G. Peripheral Vascular Ageing in Pulmonary Arterial Hypertension as Assessed by Common Carotid Artery Intima Thickness and Intima/Media Thickness Ratio: An Investigation Using Non-Invasive High-Resolution Ultrasound. *Heart Lung Circ.* **2023**, *32*, 338–347. [CrossRef]
26. Li, Z.; Du, L.; Wang, F.; Luo, X. Assessment of the arterial stiffness in patients with acute ischemic stroke using longitudinal elasticity modulus measurements obtained with Shear Wave Elastography. *Med. Ultrason.* **2016**, *18*, 182–189. [CrossRef] [PubMed]
27. Bjällmark, A.; Lind, B.; Peolsson, M.; Shahgaldi, K.; Brodin, L.-Å.; Nowak, J. Ultrasonographic strain imaging is superior to conventional non-invasive measures of vascular stiffness in the detection of age-dependent differences in the mechanical properties of the common carotid artery. *Eur. J. Echocardiogr.* **2010**, *11*, 630–636. [CrossRef] [PubMed]
28. Kang, K.-W.; Chang, H.-J.; Kim, Y.-J.; Choi, B.-W.; Lee, H.S.; Yang, W.-I.; Shim, C.-Y.; Ha, J.; Chung, N. Cardiac magnetic resonance imaging-derived pulmonary artery distensibility index correlates with pulmonary artery stiffness and predicts functional capacity in patients with pulmonary arterial hypertension. *Circ. J. Off. J. Jpn. Circ. Soc.* **2011**, *75*, 2244–2251. [CrossRef] [PubMed]
29. Ha, H.; Ziegler, M.; Welander, M.; Bjarnegård, N.; Carlhäll, C.-J.; Lindenberger, M.; Länne, T.; Ebbers, T.; Dyverfeldt, P. Age-Related Vascular Changes Affect Turbulence in Aortic Blood Flow. *Front. Physiol.* **2018**, *9*, 36. [CrossRef] [PubMed]
30. Sorelli, M.; Perrella, A.; Bocchi, L. Detecting Vascular Age Using the Analysis of Peripheral Pulse. *IEEE Trans. Biomed. Eng.* **2018**, *65*, 2742–2750. [CrossRef]
31. Rogers, E.M.; Banks, N.F.; Jenkins, N.D.M. Metabolic and microvascular function assessed using near-infrared spectroscopy with vascular occlusion in women: Age differences and reliability. *Exp. Physiol.* **2023**, *108*, 123–134. [CrossRef]
32. Melo e Silva, F.V.; Almonfrey, F.B.; Freitas, C.M.N.D.; Fonte, F.K.; Sepulvida, M.B.D.C.; Almada-Filho, C.D.M.; Cendoroglo, M.S.; Quadrado, E.B.; Amodeo, C.; Povoa, R.; et al. Association of Body Composition with Arterial Stiffness in Long-lived People. *Arq. Bras. Cardiol.* **2021**, *117*, 457–462. [CrossRef]
33. Perrault, R.; Omelchenko, A.; Taylor, C.G.; Zahradka, P. Establishing the interchangeability of arterial stiffness but not endothelial function parameters in healthy individuals. *BMC Cardiovasc. Disord.* **2019**, *19*, 190. [CrossRef]
34. Markakis, K.; Pagonas, N.; Georgianou, E.; Zgoura, P.; Rohn, B.J.; Bertram, S.; Seidel, M.; Bettag, S.; Trappe, H.-J.; Babel, N.; et al. Feasibility of non-invasive measurement of central blood pressure and arterial stiffness in shock. *Eur. J. Clin. Investig.* **2021**, *51*, e13587. [CrossRef]
35. Costa, J.J.O.A.; Cunha, R.D.C.A.; Alves Filho, A.D.A.O.; Bessa, L.R.; de Lima, R.L.S.; dos Reis Silv, A.; e Souza, B.A.; de Almeida Viterbo, C.; Requião, M.B.; Brustolim, D.; et al. Analysis of Vascular Aging in Arterial Hypertension—Population-based Study: Preliminary Results. *Artery Res.* **2019**, *25*, 131–138. [CrossRef]
36. Sridhar, Y.; Naidu, M.U.R.; Usharani, P.; Raju, Y.S.N. Non-invasive evaluation of arterial stiffness in patients with increased risk of cardiovascular morbidity: A cross-sectional study. *Indian J. Pharmacol.* **2007**, *39*, 294. [CrossRef]
37. Komine, H.; Asai, Y.; Yokoi, T.; Yoshizawa, M. Non-invasive assessment of arterial stiffness using oscillometric blood pressure measurement. *Biomed. Eng. OnLine* **2012**, *11*, 6. [CrossRef] [PubMed]
38. Hoffmann, F.; Möstl, S.; Luchitskaya, E.; Funtova, I.; Jordan, J.; Baevsky, R.; Tank, J. An oscillometric approach in assessing early vascular ageing biomarkers following long-term space flights. *Int. J. Cardiol. Hypertens.* **2019**, *2*, 100013. [CrossRef] [PubMed]
39. Juganaru, I.; Luca, C.; Dobrescu, A.-I.; Voinescu, O.; Puiu, M.; Farcaş, S.; Andreescu, N.; Iurciuc, M. A Non-invasive, Easy to Use Medical Device for Arterial Stiffness. *Rev. Chim.* **2019**, *70*, 642–645. [CrossRef]
40. Osman, M.W.; Nath, M.; Khalil, A.; Webb, D.R.; Robinson, T.G.; Mousa, H.A. Longitudinal study to assess changes in arterial stiffness and cardiac output parameters among low-risk pregnant women. *Pregnancy Hypertens.* **2017**, *10*, 256–261. [CrossRef] [PubMed]
41. Kostis, J.B.; Lin, C.P.; Dobrzynski, J.M.; Kostis, W.J.; Ambrosio, M.; Cabrera, J. Prediction of stroke using an algorithm to estimate arterial stiffness. *Int. J. Cardiol. Cardiovasc. Risk Prev.* **2021**, *11*, 200114. [CrossRef] [PubMed]

42. Negoita, M.; Abdullateef, S.; Hughes, A.D.; Parker, K.H.; Khir, A.W. Semiautomatic Vendor Independent Software for Assessment of Local Arterial Stiffness. In Proceedings of the 2018 Computing in Cardiology Conference, Maastricht, The Neterlands, 23–26 September 2018.
43. Obeid, A.N.; Barnett, N.J.; Dougherty, G.; Ward, G. A critical review of laser Doppler flowmetry. *J. Med. Eng. Technol.* **1990**, *14*, 178–181. [CrossRef]
44. Hong, J.; Nandi, M.; Charlton, P.H.; Alastruey, J. Noninvasive hemodynamic indices of vascular aging: An in silico assessment. *Am. J. Physiol.-Heart Circ. Physiol.* **2023**, *325*, H1290–H1303. [CrossRef]
45. Kyriacou, P.A.; Chatterjee, S. 2—The origin of photoplethysmography. In *Photoplethysmography*; Allen, J., Kyriacou, P., Eds.; Academic Press: Cambridge, MA, USA, 2022; pp. 17–43. [CrossRef]
46. Charlton, P.H.; Kyriacou, P.A.; Mant, J.; Marozas, V.; Chowienczyk, P.; Alastruey, J. Wearable Photoplethysmography for Cardiovascular Monitoring. *Proc. IEEE* **2022**, *110*, 355–381. [CrossRef]
47. May, J.M.; Mejía-Mejía, E.; Nomoni, M.; Budidha, K.; Choi, C.; Kyriacou, P.A. Effects of Contact Pressure in Reflectance Photoplethysmography in an In Vitro Tissue-Vessel Phantom. *Sensors* **2021**, *21*, 8421. [CrossRef]
48. Wang, M.; Li, Z.; Zhang, Q.; Wang, G. Removal of Motion Artifacts in Photoplethysmograph Sensors during Intensive Exercise for Accurate Heart Rate Calculation Based on Frequency Estimation and Notch Filtering. *Sensors* **2019**, *19*, 3312. [CrossRef] [PubMed]
49. Allen, J.; Murray, A. Similarity in bilateral photoplethysmographic peripheral pulse wave characteristics at the ears, thumbs and toes. *Physiol. Meas.* **2000**, *21*, 369. [CrossRef] [PubMed]
50. Bentham, M.; Stansby, G.; Allen, J. Innovative Multi-Site Photoplethysmography Analysis for Quantifying Pulse Amplitude and Timing Variability Characteristics in Peripheral Arterial Disease. *Diseases* **2018**, *6*, 81. [CrossRef] [PubMed]
51. Brillante, D.G.; O'sullivan, A.J.; Howes, L.G. Arterial stiffness indices in healthy volunteers using non-invasive digital photoplethysmography. *Blood Press.* **2008**, *17*, 116–123. [CrossRef] [PubMed]
52. Jannasz, I.; Sondej, T.; Targowski, T.; Mańczak, M.; Obiała, K.; Dobrowolski, A.P.; Olszewski, R. Relationship between the Central and Regional Pulse Wave Velocity in the Assessment of Arterial Stiffness Depending on Gender in the Geriatric Population. *Sensors* **2023**, *23*, 5823. [CrossRef] [PubMed]
53. Tąpolska, M.; Spałek, M.; Szybowicz, U.; Domin, R.; Owsik, K.; Sochacka, K.; Skrypnik, D.; Bogdański, P.; Owecki, M. Arterial Stiffness Parameters Correlate with Estimated Cardiovascular Risk in Humans: A Clinical Study. *Int. J. Environ. Res. Public. Health* **2019**, *16*, 2547. [CrossRef] [PubMed]
54. Tanaka, G.; Yamakoshi, K.; Sawada, Y.; Matsumura, K.; Maeda, K.; Kato, Y.; Horiguchi, M.; Ohguro, H. A novel photoplethysmography technique to derive normalized arterial stiffness as a blood pressure independent measure in the finger vascular bed. *Physiol. Meas.* **2011**, *32*, 1869. [CrossRef]
55. von Wowern, E.; Östling, G.; Nilsson, P.M.; Olofsson, P. Digital Photoplethysmography for Assessment of Arterial Stiffness: Repeatability and Comparison with Applanation Tonometry. *PLoS ONE* **2015**, *10*, e0135659. [CrossRef]
56. Djurić, B.; Žikić, K.; Nestorović, Z.; Lepojević-Stefanović, D.; Milošević, N.; Žikić, D. Using the photoplethysmography method to monitor age-related changes in the cardiovascular system. *Front. Physiol.* **2023**, *14*, 1191272. [CrossRef]
57. Huotari, M.; Yliaska, N.; Lantto, V.; Määttä, K.; Kostamovaara, J. Aortic and arterial stiffness determination by photoplethysmographic technique. *Procedia Chem.* **2009**, *1*, 1243–1246. [CrossRef]
58. Zekavat, S.M.; Aragam, K.; Emdin, C.; Khera, A.V.; Klarin, D.; Zhao, H.; Natarajan, P. Genetic Association of Finger Photoplethysmography-Derived Arterial Stiffness Index With Blood Pressure and Coronary Artery Disease. *Arterioscler. Thromb. Vasc. Biol.* **2019**, *39*, 1253–1261. [CrossRef] [PubMed]
59. Sharkey, E.J.; Maria, C.D.; Klinge, A.; Murray, A.; Zheng, D.; O'Sullivan, J.; Allen, J. Innovative multi-site photoplethysmography measurement and analysis demonstrating increased arterial stiffness in paediatric heart transplant recipients. *Physiol. Meas.* **2018**, *39*, 074007. [CrossRef] [PubMed]
60. Wei, C.-C. Developing an Effective Arterial Stiffness Monitoring System Using the Spring Constant Method and Photoplethysmography. *IEEE Trans. Biomed. Eng.* **2013**, *60*, 151–154. [CrossRef] [PubMed]
61. Park, Y.-J. Association between blood glucose levels and arterial stiffness marker: Comparing the second derivative of photoplethysmogram and cardio-ankle vascular index scores. *Front. Endocrinol.* **2023**, *14*, 1237282. [CrossRef] [PubMed]
62. Bortolotto, L.A.; Blacher, J.; Kondo, T.; Takazawa, K.; Safar, M.E. Assessment of vascular aging and atherosclerosis in hypertensive subjects: Second derivative of photoplethysmogram versus pulse wave velocity. *Am. J. Hypertens.* **2000**, *13*, 165–171. [CrossRef] [PubMed]
63. Korneeva, O.; Drapkina, O. PP.33.08: Arterial stiffness parameters in obese hypertensive patients using different statin-based regimens and fixed antihypertensive combination. *J. Hypertens.* **2015**, *33*, e431. [CrossRef]
64. Drapkina, O.; Ivashkin, V. Improving in arterial stiffness parameters by pulse wave analysis in obese patients. *Atherosclerosis* **2014**, *235*, e156–e157. [CrossRef]
65. Drapkina, O. Rosuvastatin significantly improves arterial stiffness parameters by pulse wave analysis in high risk patients with arterial hypertension and dyslipidemia. *Atherosclerosis* **2014**, *235*, e95. [CrossRef]
66. von Wowern, E.; Källén, K.; Olofsson, P. Arterial stiffness in normal pregnancy as assessed by digital pulse wave analysis by photoplethysmography—A longitudinal study. *Pregnancy Hypertens.* **2019**, *15*, 51–56. [CrossRef]
67. Lefferts, W.K.; Lefferts, E.C.; Hibner, B.A.; Smith, K.J.; Fernhall, B. Impact of acute changes in blood pressure and arterial stiffness on cerebral pulsatile haemodynamics in young and middle-aged adults. *Exp. Physiol.* **2021**, *106*, 1643–1653. [CrossRef]

68. Salvi, P.; Magnani, E.; Valbusa, F.; Agnoletti, D.; Alecu, C.; Joly, L.; Benetos, A. Comparative study of methodologies for pulse wave velocity estimation. *J. Hum. Hypertens.* **2008**, *22*, 669–677. [CrossRef] [PubMed]
69. Djeldjli, D.; Bousefsaf, F.; Maaoui, C.; Bereksi-Reguig, F.; Pruski, A. Remote estimation of pulse wave features related to arterial stiffness and blood pressure using a camera. *Biomed. Signal Process. Control* **2021**, *64*, 102242. [CrossRef]
70. Kock, K.D.S.; da Silva, J.B.F.; Marques, J.L.B. Comparison of the ankle-brachial index with parameters of stiffness and peripheral arterial resistance assessed by photoplethysmography in elderly patients. *J. Vasc. Bras.* **2019**, *18*, e20180084. [CrossRef]
71. Murakami, T.; Asai, K.; Kadono, Y.; Nishida, T.; Nakamura, H.; Kishima, H. Assessment of Arterial Stiffness Index Calculated from Accelerated Photoplethysmography. *Artery Res.* **2019**, *25*, 37–40. [CrossRef]
72. Clarenbach, C.F.; Stoewhas, A.-C.; van Gestel, A.J.; Latshang, T.D.; Lo Cascio, C.M.; Bloch, K.E.; Kohler, M. Comparison of photoplethysmographic and arterial tonometry-derived indices of arterial stiffness. *Hypertens. Res.* **2012**, *35*, 228–233. [CrossRef] [PubMed]
73. Dall'Olio, L.; Curti, N.; Remondini, D.; Safi Harb, Y.; Asselbergs, F.W.; Castellani, G.; Uh, H.-W. Prediction of vascular aging based on smartphone acquired PPG signals. *Sci. Rep.* **2020**, *10*, 19756. [CrossRef] [PubMed]
74. Shin, H.; Noh, G.; Choi, B.-M. Photoplethysmogram based vascular aging assessment using the deep convolutional neural network. *Sci. Rep.* **2022**, *12*, 11377. [CrossRef]
75. Park, J.; Shin, H. Vascular Aging Estimation Based on Artificial Neural Network Using Photoplethysmogram Waveform Decomposition: Retrospective Cohort Study. *JMIR Med. Inform.* **2022**, *10*, e33439. [CrossRef]
76. Olmedo-Aguirre, J.O.; Reyes-Campos, J.; Alor-Hernández, G.; Machorro-Cano, I.; Rodríguez-Mazahua, L.; Sánchez-Cervantes, J.L. Remote Healthcare for Elderly People Using Wearables: A Review. *Biosensors* **2022**, *12*, 73. [CrossRef]
77. Reeves, C.; Islam, A.; Gentry, T. The State of Health and Care of Older People. 2023. Available online: https://www.ageuk.org.uk/globalassets/age-uk/documents/reports-and-publications/reports-and-briefings/health--wellbeing/age-uk-briefing-state-of-health-and-care-july-2023-abridged-version.pdf (accessed on 30 November 2023).
78. Lian, C.; Yang, Y.; Yu, X.; Sun, H.; Zhao, Y.; Zhang, G.; Li, W.J. Robust and Remote Photoplethysmography Based on Smartphone Imaging of the Human Palm. *IEEE Trans. Instrum. Meas.* **2023**, *72*, 5012611. [CrossRef]
79. Zhao, L.; Zhang, X.; Niu, X.; Sun, J.; Geng, R.; Li, Q.; Zhu, X.; Dai, Z. Remote photoplethysmography (rPPG) based learning fatigue detection. *Appl. Intell.* **2023**, *53*, 27951–27965. [CrossRef]
80. Casalino, G.; Castellano, G.; Zaza, G. Evaluating the robustness of a contact-less mHealth solution for personal and remote monitoring of blood oxygen saturation. *J. Ambient Intell. Humaniz. Comput.* **2023**, *14*, 8871–8880. [CrossRef] [PubMed]
81. Perpetuini, D.; Chiarelli, A.M.; Maddiona, L.; Rinella, S.; Bianco, F.; Bucciarelli, V.; Gallina, S.; Perciavalle, V.; Vinciguerra, V.; Merla, A.; et al. Multi-Site Photoplethysmographic and Electrocardiographic System for Arterial Stiffness and Cardiovascular Status Assessment. *Sensors* **2019**, *19*, 5570. [CrossRef] [PubMed]
82. Charlton, P.H.; Paliakaitė, B.; Pilt, K.; Bachler, M.; Zanelli, S.; Kulin, D.; Allen, J.; Hallab, M.; Bianchini, E.; Mayer, C.C.; et al. Assessing hemodynamics from the photoplethysmogram to gain insights into vascular age: A review from VascAgeNet. *Am. J. Physiol.-Heart Circ. Physiol.* **2022**, *322*, H493–H522. [CrossRef]
83. Hydren, J.R.; Broxterman, R.M.; Trinity, J.D.; Gifford, J.R.; Kwon, O.S.; Kithas, A.C.; Richardson, R.S. Delineating the age-related attenuation of vascular function: Evidence supporting the efficacy of the single passive leg movement as a screening tool. *J. Appl. Physiol.* **2019**, *126*, 1525–1532. [CrossRef]
84. Gifford, J.R.; Richardson, R.S. CORP: Ultrasound assessment of vascular function with the passive leg movement technique. *J. Appl. Physiol.* **2017**, *123*, 1708–1720. [CrossRef]

sensors

MDPI

Review

Review of Three-Dimensional Handheld Photoacoustic and Ultrasound Imaging Systems and Their Applications

Changyeop Lee [1], Chulhong Kim [1,*] and Byullee Park [2,*]

[1] Department of Electrical Engineering, Convergence IT Engineering, Mechanical Engineering, Medical Science and Engineering, Graduate School of Artificial Intelligence, and Medical Device Innovation Center, Pohang University of Science and Technology, Pohang 37673, Republic of Korea; ckdduq0801@postech.ac.kr

[2] Department of Biophysics, Institute of Quantum Biophysics, Sungkyunkwan University, Suwon 16419, Republic of Korea

* Correspondence: chulhong@postech.edu (C.K.); byullee@skku.edu (B.P.)

Abstract: Photoacoustic (PA) imaging is a non-invasive biomedical imaging technique that combines the benefits of optics and acoustics to provide high-resolution structural and functional information. This review highlights the emergence of three-dimensional handheld PA imaging systems as a promising approach for various biomedical applications. These systems are classified into four techniques: direct imaging with 2D ultrasound (US) arrays, mechanical-scanning-based imaging with 1D US arrays, mirror-scanning-based imaging, and freehand-scanning-based imaging. A comprehensive overview of recent research in each imaging technique is provided, and potential solutions for system limitations are discussed. This review will serve as a valuable resource for researchers and practitioners interested in advancements and opportunities in three-dimensional handheld PA imaging technology.

Keywords: photoacoustic imaging; ultrasound imaging; 3D handheld; clinical applications; biomedical studies

Citation: Lee, C.; Kim, C.; Park, B. Review of Three-Dimensional Handheld Photoacoustic and Ultrasound Imaging Systems and Their Applications. *Sensors* **2023**, *23*, 8149. https://doi.org/10.3390/s23198149

Academic Editor: Evgeny Katz

Received: 1 September 2023
Revised: 25 September 2023
Accepted: 25 September 2023
Published: 28 September 2023

1. Introduction

Photoacoustic (PA) imaging is a non-invasive biomedical imaging modality that has been gaining increasing attention due to its unique ability to provide high-resolution structural and functional information on various endogenous light absorbers without needing exogenous contrast agents [1–5]. This imaging technique is based on the PA effect [6], which occurs when a short-pulse laser irradiates a light-absorbing tissue, causing local thermal expansion and extraction to occur and subsequently generating ultrasound (US) waves. Collected US signals are reconstructed to PA images that combine beneficial features of both optics and acoustics, including high spatial resolution, deep tissue penetration, and high optical contrasts. Further, PA images can visualize multi-scale objects from cells to organs using the same endogenous light absorbers [7,8]. Moreover, compared to traditional medical imaging modalities such as magnetic resonance imaging (MRI), computed tomography (CT), and positron emission tomography (PET), PA imaging systems offer several advantages, including simpler system configuration, relatively lower cost, and the use of non-ionizing radiation [9,10]. These benefits have resulted in the emergence of PA imaging as a promising technique for various biomedical applications, including imaging of cancers, peripheral diseases, skin diseases, and hemorrhagic and ischemic diseases [11–22].

Two-dimensional (2D) cross-sectional PA imaging systems are widely used due to their relatively simple implementation by adding a laser to a typical US imaging system. Moreover, PA images can be acquired at the same location as conventional US images, thus facilitating clinical quantification analysis [23–27]. However, despite the advantages of 2D imaging, poor reproducibility performance can arise depending on the operator's proficiency and system sensitivity, which can adversely affect accurate clinical analysis [28]. To address these issues, the development of three-dimensional (3D) PA imaging systems utilizing various scanning methods such as direct, mechanical, and mirror-based scanning has gained momentum. Direct scanning employs 2D hemispherical or matrix-shaped array US transducers (USTs) for real-time acquisition of volumetric PA images [29,30]. Mechanical scanning combines 1D array USTs of various shapes with a motorized stage to obtain large-area PA images, albeit at a relatively slower pace. Mirror-scanning-based 3D PA imaging utilizes single-element USTs with microelectromechanical systems (MEMS) or galvanometer scanners (GS), enabling high-resolution images with fast imaging speed, although it has a limited range of acquisition [31,32].

While the aforementioned fixed 3D PA imaging systems have demonstrated promising results, the inherent fixed nature of these systems can potentially limit the acquisition of PA images due to patient-specific factors or the specific anatomical location of the pathology under consideration [9,33–35]. Consequently, to expedite clinical translation, it becomes imperative to develop versatile 3D handheld PA imaging systems. Here, we aim to provide a detailed review of 3D handheld PA imaging systems. Our strategy involves classifying these systems into four scanning techniques—direct [30,36–57], mechanical [10,58–62], mirror [31,32,63–67], and freehand [50,67–70] scanning—and providing a comprehensive summary of the latest research corresponding to each technique. Freehand scanning involves image reconstruction through image processing techniques with or without additional aids such as position sensors and optical patterns. It is not limited by the UST type. Further, we discuss potential solutions to overcome 3D handheld PA imaging system limitations such as motion artifacts, anisotropic spatial resolution, and limited view artifacts. We believe that our review will provide valuable insights to researchers and practitioners in the field of 3D handheld PA imaging to uncover the latest developments and opportunities to advance this technology.

2. Classification of 3D Handheld PA Imaging Systems into Four Scanning Techniques

Three-dimensional handheld PA imaging systems can be categorized into four distinct groups based on the scanning mechanism employed: direct scanning, mechanical scanning, mirror-based scanning, and freehand scanning (Figure 1). Each of these systems possesses unique characteristics and limitations. Therefore, it is essential for researchers to comprehensively understand general features associated with each system to facilitate further advancements in this field.

Direct scanning involves utilization of 2D array USTs with various shapes such as hemispherical and matrix configurations [30,40,57]. These 2D array USTs enable the acquisition of 3D image data, facilitating direct generation of real-time volumetric PA images. Imaging systems based on 2D array USTs offer isotropic spatial resolutions in both lateral and elevational directions thanks to the arrangement of their transducer elements. Furthermore, a 2D hemispherical array UST provides optimal angular coverage for receiving spherical spreading PA signals [71]. This mitigates the effects of a limited view, thereby enhancing the spatial resolutions of obtained images. It is crucial to maintain coaxiality between US and light beams to achieve good signal-to-noise ratios (SNRs). To accomplish this, some imaging systems based on a 2D hemispherical array UST or certain matrix array USTs incorporate a hole in the center of the UST array, which allows light to be directed towards the US imaging plane [39,46,54].

Mechanical-scanning-based 3D handheld PA imaging systems mainly use 1D array USTs [58,59]. With the help of mechanical movements, 1D array USTs can collect scan data in the scanning direction. Three-dimensional image reconstruction is typically performed by stacking collected data in the scanning direction. Since 1D array USTs provide array elements in the lateral direction, beamforming can be implemented in the lateral direction. However, they do not provide array elements in the elevation direction. Thus, beamforming reconstruction cannot be performed, thereby providing anisotropic spatial resolutions. Although the synthetic aperture focusing technique (SAFT) can be applied as an alternative using scan data in the elevational direction [72], its effectiveness is low due to a tight elevation beam focus fixed by the US lens. Fiber bundles (FBs) are routinely used for light transmission. They are obliquely attached to the side of USTs using an adapter. To cross the US imaging plane and the light area irradiated obliquely, they typically use a stand-off such as water or a gel pad between the bottom of USTs and image targets [9,10,58].

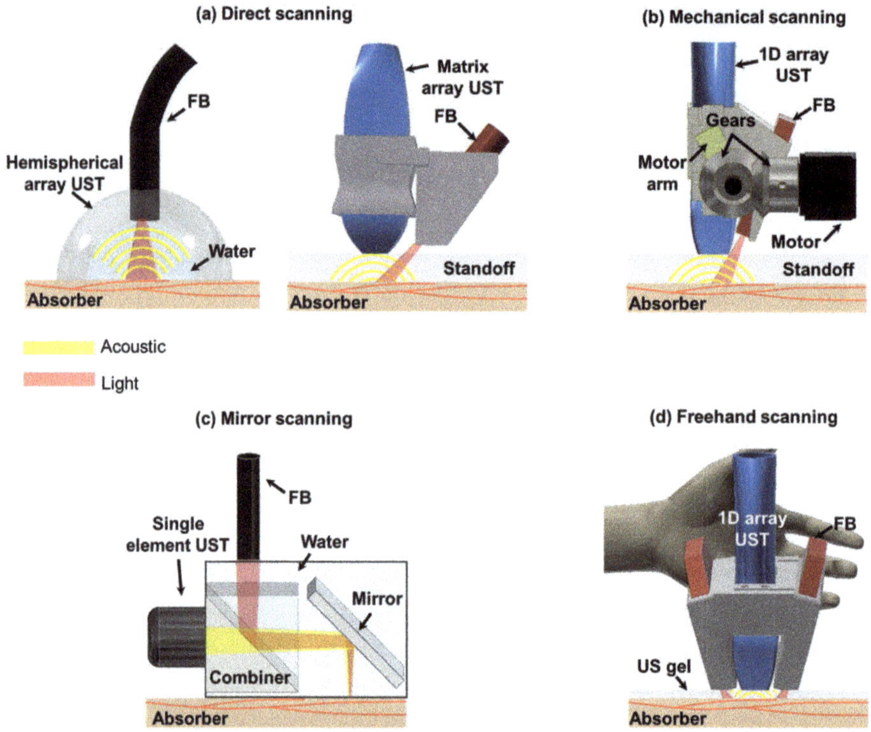

Figure 1. Schematics of 3D handheld PA imaging systems. (**a**) Direct scanning with 2D array UST-based 3D handheld PA imaging system. (**b**) Mechanical scanning with 1D array UST-based 3D handheld PA imaging system. (**c**) Mirror scanning with single-element UST-based 3D handheld PA imaging system. (**d**) Freehand scanning with 1D array UST-based 3D handheld PA imaging system. PA, photoacoustic; UST, ultrasound transducer; FB, fibers; LS, linear stage.

Mirror-scanning-based 3D handheld PA imaging systems use single-element USTs with different mirrors (e.g., MEMS or GS) to ensure high imaging speed [31,32]. Spatial resolutions of these systems are generally superior to array-based imaging systems because they use a single element with high frequency USTs. Some of them can provide tight optical focusing to provide better lateral resolutions [73]. Conversely, they can provide shallower penetration depth than other systems. In particular, penetration depths of these systems providing optical focusing are limited to 1 mm, the mean optical path length of the biological tissue [74]. Thus, these systems are disadvantageous to use for clinical imaging [63]. To ensure coaxial configuration between light and acoustic beams, they typically use beam combiners [64,67].

Freehand scanning can be implemented with all UST types. Additional devices such as a global positioning system (GPS) and optical tracker are attached to UST bodies for recording the scanning location [69,70]. With scan data and the information of the scanning location, 3D image reconstruction is implemented. In addition, by analyzing similarity between successively acquired freehand scan images, position calibration can be performed without additional hardware units [50]. Furthermore, they provide flexibility to combine with different scanning-based 3D PA imaging systems [67]. Therefore, spatial resolutions, penetration depths, and coaxial configurations might vary depending on the UST type and scanning system.

These four scanning mechanisms provide useful criteria for distinguishing 3D handheld PA imaging systems. Their respective characteristics are summarized in Table 1.

Table 1. General features of 3D clinical handheld imaging systems.

Scanning Mechanism	Direct [30,39]	Mechanical [10,58]	Mirror [32,64]	Freehand
UST type	2D array	1D array	Single element	All types
Lateral resolution (μm)	200	592–799	5–12	*
Penetration depth (mm)	<30	<30	<1	*
Coaxial configuration unit	Hole	Standoff	Beam combiner	*

*: Depends on types of USTs and combined scan systems. UST, ultrasound transducer.

2.1. Three-Dimensional Handheld PA Imaging Systems Using Direct Scanning

Direct 3D scanning is typically achieved using hemispherical or matrix-shaped 2D array USTs. In PA imaging, the PA wave propagates in a three-dimensional direction, which makes hemispherical 2D array USTs particularly effective for wide angular coverage, minimizing limited view artifacts and ensuring high SNRs [71]. Meanwhile, matrix-shaped 2D array USTs can be custom fabricated into smaller sizes, making them highly suitable for handheld applications [54,57]. Furthermore, their unique features make them suitable for use as wearable devices [55].

Dea'n-Ben et al. [39] have introduced a 3D multispectral PA imaging system based on a 2D hemispherical array UST and visualized human breast with the system. They used an optical parametric oscillator (OPO) laser, a hemispherical shaped UST, and a custom designed data acquisition (DAQ) system (Figure 2a(i)). The OPO laser emitted short pulses (<10 ns) and was rapidly tuned over a wavelength range of 690 to 900 nm. The laser fluence and average power were less than 15 mJ/cm^2 and 200 mW/cm^2, respectively. The hemispherical US TR was composed of 256 piezocomposite elements arranged on a 40 mm radius hemispherical surface that covered a solid angle of 90°. The element was about 3×3 mm^2. It provided 4 MHz of a center frequency with a 100% bandwidth. At the center of the TR, an 8 mm cylindrical cavity was formed. Light was delivered through the cavity using an FB. Data acquisition was conducted at 40 Msamples/s with the DAQ system. Single-wavelength volume imaging was performed at a rate of 10 Hz. For online image reconstruction, a back-projection algorithm accelerated with a graphics processing unit (GPU) was implemented. For accurate imaging reconstruction, 3D model-based image reconstruction was conducted offline. The spatial resolution was measured by imaging spherical-shaped geometry. It was about 0.2 mm. Spectral unmixing was performed using the least-square method. Unmixed PA oxy-hemoglobin (HbO2), deoxy-hemoglobin (Hb), and melanin images of human breast are shown in Figure 2a(ii).

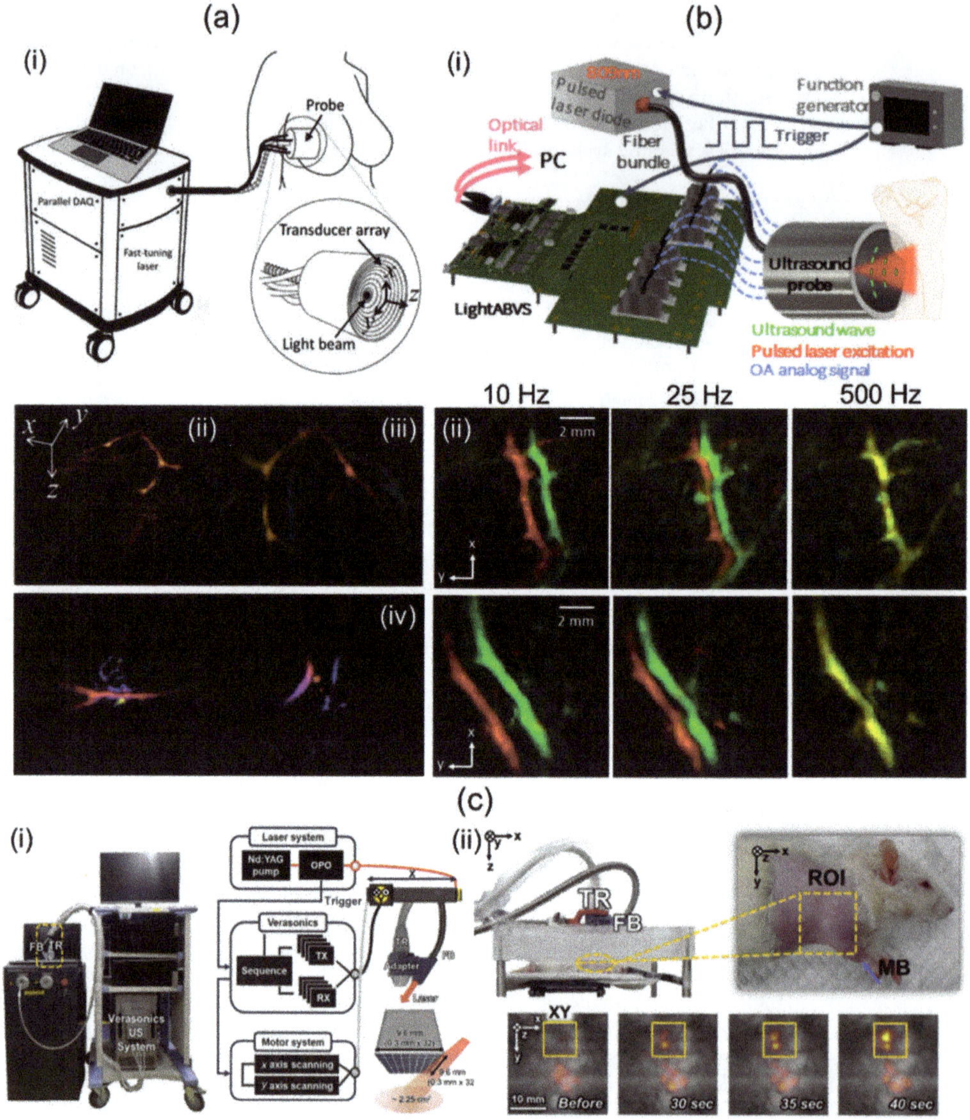

Figure 2. Direct-scanning-based 3D handheld PA imaging systems and their applications. (**a**) (**i**) Schematic diagram of a 2D hemispherical array transducer-based direct 3D imaging system and its clinical application. PA (**ii**) amplitude (**iii**) depth-encoded and (**iv**) sO_2 images of human breasts. (**b**) (**i**) A cost-effective direct 3D handheld PA imaging system based on a 2D hemispherical array probe. (**ii**) In vivo human wrist PA MAP images acquired at various acquisition rates. Top and bottom rows indicate different regions around the wrist. (**c**) (**i**) Conventional 2D matrix array UST-based 3D PA/US imaging system. (**ii**) Photograph of the rat and in vivo PA/US MAP images after MB injection. ROI indicates the SLN area. PA, photoacoustic; US, ultrasound; MAP, maximum amplitude projection; SLN, sentinel lymph node; MB, methylene blue; sO_2, oxygen saturation. Reprinted with permission from references [35,39,75].

Özsoy et al. [75] have demonstrated a cost-effective and compact 3D PA imaging system based on a 2D hemispherical array UST. Their study was performed using a low-cost diode laser, the hemispherical array UST, and an optical-link-based US acquisition platform [76] (Figure 2b(i)). The diode laser delivered a single-wavelength (809 nm) pulse with a length of about 40 ns. The lasers' pulse repetition frequency (PRF) was up to 500 Hz. The output energy was measured to be about 0.65 mJ. The laser system was operated by a power supply and synchronized by a function generator. The size of the laser was $9 \times 5.6 \times 3.4$ cm^3, which was sufficiently smaller than OPO-based laser systems. The hemispherical array UST consisted of 256 piezoelectric composite elements with a 4 MHz center frequency and 100% bandwidth. It provided a solid angle of 0.59 π°. A total of 480 single-mode fibers were inserted through the central hole of the transducer and used to deliver the beam. The custom DAQ system had a total cost of approximately EUR 9.4. It offered 192 channels. Its measurements were $18 \times 22.6 \times 10$ cm^3. The DAQ system offered a 40 MHz sampling frequency with a 12-bit resolution. A total of 1000 sample data were collected for each channel, acquiring a region of 37.5 mm. Data were transmitted to a PC via 100G Ethernet with two 100 G Fireflies. Image reconstruction was performed with back projection [77]. The FOV of the reconstructed volume was $10 \times 10 \times 10$ mm^3. In vivo human wrist imaging was performed by freely scanning along a random path using the developed system. Two consecutively acquired PA images were overlaid in red and green as shown in Figure 2b(ii). The top and bottom rows represent different imaged regions around the wrist. As the image acquisition rate increases, the accuracy of spectral unmixing is expected to improve as two successive images cannot be distinguished.

Liu et al. [57] have presented a compact 3D handheld PA imaging system based on a custom-designed 2D matrix array UST. Their research was conducted using a custom fiber-connected laser, the custom fabricated 2D matrix array probe, and a DAQ. The laser delivering 10.7 mJ of pulse energy at the fiber output tip was $27 \times 8 \times 6$ cm^3 in size with a weight of 2 kg. The probe contained 72 piezoelectric elements with a size, center frequency, and bandwidth of 1×1 mm^2, 2.25 MHz, and 65%, respectively. US elements were arranged to make a central hole (diameter: 2 mm). The light was delivered through the hole after passing through a collimator (NA: 0.23 in the air) and a micro condenser lens (diameter: 6 mm, f: 6 mm). The light was then homogenized with an optical diffusing gel pad to reshape a Gaussian-distributed beam in the imaging FOV. The dimensions and weight of the entire probe were <10 cm^3 and 44 g, respectively. Then, 72-channel data for each volume were digitized by the DAQ, providing 128-channel 12-bit 1024 samples at 40 Msamples/s. Data acquisition timing was controlled by a programmed micro-control unit (MCU). Acquired volume data were reconstructed by the fast phase shift migration (PSM) method [56], which enabled real-time imaging. Spatial resolution quantification was implemented by imaging human hairs. Lateral and axial resolutions were measured to be 0.8 and 0.73 mm, respectively. The system took 0.1 sec of volumetric imaging time to scan an FOV of $10 \times 10 \times 10$ mm^3. The handheld feasibility of the system was demonstrated by imaging the cephalic vein of a human arm.

Although 3D handheld PA/US imaging using a conventional 2D matrix array UST has not been reported, 3D handheld operation is readily possible using stand-off or optical systems. Therefore, studies using 2D matrix array USTs will potentially lead to the development of 3D handheld PA/US imaging systems. Representatively, Kim et al. [35] have demonstrated a conventional 2D matrix array UST-based 3D PA/US imaging system. They used tunable laser, the 2D matrix array UST, an FB, a US imaging platform, and two motorized stages (Figure 2c(i)). The laser wavelength range and PRF were 660–1320 nm and 10 Hz, respectively. The 2D matrix array UST provides 1024 US elements with a center frequency of 3.3 MHz. For light transmission, FB was attached to the side of the 2D matrix array UST obliquely, and light was delivered to the US imaging plane. The US imaging platform offers 256 receive channels. Thus, four laser shots were required for one PA image. To obtain wide-field 3D images, they combined 2D matrix array UST with two motorized stages and implemented 2D raster scanning. For offline image processing, 3D filtered back projection and enveloped detection were conducted. To evaluate system performance, a 54 × 18 mm^2 phantom containing 90 μm thick black absorbers was imaged with a total scan time of 30 min at a raster scan step size of 0.9 mm. Quantified axial and lateral resolutions were 0.76 and 2.8 mm, respectively. Further, they implemented in vivo rat sentinel lymph node (SLN) imaging after methylene blue injection (Figure 2c(ii)). Wang et al. [78] have showcased a 3D PA/US imaging system based on a conventional 2D matrix array UST. In their study, a tunable dye laser, the 2D matrix array UST, and a custom-built DAQ were used. The laser's PRF and pulse duration were 10 Hz and 6.5 ns, respectively. The laser was coupled with bifurcated FB. The matrix array UST has 2500 US elements with a nominal bandwidth of 2–7 MHz. Thirty-six laser shots were required for the acquisition of one PA image. Image data transferred by the DAQ were then reconstructed with a 3D back projection algorithm [79] on a four-core central processing unit (CPU). This took 3 h. The reconstructed PA image's volume was 2 × 2 × 2 cm^3. For the benchmark of their system, gelatin phantom containing a human hair was imaged. Measured profiles in axial, lateral, and elevational directions were 0.84, 0.69, and 0.90 mm, respectively. In addition, in vivo mouse SLN imaging with methylene blue was performed for further verification (Table 2).

Table 2. Specifications of direct-scanning-based 3D handheld PA/US imaging systems.

Author		Dea'n-Ben et al. [39]	Özsoy et al. [75]	Liu et al. [57]	Kim et al. [35]	Wang et al. [78]
Laser	Wavelength	660–900 nm	809 nm	-	660–1320 nm	650 nm
	Pulse duration	<10 ns	40 ns	-	-	6.5 ns
	PRF	-	~500 Hz	-	10 Hz	10 Hz
	Light delivery	Fiber	Fiber	Fiber	Fiber	Fiber
Probe	Scanning aids	-	-	2 motor stages	-	-
	Scanning type	Direct	Direct	Direct	Direct + mechanical	Direct
	UST type	2D hemispherical	2D hemispherical	2D matrix	2D matrix	2D matrix
	# of elements	256	256	72	1024	2500
	Center frequency	4 MHz	4 MHz	2.25 MHz	3.3 MHz	-
	Bandwidth	100%	100%	65%	-	-
	Resolution	S: 200 μm	-	A, L: 730, 800 μm	A, L: 760, 2800 μm	A, L, E: 840, 690, 900 μm
	Scan time	0.1 s	0.002 s	0.1 s	-	3 h
	FOV	-	10 × 10 × 10 mm^3	10 × 10 × 10 mm^3	10 × 10 mm^2	20 × 20 × 20 mm^3
	Weight	-	-	44 g	-	-
	Dimension	-	-	<10 cm^3	-	-
Data acquisition/processing system	Platform	Custom DAQ	Custom DAQ	SonixDAQ	Verasonics	Custom DAQ
	Sampling rate	40 MHz	40 MHz	40 MHz	-	-

S, A, L, E: spatial, axial, lateral, and elevational resolutions, respectively.

While 2D array USTs with a hemispherical shape offer advantages for PA imaging, their high cost, complex electronics, and high computational demands are cited as disadvantages. Additionally, their large element pitch sizes render them unsuitable for US imaging, limiting their use in clinical research. Conversely, matrix-shaped 2D array USTs are highly portable due to their light weight and small size. However, their suboptimal angular coverage for PA signal reception results in poor spatial resolution and limited view artifacts.

2.2. Three-Dimensional Handheld PA Imaging Systems Using Mechanical Scanning

Several 3D handheld PA imaging systems utilizing mechanical scanning methods and various types of USTs have been developed. Typically, these systems employ one- to three-axis motorized stages for 3D scanning. They are relatively inexpensive, requiring simple electronic devices. In this chapter, we will introduce two such systems: one based on a 1D linear array UST and another based on a single-element UST [10,58,59,62].

Lee et al. [58] have developed a 1D linear array UST-based 3D handheld PA/US imaging system that uses a lightweight and compact motor whose scanning method is controlled by a non-linear scanning mechanism. In their study, they used a portable OPO laser with fast pulse tuning, a conventional clinical US machine, the 1D linear array UST, and a custom-built scanner (Figure 3a(i)). The laser delivered light at pulse energies of 7.5 and 6.1 mJ/cm^2 at 797 and 850 nm, respectively, with a 10 Hz PRF. With 64 receive channels, the US machine was an FDA-approved US system that offered a programmable environment. The UST had an 8.5 MHz central frequency and a 62% bandwidth at −6 dB. The scanner included the guide rail, handle, guide rod, plate, connector, drain hole, plastic cap, FB, holder, UST, motor, arm, water, water tank, and membrane. The FB was integrated into the holder with the UST. It was inserted at 15° for efficient light transmission. The motor arm moved the holder to drive the UST in the scanning direction after receiving a laser trigger signal. The movement was managed with a non-linear scanning method (i.e., the scotch yoke mechanism) that changed linear motion into rotational movement and vice versa, allowing the use of a small, lightweight motor (170 g) instead of bulky, heavy motor stages. They used a 3 cm deep water tank to avoid PA reflection artifacts within 3 cm of the axial region. Polyvinyl chloride (PVC) membranes with a thickness of 0.2 mm were attached to the water tank to seal the water. Dimensions and weight of the scanner were 100 × 80 × 100 mm^3 and 950 g, respectively. The maximum FOV was 40 × 38 mm^2, and the maximum scan time for single-wavelength imaging was 20.0 s. Acquired volume data were reconstructed by Fourier-domain beamforming [80]. They measured spatial resolutions by quantifying the full width at half maximum (FWHM) of black threads having a thickness of 90 μm. The calculated axial and lateral resolutions were 191 μm and 799 μm, respectively. To confirm the usefulness of the handheld system for human body imaging, they implemented human in vivo imaging of the wrist, neck, and thigh (Figure 3a(ii)).

Using the developed 3D handheld imaging system, Park et al. [59] have conducted a clinical study of various types of cutaneous melanomas. They used multiple wavelengths of 700, 756, 797, 866, and 900 nm. The FOV was scaled to 31 × 38 mm^2, resulting in a scan time of 57 sec. They recruited six patients who had in situ, nodular, acral lentiginous, and metastasis types of melanomas. US/PA/PA unmixed melanin images of metastasized melanoma were acquired as shown in Figure 3b(i). They compared unmixed PA melanin depth and histopathological depth and confirmed that the two depths agreed well within a mean absolute error of 0.36 mm (Figure 3b(ii)). Further, they measured a maximum PA penetration depth of 9.1 mm in nodular melanoma.

Lee et al. [10] have showcased updated 3D clinical handheld PA/US imaging scanner and systems. The 3D handheld imager weighs 600 g and measures 70 × 62 × 110 mm^3, increasing handheld usability (Figure 3c(i)). The updated system improved SNRs of the system by an average of 11 dB over the previous system using a transparent solid US gel pad with similar attenuation coefficients to water. The previous system could not immediately check 3D image results, making it difficult to filter out motion-contaminated images. Through imaging system updates, PA/US maximum amplitude projection (MAP) preview images considered as 3D images were provided online on the US machine to help select data. Offline 3D panoramic scanning was performed to provide super wide-field images of the human body (Figure 3c(ii)). An updated 3D PHOVIS, providing six-degree mosaic stitching, was used for image position correction. Twelve panoramic scans were performed along the perimeter of the human neck for system evaluation (Figure 3c(iii)). The total scan time was 601.2 s, and the corrected image FOV was 129 × 120 mm^2. Hemoglobin oxygen saturation (sO_2) levels of the carotid artery and jugular vein were quantified to be 97 ± 8.2% and 84 ± 12.8%, respectively.

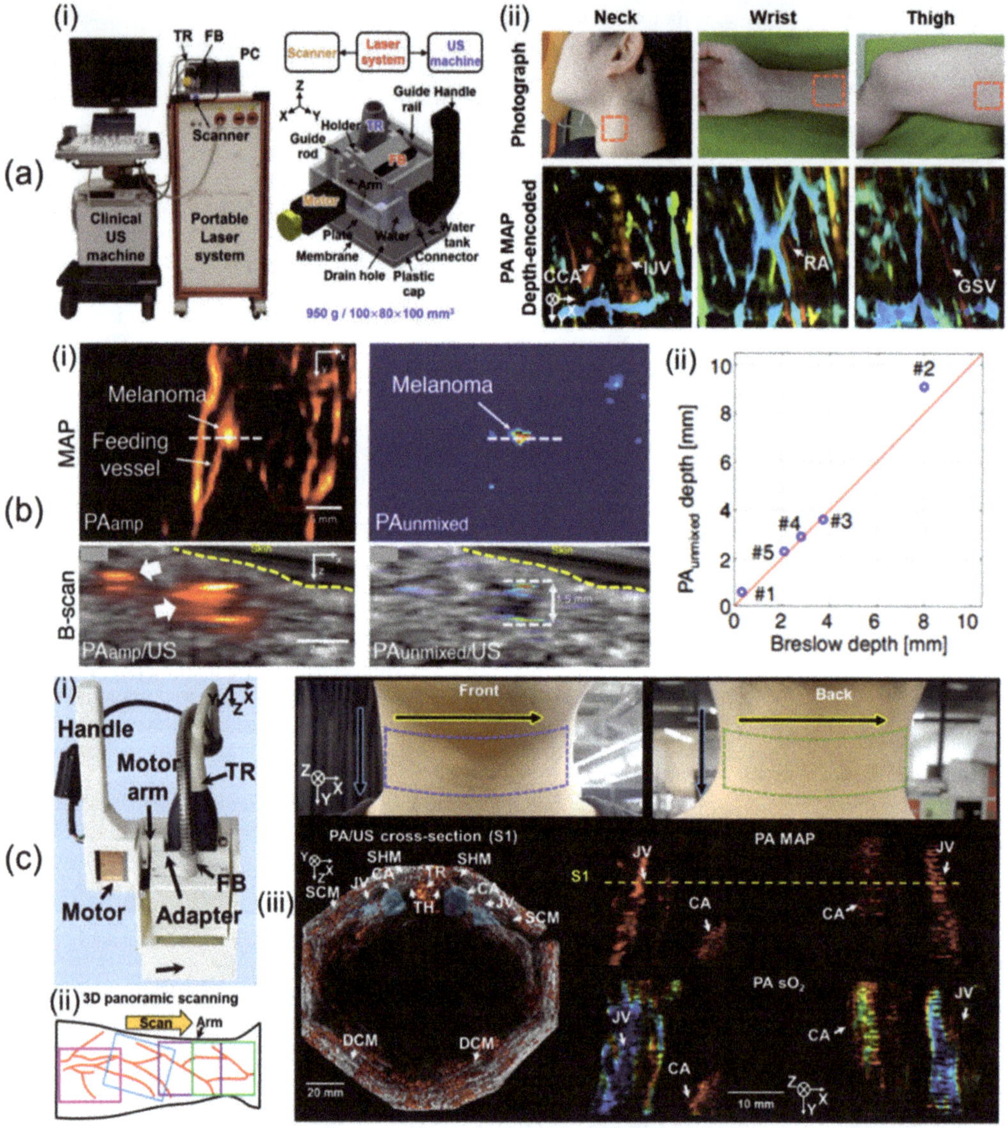

Figure 3. Three-dimensional clinical handheld PA/US imaging systems based on 1D linear array UST with mechanical scanning and their applications. (**a**) (**i**) Photograph of a mechanical-scanning-based 3D handheld PA/US imaging system using a 1D linear array UST. (**ii**) In vivo human imaging of various human bodies. (**b**) (**i**) US, PA amplitude, and unmixed PA melanin images of metastasized melanoma. (**ii**) Graph of unmixed PA melanin depth vs. histopathological depth. (**c**) (**i**) Updated 3D clinical handheld PA/US imager. (**ii**) Schematic of 3D panoramic scanning. (**iii**) Three-dimensional panoramic imaging of the human neck. PA, photoacoustic; UST, ultrasound transducer. Reprinted with permission from references [10,58,59].

Bost et al. [62] have presented a single-element UST with a two-axis linear motorized stage-based 3D handheld PA/US imaging system. In their study, a solid laser, a custom probe, and a digitizer were used. The solid laser produced 532 and 1024 nm pulses with a PRF of 1 kHz and a pulse length of 1.5 ns. The maximum pulse energies at 523 and 1024 nm were 37 and 77 μJ, respectively. The probe integrated a 35 MHz single-element UST with 100% bandwidth, a custom-designed fiber bundle, a two-axis linear stage, and a waterproof polyvinylchloride plastisol membrane for acoustic coupling. The number and geometry of fibers were effectively designed with Monte Carlo simulation. Data were acquired at a sampling rate of 200 MSamples/s. For volumetric US and PA imaging, scan times were 1.5 min and 4 min, respectively, for an area of 9.6 × 9.6 mm^2. US and PA lateral resolutions were 86 μm and 93 μm, respectively. A visualized B-scan superimposed US/PA images of a human scapula where the PA vascular network was identified. Furthermore, they acquired a volume-rendered image of a mouse with a subcutaneous tumor (Table 3).

Table 3. Specifications of mechanical-scanning-based 3D handheld PA/US imaging systems.

Author		Lee et al. [58]	Park et al. [59]	Lee et al. [10]	Bost et al. [62]
Laser	Wavelength	690–950 nm	690–950 nm	690–950 nm	532/1024 nm
	Pulse duration	5 ns	5 ns	5 ns	1.5 ns
	PRF	10 Hz	10 Hz	10 Hz	1 kHz
	Light delivery	Fiber	Fiber	Fiber	Fiber
Probe	Scanning aids	1 motor	1 motor	1 motor	2 motor stages
	Scanning type	Mechanical	Mechanical	Mechanical	Mechanical
	UST type	1D linear	1D linear	1D linear	Single element
	# of elements	128	128	128	1
	Center frequency	8.5 MHz	8.5 MHz	8.5 MHz	35 MHz
	Bandwidth	62%	62%	62%	100%
	Resolution	A, L: 191, 799 μm	A, L: 200, 1000 μm	A, L, E: 195, 592, 1976 μm	L: 93 μm
	Scan time	20 s	11.4 s	16.7 s	4 min
	FOV	40 × 38 mm^2	31 × 38 mm^2	25 × 38 mm^2	9.6 × 9.6 mm^2
	Weight	950 g	950 g	600 g	-
	Dimension	100 × 80 × 100 mm^3	100 × 80 × 100 mm^3	70 × 62 × 110 mm3	-
Data acquisition/ processing system	Platform	EC-12R	EC-12R	EC-12R	AMI US/OA platform
	Sampling rate	40 MHz	40 MHz	40 MHz	200 MHz

A, L, and E: axial, lateral, and elevational resolutions, respectively.

Three-dimensional handheld imaging systems using mechanical scanning methods can be implemented with relatively simple devices while still providing acceptable image quality. However, these systems are susceptible to motor noises and vulnerable to motion contamination. Furthermore, the inclusion of motorized parts can make the entire scanner bulky and heavy, which can make handheld operation challenging.

2.3. Three-Dimensional Handheld PA Imaging Systems Using Mirror Scanning

Mechanical scanning is limited in speed due to physical movement of the scanner [81,82]. Additionally, the use of relatively bulky and heavy motor systems adversely affects handheld operation. To overcome these problems, optical scanning (e.g., MEMS- and galvanometer-based mirror scanners) was used to develop single-element UST-based 3D PA imaging systems.

Lin et al. [64] have demonstrated a volumetric PA imaging system implemented using a two-axis MEMS and a single-element UST. For that study, they utilized a fiber laser, a custom handheld probe, and a DAQ. The laser pulse length at 532 nm was 5 nm, and the PRF was 88 kHz. The probe consisted of a single-mode fiber (SMF) for delivering the light, optical lenses for focusing laser beams into an opto-acoustic combiner, prisms for ensuring opto-acoustic coaxial alignment, the MEMS for reflecting focused laser beams and generated PA signals, a 50 MHz center frequency UST for detecting reflected PA signals, an optical correction lens for calibrating prism-induced aberration, and a motorized stage for correcting the location of the MEMS mirror (Figure 4a(i)). Laser pulses synchronized the MEMS scanner and the DAQ system. PA signal data were collected by the DAQ with a sampling frequency of 250 Msamples/s. PA volume imaging was implemented at a repetition rate of 2 Hz. The overall dimension and imaging FOV of the probe were $80 \times 115 \times 150$ mm^3 and $2.5 \times 2.0 \times 0.5$ mm^3, respectively. Resolution measurements of the system were performed in lateral and axial directions, resulting in 5 μm and 26 μm, respectively. To test the developed 3D handheld imager, a mouse ear (Figure 4a(ii)) and a human cuticle (Figure 4a(iii)) were imaged.

Park et al. [32] have developed a two-axis MEMS and single-element UST-based miniaturized system without using a mechanical stage. A diode laser, an SMF, a custom-designed 3D probe, and DAQ were used in their study. The 532 nm diode laser provided a 50 kHz PRF. It was coupled with optical fiber. The probe integrated light delivery assembly (LDA), objective lens (OL), opto-ultrasound combiner (OUC), the MEMS, an acoustic lens (AL), and a UST (Figure 4b(i)). After the light was delivered from the LDA, it was focused by the OL and delivered to the OUC containing prisms and silicon fluid. The delivered light was reflected by the MEMS scanner. It then illuminated objects that could absorb light. Generated PA signals were reflected from the MEMS and at prisms in OUC. Reflected PA signals were collected at the UST with a central frequency of 50 MHz. The AL attached to the right side of the OUC was used to increase SNRs. The DAQ was used to create trigger signals for operating the laser and the digitizer. At the same time, the DAQ generated sinusoidal and sawtooth waves to control the two-axis MEMS scanner. PA volumetric imaging was implemented at a PRF of 0.05 Hz. The probe was 17 mm in diameter with a weight of 162 g. It provided a maximum FOV of 2.8×2 mm^2. Lateral and axial resolutions were 12 μm and 30 μm, respectively. To confirm the feasibility of the system, they conducted in vivo mouse imaging in various parts such as the iris and brain, as shown in Figure 4b(ii,iii).

Zhang et al. [31] presented a 2D GS and single-element UST-based 3D handheld system (Figure 4c(i)). Their study was performed using a pulsed laser, a custom-made probe, and a DAQ card. The laser offered a maximum PRF of 10 kHz at 532 nm. It was connected to an SMF (diameter: 4 μm) via a fiber coupler (FC). The light was delivered to a collimator mounted on the side of the probe. The collimated light was scanned with the GS and focused onto a transparent glass with an OL. Generated PA signals were reflected off the 45° tilted glass (G) and delivered to the UST with a center frequency of 15 MHz. The optical focus was adjusted using the probe's XY- and Z-direction stages. The DAQ card provided a sampling rate of 200 MSamples/s. The 3D imaging took 16 sec. The imaging FOV was 2×2 mm^2. Lateral and axial resolutions were quantified with the phantom's sharp blade and carbon rod. The results were 8.9 μm and 113 μm, respectively. Using this system, they conducted in vivo rooster wattle and human lip imaging experiments, as shown in Figure 4c(ii,iii).

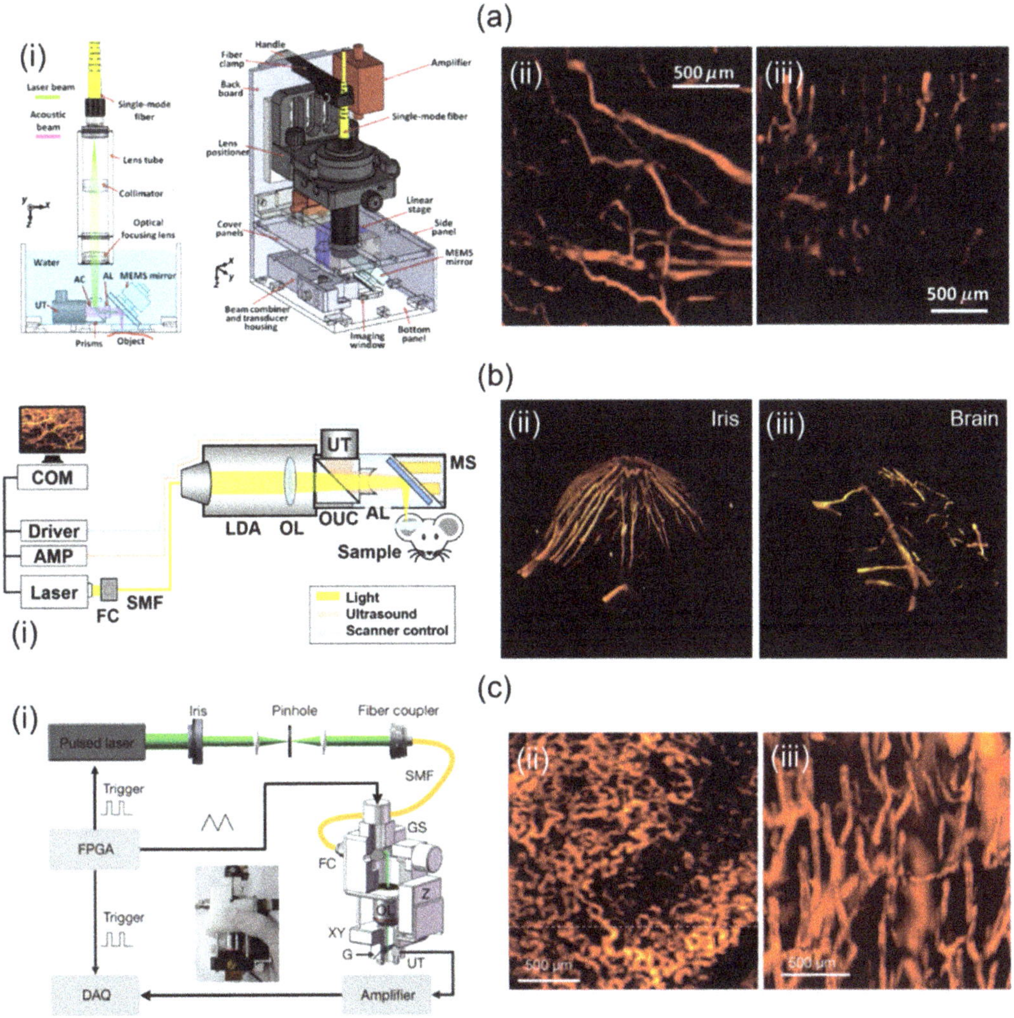

Figure 4. Mirror-scanning-based 3D handheld PA imaging systems and their applications. (**a**) (**i**) Schematic of a single-element UST and 2-axis MEMS-based 3D PA imaging system. In vivo PA vessel visualizations of (**ii**) a mouse ear and (**iii**) a human cuticle. (**b**) (**i**) Schematic diagram of a 2-axis MEMS and single-element UST-based miniaturized 3D handheld imaging system. In vivo PA volume rendered images of a mouse (**ii**) iris and (**iii**) brain. (**c**) Schematic diagram of a compact single-element UST-based 3D portable imaging system scanning with a galvanometer. In vivo PA MAP visualizations of (**ii**) the wattle of a Leghorn rooster and (**iii**) the lower lip of a human. MEMS, microelectromechanical systems; PA, photoacoustic; MAP, maximum amplitude projection. Reprinted with permission from references [31,32,64].

Qin et al. [65] have presented a dual-modality 3D handheld imaging system capable of PA and optical coherence tomography (OCT) imaging using lasers, a 2D MEMS, a miniaturized flat UST, and OCT units. For PA imaging, a pulsed laser transmitted light with a high repetition rate of 10 kHz and a pulse duration of 8 ns. For OCT imaging, a diode laser delivered light that provided a center wavelength of 839.8 nm with a FWHM of 51.8 nm. The light for PA and OCT imaging used the same optical path and was reflected by the MEMS mirror, respectively, to scan targets. The dual-modality 3D handheld scanner measured 65 × 30 × 18 mm^3, providing an effective FOV of 2 × 2 mm^2. Their lateral and axial PA resolutions were quantified to 3.7 µm and 120 µm, respectively, while lateral and axial OCT resolutions were quantified to 5.6 µm and 7.3 µm, respectively (Table 4).

Table 4. Specifications of mirror-scanning-based 3D handheld PA/US imaging systems.

Author		Lin et al. [64]	Park et al. [32]	Zhang et al. [31]	Qin et al. [65]
Laser	Wavelength	532 nm	532 nm	532 nm	532 nm
	Pulse duration	5 ns	-	-	8 ns
	PRF	88 kHz	50 kHz	10 kHz	10 kHz
	Light delivery	Fiber	Fiber	Fiber	Fiber
Probe	Scanning aids	2D MEMS	2D MEMS	2D galvo	2D MEMS
	Scanning type	Mirror	Mirror	Mirror	Mirror
	UST type	Single element	Single element	Single element	Single element
	# of elements	1	1	1	1
	Center frequency	50 MHz	50 MHz	15 MHz	10 MHz
	Bandwidth	-	-	-	60%
	Resolution	A, L: 26, 5 µm	A, L: 30, 12 µm	A, L: 113, 9 µm	A, L: 120, 3.7 µm
	Scan time	0.5 s	20 s	16 s	-
	FOV	2.5 × 2.0 × 0.5 mm^3	2.8 × 2 mm^2	2 × 2 mm^2	2 × 2 mm^2
	Weight	-	162 g	-	-
	Dimension	80 × 115 × 150 mm^3	12 cm	-	65 × 30 × 18 mm^3
Data acquisition/ processing system	Platform	ATS9350	NI PCIe-6321	A DAQ card	NI PCI-5122
	Sampling rate	250 MHz	-	200 MHz	100 MHz

A and L: axial and lateral resolutions, respectively.

Optical-scanning-based 3D PA probes provide high-quality PA images with fast acquisition and high handheld usability by eliminating motors and associated systems. However, theses probes have limitations such as small FOVs and shallow penetration depths, which are unfavorable for clinical PA imaging.

2.4. Three-Dimensional Handheld PA Imaging Systems Using Freehand Scanning

Three-dimensional handheld PA imaging systems utilizing freehand scanning methods have been developed using various types of USTs. This section aims to showcase a range of 3D freehand scanning systems that utilize different aids, including a GPS, an optical pattern, and an optical tracker. Additionally, we will explore freehand scanning imaging systems combined with other scanning-based imaging systems such as mirror-scanning-based and direct-scanning-based imaging systems.

Jiang et al. [69] have proposed a 3D freehand PA tomography (PAT) imaging system using a GPS sensor. For their research, they utilized an OPO laser and a laser providing high power, a linear UST, a DAQ card, and a 3D GPS (Figure 5a(i)). The PRF and wavelength ranges of the OPO laser were 10 Hz and 690–950 nm, respectively. The high-power laser delivered a 1064 nm wavelength at a 10 Hz PRF. The energy fluence was less than 20 mJ/cm^2. The linear UST provided a center frequency of 7.5 MHz with 73% bandwidth. The DAQ system sampled data on 128 channels at a rate of 40 Msamples/s. The GPS consisted of a system electronics unit (SEU), a standard sensor (SS), and an electromagnetic field source (EFS). The SS was attached to the US probe as shown in Figure 5a(i). It collected six degrees of freedom (DOF) data at a sampling rate of 120 Hz for each B-mode image. The 6-DOF data contained three translational (x, y, z) and three rotational (α, β, γ) data. Obtained 6-DOF data were transmitted to a PC via the SEU. Imaging positions were accurately tracked within a magnetic field generated by the EFS. GPS calculation errors for translational and rotational motions were 0.2% and 0.139%, respectively. They used the sensor within a translation resolution of 0.0254 mm and an angular resolution of 0.002° for 3D PAT. Free-scan data were acquired at scan speeds of 60–240 mm/min with scan intervals of 0.1–0.4 mm. After PA data collection, a back-projection beamformer was applied for 2D image reconstruction. Three-dimensional PAT was implemented as pixel nearest-neighbor-based fast-dot projection using 6-DOF coordinate data [83,84]. Measured lateral and elevational resolutions were 237.4 and 333.4 μm, respectively. The application of free-scan 3D PAT on a human wrist is shown in Figure 5a(ii).

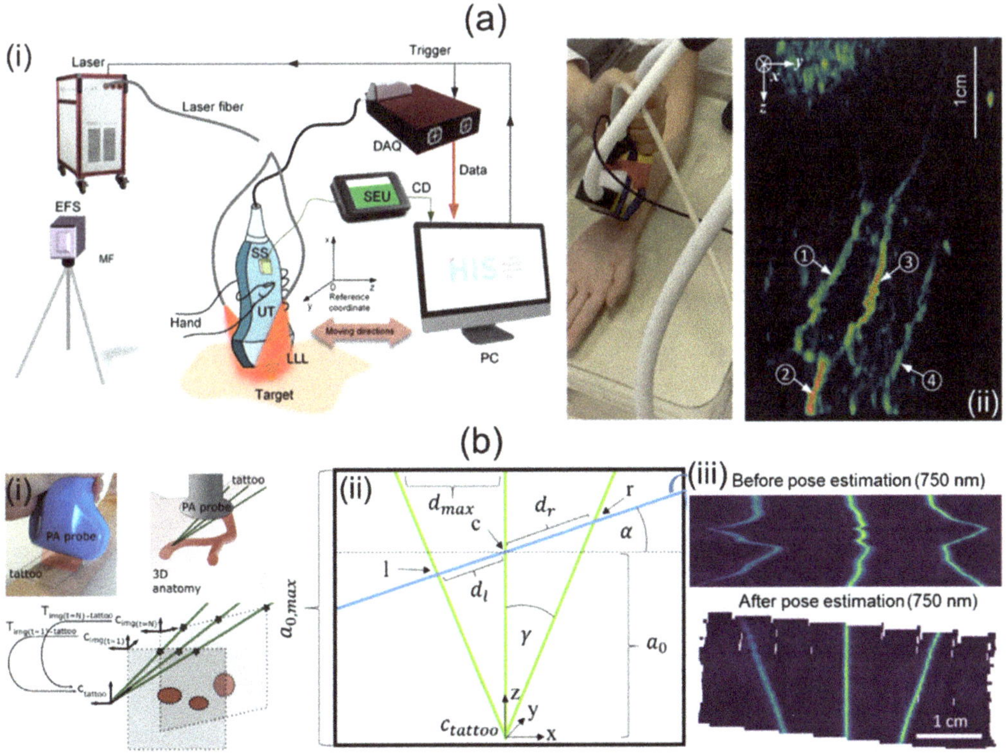

Figure 5. Freehand-scanning-based 3D handheld PA imaging systems and their applications. (**a**) (**i**) Schematic description of a free-scanning-based 3D handheld PA imaging system using a GPS sensor. (**ii**) In vivo 3D free-scan imaging of a human arm. (**b**) (**i**) Conceptual explanation of a 3D freehand scan imaging system using an optical tattoo. (**ii**) Schematic geometry for the tattoo. l, c, and r represent intersections of a free-scan image and the optical tattoo. (**iii**) Original and repositioned PA images acquired by freehand scanning. PA, photoacoustic; UT, ultrasound transducer; GPS, global positioning system. Reprinted with permission from references [69,70].

Holzwarth et al. [70] have presented a 3D handheld imaging system using freehand scanning with a 1D concave array UST. They used a tunable ND:YAG laser, a multispectral opto-acoustic tomography (MSOT) Acuity Echo research system, and a custom optical pattern. The laser provided a spectral range of 660–1300 nm, with pulse energy, durations, and an emission rate of 30 mJ, 4–10 ns, and 25 Hz, respectively. The concave UST was

80 mm in diameter. It provided a central frequency of 4 MHz with 256 individual elements. The optical pattern was engraved on a transparent foil and filled with cyan ink. The pattern was attached to regions of interest (ROIs). Free-scan data were acquired within these ROIs. As shown in Figure 5b(i), the trident optical pattern representing the three green lines gave three points corresponding to each probe position in each free scan. They obtained a transformation matrix using probe positions and geometry (Figure 5b(ii)). Based on 2D images and the transformation matrix, 3D volume compounding was implemented [85]. Three-dimensional free-scan images before and after position calibration are shown in Figure 5b(iii).

Fournelle et al. [68] have demonstrated a 3D freehand scan imaging system using an optical tracker and a commercial 1D linear array UST. This study was performed using a solid-state ND:YAG laser, a wavelength tunable OPO laser, the linear array UST, a custom US platform, an optical tracking system, and an FB. Frame rates for the ND:YAG and OPO lasers were 20 and 10 frames/sec, respectively. Light emitted from these laser systems was transmitted through the FB. Due to the opening angle (22°) of the FB, the transmitted light formed a rectangular shape measured 2×20 mm^2. The laser fluence was 10 mJ/cm^2 at 1064 nm. The center frequency and pitch size of the UST were 7.5 MHz and 300 μm, respectively. The US platform digitized 128-channel data at a sampling rate of 80 MSamples/s. The optical tracker provided the position and orientation of each B-mode image with a root-mean-square error (0.3 mm3). An adaptive delay-and-sum beamformer was applied for reconstructing B-mode US and PA images. The reconstruction was accelerated with multi-core processors and parallel graphics processors. Reconstructed images were placed in 3D space using position and orientation information obtained from the optical tracker and corrected with a calibration matrix. Resolution characterization of the system was achieved by imaging the tip of a steel needle. Measured FWHMs in lateral and elevational directions were 600 μm and 1100 μm, respectively. To confirm the availability of the 3D freehand approach for humans, PA/US imaging was performed for a human hand. The total number of image frames acquired for the volume was 150, resulting in a scan time of 15 s.

Chen et al. [67] have showcased a 3D freehand scanning probe based on a resonant-galvo scanner and a single-element UST. The resonant mirror in the probe generated a periodic magnetic field to rotate the reflector at a rate of 1228 Hz. The galvo scanner was attached to the resonant mirror for slow-axis scanning. Taking advantage of the system's high scanning speed (C-scan rate, 5–10 Hz), they applied simultaneous localization and mapping called SLAM. The SLAM performed feature point extraction between consecutive scan images using the speeded-up robust features (SURF) method and scale-invariant feature transform (SIFT) method [86–89]. Points were then used to calibrate positions of consecutive images. They demonstrated an extended FOV image of the mouse brain with the developed system and processing method. The expanded FOV was 8.3–13 times larger than the original FOV of ~1.7×5 mm^2.

Knauer et al. [50] have presented a free-scan method using a hemispherical 3D hand-held scanner to extend a limited FOV using acquired volumetric images. Acquired volume images with the hemispherical probe were spatially compounded. Fourier-based orientation and position correction were then applied. For translation (t_x, t_y, t_z) and rotation (θ_x, θ_y, θ_z) corrections, a phase correlation [90] and PROPELLER [91] were used, respectively. Volume areas that overlapped with each other by less than approximately 85% were used to apply the position correction algorithm. To validate their algorithm, they imaged a human palm. The extended volume FOV was $50 \times 70 \times 15$ mm^3, which was larger than the single volume image FOV (Table 5).

Table 5. Specifications of freehand-scanning-based 3D handheld PA/US imaging systems.

Author		Jiang et al. [69]	Holzwarth et al. [70]	Fournelle et al. [68]	Chen et al. [67]	Knauer et al. [50]
Laser	Wavelength	690 nm	660–1300 nm	532/1024 nm	532/588 nm	680–950 nm
	Pulse duration	-	4–10 ns	3–10 ns	-	-
	PRF	10 Hz	25 Hz	10–20 Hz	500 kHz	10 Hz
	Light delivery	Fiber	Fiber	Fiber	Fiber	Fiber
Probe	Scanning aids	GPS sensor	Optical pattern	Optical tracker	Resonant galvo	-
	Scanning type	Freehand	Freehand	Freehand	Freehand + mirror	Freehand + direct
	UST type	1D linear	1D concave	1D linear	Single element	2D hemispherical
	# Of elements	128	256	128	1	256
	Center frequency	7.5 MHz	4 MHz	7.5 MHz	-	4 MHz
	Bandwidth	73%	-	-	-	100%
	Resolution	L, E: 237, 333 μm	-	L, E: 600, 1100 μm	A, L: 39, 6 μm	S: 200 μm
	Scan time	240 mm/min	-	15 s	0.1–0.2 s	-
	FOV	45 × 38 × 38 mm^3	-	-	1.7 × 5 mm^2	50 × 70 × 15 mm^3
	Weight	-	-	-	158 g	-
	Dimension	-	-	-	59 × 30 × 44 mm^3	-
Data acquisition/ processing system	Platform	A DAQ card	MSOT Acuity Echo	DiPhAS	-	Custom DAQ
	Sampling rate	40 MHz	-	80 MHz	-	40 MHz

S, A, L, and E: spatial, axial, lateral, and elevational resolutions, respectively.

Freehand-scanning-based 3D PA imaging systems are implemented with many types of UST. They also offer flexibility to combine with mirror- or direct-scanning-based 3D PA imaging systems. Combined scanning methods can overcome limitations of small FOV systems. However, they can be severely affected by various motion artifacts, making spectroscopic PA analysis difficult.

3. Discussion and Conclusions

In Section 2 of this paper, we classified and provided detailed explanations of four scanning techniques (i.e., direct, mechanical, mirror-based, and freehand scanning) for 3D handheld PA imaging. Subsequently, in Sections 2.1–2.4 we introduced recent research on 3D handheld PA imaging corresponding to these four scanning techniques. As mentioned earlier, 3D handheld PA imaging holds great potential for widespread applications, including both preclinical and clinical applications. Despite its active development and utilization, three major limitations need to be overcome: motion artifacts, anisotropic spatial resolution, and limited view artifacts. In the following sections, we will delve into these three limitations in greater detail.

Motion artifacts can occur in all types of scanning-based 3D imaging systems. Artifacts can degrade structural information. They can also deteriorate functional information such as sO_2 due to pixel-by-pixel spectral unmixing calculations [60]. For clinical translation, 3D imaging systems should mitigate motion contaminations to ensure accurate imaging and quantifications. Mozaffarzadeh et al. [92] have showcased motion-corrected freehand scanned PA images using the modality independent neighborhood descriptor (MIND), which is based on self-similarity [93]. By applying the MIND algorithm, they corrected a motion-contaminated phantom image as shown in Figure 6a(i). After correction, the structural similarity index (SSIM) was greatly enhanced (Figure 6a(ii)). Yoon et al. [60] have proposed a 3D motion-correction method using a motor-scanning-based 3D imaging system. They applied US motion compensations by maximizing structural similarities of subsequently acquired US skin profiles. PA motion correction was implemented using the corrected US information. The effectiveness of motion compensation was verified through in vivo human wrist imaging (Figure 6b(i)). Structural similarities were measured by quantifying cross-correlations of each method. Significant improvements were confirmed after motion corrections. Further, inaccuracy of the spectral unmixing was greatly reduced after calibrations (Figure 6b(ii)).

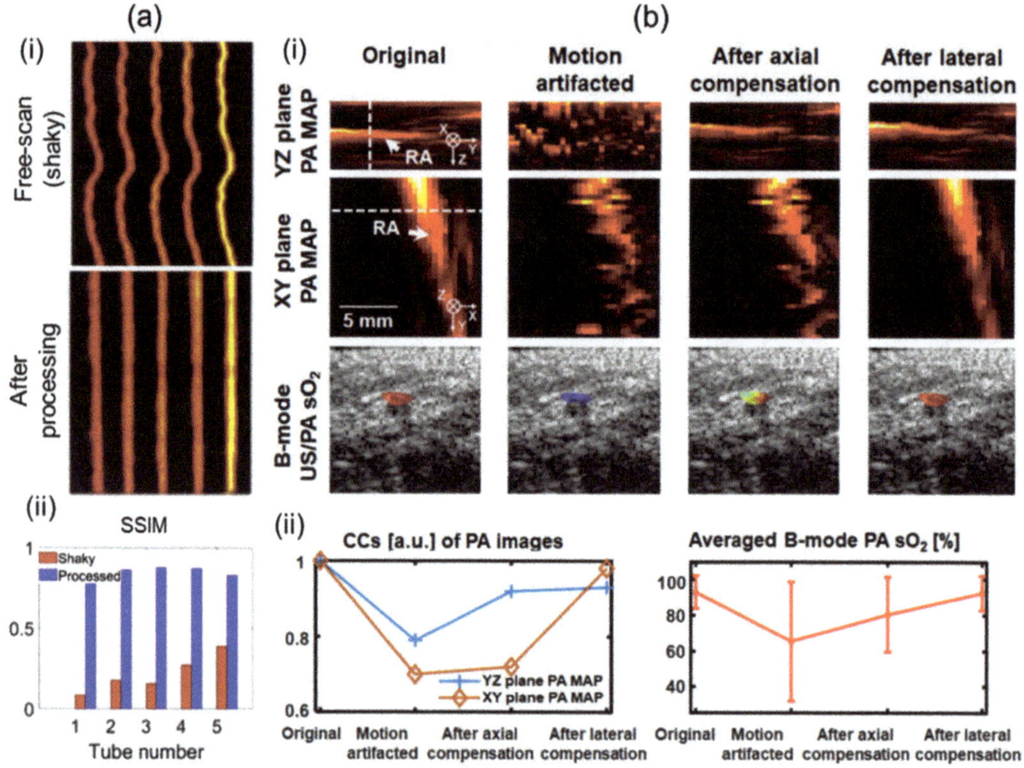

Figure 6. Motion compensations applied to 3D handheld imaging systems. (**a**) (**i**) Before and after motion-corrected PA MAP images. Motion contamination occurred in the freehand scan. (**ii**) Comparison of SSIM before and after motion correction. (**b**) (**i**) PA MAP and B-mode US/PA sO_2 images before and after motion compensation. (**ii**) Motion compensation validation charts. Degrees of recovery of structural and functional information were measured. PA, photoacoustic; MAP, maximum amplitude projection; US, ultrasound; SSIM, the structural similarity index; sO_2, hemoglobin oxygen saturation. Reprinted with permission from references [60,92].

It is known that 1D linear array UST-based 3D imaging systems suffer from low elevation resolution due to their narrow elevation beamwidth and limited view artifacts caused by a narrow aperture size. To overcome poor elevational resolution, Wang et al. [94] have proposed a system using stainless steel at the focal point of a 1D linear UST as shown in Figure 7a(i). The system diffracted PA waves to create wide signal receptions in the elevational direction [95]. They effectively applied the beamformer in the elevation direction with sufficiently overlapping signal areas. Results of depth-encoded tube phantom images before and after slit application are shown in Figure 7a(ii,iii). The elevation resolution of the system with slit was 640 μm, which was superior to the 1500 μm elevation resolution of the system without the slit.

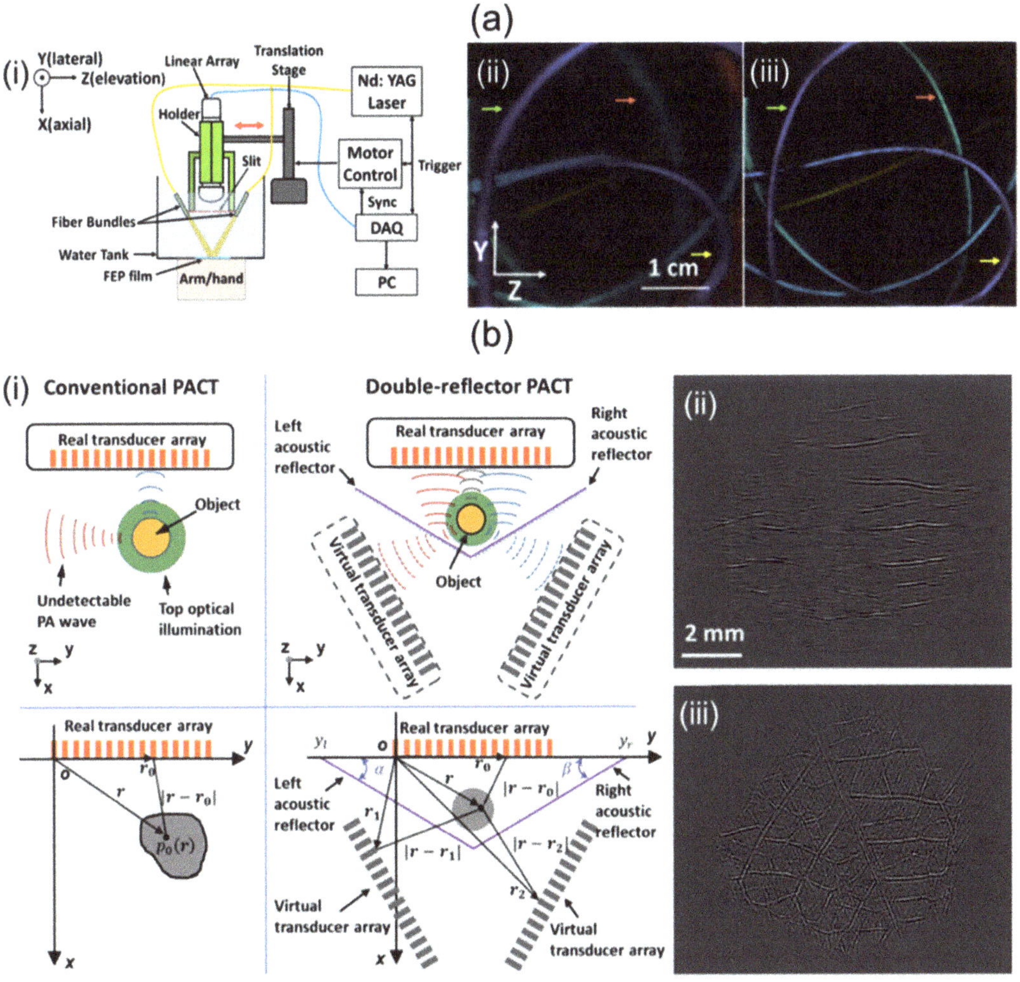

Figure 7. Potential ways to improve performance of 1D linear array UST-based 3D portable imaging systems. (**a**) (**i**) Schematic explanation of a slit-enabled PACT system. (**ii**) Conventional and (**iii**) slit-applied tube phantom depth-encoded PA images. (**b**) (**i**) Conceptual schematic of conventional and double-reflector PACT. PA leaf skeleton images of (**ii**) conventional and (**iii**) double-reflector application. The double-reflector PACT restored the vascular network of the leaf skeleton. PACT, photoacoustic computed tomography; UST, ultrasound transducer. Reprinted with permission from references [94,96].

To mitigate limited-view artifacts, solid angle coverage of a UST should be at least $>\pi$. Li et al. [96] have demonstrated a 1D linear array UST-based imaging system that applies two planar acoustic reflectors to virtually increase the detection view angle. The two reflectors were located at an angle of 120° relative to each other (Figure 7b(i)). The system using an increased detection view angle of 240° successfully recovered the vascular network image of the leaf skeleton (Figure 7b(iii)) compared to the original one (Figure 7b(ii)).

Recent advancements in silicon-photonics technology have led to the development of on-chip optical US sensors [97,98]. These sensors offer several advantages over traditional piezoelectric US sensors. Unlike piezoelectric sensors, which often sacrifice sensitivity when miniaturized, optical US sensors exhibit high sensitivity and broadband detection capabilities while maintaining a compact size. Furthermore, there have been recent breakthroughs in achieving parallel interrogation of these on-chip optical US sensors [99]. These innovations open up exciting possibilities for the creation of ultracompact 3D handheld imaging systems. In addition, multispectral imaging, real-time capabilities, hybrid modalities [100], and AI integration [101] hold great promise for improving future 3D handheld PA imaging systems. These systems can unlock new opportunities for patient care, early disease detection and image-guided interventions [102], ultimately enhancing our understanding and management of various medical conditions.

Author Contributions: Conceptualization, C.L. and B.P.; investigation, C.L.; writing—original draft preparation, C.L.; writing—review and editing, C.L., B.P. and C.K.; supervision, B.P. and C.K.; project administration, C.K.; funding acquisition, C.K. All authors have read and agreed to the published version of the manuscript.

Funding: This research was supported by the Basic Science Research Program through the National Research Foundation of Korea (NRF) funded by the Ministry of Education (MOE) (2020R1A6A1A03047902). It was also partly supported by the national R&D Program through the NRF funded by the Ministry of Science and ICT (MSIT) (2021M3C1C3097624, RS-2023-00210682). It was also supported by an NRF grant funded by MSIT (NRF-2023R1A2C3004880). It was in part supported by grants (1711137875, RS-2020-KD000008) from the Korea Medical Device Development Fund funded by the Ministry of Trade, Industry and Energy (MOTIE). It was also in part supported by the BK21 FOUR project.

Conflicts of Interest: Chulhong Kim has financial interest in OPTICHO. However, OPTICHO did not support this work.

References

1. Chen, Q.; Qin, W.; Qi, W.; Xi, L. Progress of clinical translation of handheld and semi-handheld photoacoustic imaging. *Photoacoustics* **2021**, *22*, 100264. [CrossRef]
2. Choi, W.; Park, B.; Choi, S.; Oh, D.; Kim, J.; Kim, C. Recent Advances in Contrast-Enhanced Photoacoustic Imaging: Overcoming the Physical and Practical Challenges. *Chem. Rev.* **2023**, *123*, 7379–7419. [CrossRef]
3. Cao, R.; Zhao, J.; Li, L.; Du, L.; Zhang, Y.; Luo, Y.; Jiang, L.; Davis, S.; Zhou, Q.; de la Zerda, A.; et al. Optical-resolution photoacoustic microscopy with a needle-shaped beam. *Nat. Photonics* **2022**, *17*, 89–95. [CrossRef]
4. Liu, C.; Wang, L. Functional photoacoustic microscopy of hemodynamics: A review. *Biomed. Eng. Lett.* **2022**, *12*, 97–124. [CrossRef]
5. Zhang, Z.; Mu, G.; Wang, E.; Cui, D.; Yang, F.; Wang, Z.; Yang, S.; Shi, Y. Photoacoustic imaging of tumor vascular involvement and surgical margin pathology for feedback-guided intraoperative tumor resection. *Appl. Phys. Lett.* **2022**, *121*, 193702. [CrossRef]
6. Bell, A.G. Upon the production and reproduction of sound by light. *J. Soc. Telegr. Eng.* **1880**, *9*, 404–426. [CrossRef]
7. Yao, J.; Kaberniuk, A.A.; Li, L.; Shcherbakova, D.M.; Zhang, R.; Wang, L.; Li, G.; Verkhusha, V.V.; Wang, L.V. Multiscale photoacoustic tomography using reversibly switchable bacterial phytochrome as a near-infrared photochromic probe. *Nat. Methods* **2015**, *13*, 67–73. [CrossRef]
8. Wang, L.V.; Yao, J. A practical guide to photoacoustic tomography in the life sciences. *Nat. Methods* **2016**, *13*, 627–638. [CrossRef]
9. Choi, W.; Park, E.-Y.; Jeon, S.; Yang, Y.; Park, B.; Ahn, J.; Cho, S.; Lee, C.; Seo, D.-K.; Cho, J.-H.; et al. Three-dimensional multistructural quantitative photoacoustic and US imaging of human feet in vivo. *Radiology* **2022**, *303*, 467–473. [CrossRef]
10. Lee, C.; Cho, S.; Lee, D.; Lee, J.; Park, J.-I.; Kim, H.-J.; Park, S.H.; Choi, W.; Kim, U.; Kim, C. Panoramic Volumetric Clinical Handheld Photoacoustic and Ultrasound Imaging. *Photoacoustics* **2023**, *31*, 100512. [CrossRef]
11. Neuschler, E.I.; Butler, R.; Young, C.A.; Barke, L.D.; Bertrand, M.L.; Böhm-Vélez, M.; Destounis, S.; Donlan, P.; Grobmyer, S.R.; Katzen, J.; et al. A pivotal study of optoacoustic imaging to diagnose benign and malignant breast masses: A new evaluation tool for radiologists. *Radiology* **2018**, *287*, 398–412. [CrossRef]
12. Yang, X.; Chen, Y.-H.; Xia, F.; Sawan, M. Photoacoustic imaging for monitoring of stroke diseases: A review. *Photoacoustics* **2021**, *23*, 100287. [CrossRef]
13. Mantri, Y.; Dorobek, T.R.; Tsujimoto, J.; Penny, W.F.; Garimella, P.S.; Jokerst, J.V. Monitoring peripheral hemodynamic response to changes in blood pressure via photoacoustic imaging. *Photoacoustics* **2022**, *26*, 100345. [CrossRef]
14. Menozzi, L.; Del Águila, Á.; Vu, T.; Ma, C.; Yang, W.; Yao, J. Three-dimensional non-invasive brain imaging of ischemic stroke by integrated photoacoustic, ultrasound and angiographic tomography (PAUSAT). *Photoacoustics* **2023**, *29*, 100444. [CrossRef]

15. Li, H.; Zhu, Y.; Luo, N.; Tian, C. In vivo monitoring of hemodynamic changes in ischemic stroke using photoacoustic tomography. *J. Biophotonics* **2023**, e202300235. [CrossRef]
16. Li, J.; Chen, Y.; Ye, W.; Zhang, M.; Zhu, J.; Zhi, W.; Cheng, Q. Molecular breast cancer subtype identification using photoacoustic spectral analysis and machine learning at the biomacromolecular level. *Photoacoustics* **2023**, *30*, 100483. [CrossRef]
17. Lin, L.; Wang, L.V. The emerging role of photoacoustic imaging in clinical oncology. *Nat. Rev. Clin. Oncol.* **2022**, *19*, 365–384. [CrossRef]
18. Xing, B.; He, Z.; Zhou, F.; Zhao, Y.; Shan, T. Automatic force-controlled 3D photoacoustic system for human peripheral vascular imaging. *Biomed. Opt. Express* **2023**, *14*, 987–1002. [CrossRef]
19. Zhang, M.; Wen, L.; Zhou, C.; Pan, J.; Wu, S.; Wang, P.; Zhang, H.; Chen, P.; Chen, Q.; Wang, X.; et al. Identification of different types of tumors based on photoacoustic spectral analysis: Preclinical feasibility studies on skin tumors. *J. Biomed. Opt.* **2023**, *28*, 065004. [CrossRef]
20. Kim, J.; Park, B.; Ha, J.; Steinberg, I.; Hooper, S.M.; Jeong, C.; Park, E.-Y.; Choi, W.; Liang, T.; Bae, J.S.; et al. Multiparametric Photoacoustic Analysis of Human Thyroid Cancers in vivo photoacoustic Analysis of Human Thyroid Cancers. *Cancer Res.* **2021**, *81*, 4849–4860. [CrossRef]
21. Park, B.; Kim, C.; Kim, J. Recent Advances in Ultrasound and Photoacoustic Analysis for Thyroid Cancer Diagnosis. *Adv. Phys. Res.* **2023**, *2*, 2200070. [CrossRef]
22. Kim, J.; Kim, Y.; Park, B.; Seo, H.-M.; Bang, C.; Park, G.; Park, Y.; Rhie, J.; Lee, J.; Kim, C. Multispectral *ex vivo* photoacoustic imaging of cutaneous melanoma for better selection of the excision margin. *Br. J. Dermatol.* **2018**, *179*, 780–782. [CrossRef] [PubMed]
23. Yang, M.; Zhao, L.; He, X.; Su, N.; Zhao, C.; Tang, H.; Hong, T.; Li, W.; Yang, F.; Lin, L.; et al. Photoacoustic/ultrasound dual imaging of human thyroid cancers: An initial clinical study. *Biomed. Opt. Express* **2017**, *8*, 3449–3457. [CrossRef] [PubMed]
24. Garcia-Uribe, A.; Erpelding, T.N.; Krumholz, A.; Ke, H.; Maslov, K.; Appleton, C.; Margenthaler, J.A.; Wang, L.V. Dual-modality photoacoustic and ultrasound imaging system for noninvasive sentinel lymph node detection in patients with breast cancer. *Sci. Rep.* **2015**, *5*, 15748. [CrossRef] [PubMed]
25. Berg, P.J.v.D.; Daoudi, K.; Moens, H.J.B.; Steenbergen, W. Feasibility of photoacoustic/ultrasound imaging of synovitis in finger joints using a point-of-care system. *Photoacoustics* **2017**, *8*, 8–14. [CrossRef]
26. Dima, A.; Ntziachristos, V. Non-invasive carotid imaging using optoacoustic tomography. *Opt. Express* **2012**, *20*, 25044–25057. [CrossRef]
27. Dima, A.; Ntziachristos, V. In-vivo handheld optoacoustic tomography of the human thyroid. *Photoacoustics* **2016**, *4*, 65–69. [CrossRef]
28. Fenster, A.; Downey, D.B. 3-D ultrasound imaging: A review. *IEEE Eng. Med. Biol. Mag.* **1996**, *15*, 41–51. [CrossRef]
29. Yang, J.; Choi, S.; Kim, C. Practical review on photoacoustic computed tomography using curved ultrasound array transducer. *Biomed. Eng. Lett.* **2021**, *12*, 19–35. [CrossRef]
30. Deán-Ben, X.L.; Razansky, D. Portable spherical array probe for volumetric real-time optoacoustic imaging at centimeter-scale depths. *Opt. Express* **2013**, *21*, 28062–28071. [CrossRef]
31. Zhang, W.; Ma, H.; Cheng, Z.; Wang, Z.; Zhang, L.; Yang, S. Miniaturized photoacoustic probe for in vivo imaging of subcutaneous microvessels within human skin. *Quant. Imaging Med. Surg.* **2019**, *9*, 807. [CrossRef]
32. Park, K.; Kim, J.Y.; Lee, C.; Jeon, S.; Lim, G.; Kim, C. Handheld Photoacoustic Microscopy Probe. *Sci. Rep.* **2017**, *7*, 13359. [CrossRef]
33. Moothanchery, M.; Dev, K.; Balasundaram, G.; Bi, R.; Olivo, M. Acoustic resolution photoacoustic microscopy based on microelectromechanical systems scanner. *J. Biophotonics* **2019**, *13*, e201960127. [CrossRef] [PubMed]
34. Toi, M.; Asao, Y.; Matsumoto, Y.; Sekiguchi, H.; Yoshikawa, A.; Takada, M.; Kataoka, M.; Endo, T.; Kawaguchi-Sakita, N.; Kawashima, M.; et al. Visualization of tumor-related blood vessels in human breast by photoacoustic imaging system with a hemispherical detector array. *Sci. Rep.* **2017**, *7*, 41970. [CrossRef]
35. Kim, W.; Choi, W.; Ahn, J.; Lee, C.; Kim, C. Wide-field three-dimensional photoacoustic/ultrasound scanner using a two-dimensional matrix transducer array. *Opt. Lett.* **2023**, *48*, 343–346. [CrossRef]
36. Neuschmelting, V.; Burton, N.C.; Lockau, H.; Urich, A.; Harmsen, S.; Ntziachristos, V.; Kircher, M.F. Performance of a Multispectral Optoacoustic Tomography (MSOT) System equipped with 2D vs. 3D Handheld Probes for Potential Clinical Translation. *Photoacoustics* **2015**, *4*, 1–10. [CrossRef] [PubMed]
37. Dean-Ben, X.L.; Ozbek, A.; Razansky, D. Volumetric real-time tracking of peripheral human vasculature with GPU-accelerated three-dimensional optoacoustic tomography. *IEEE Trans. Med. Imaging* **2013**, *32*, 2050–2055. [CrossRef] [PubMed]
38. Deán-Ben, X.L.; Bay, E.; Razansky, D. Functional optoacoustic imaging of moving objects using microsecond-delay acquisition of multispectral three-dimensional tomographic data. *Sci. Rep.* **2014**, *4*, srep05878. [CrossRef] [PubMed]
39. Deán-Ben, X.L.; Fehm, T.F.; Gostic, M.; Razansky, D. Volumetric hand-held optoacoustic angiography as a tool for real-time screening of dense breast. *J. Biophotonics* **2015**, *9*, 253–259. [CrossRef] [PubMed]
40. Ford, S.J.; Bigliardi, P.L.; Sardella, T.C.; Urich, A.; Burton, N.C.; Kacprowicz, M.; Bigliardi, M.; Olivo, M.; Razansky, D. Structural and functional analysis of intact hair follicles and pilosebaceous units by volumetric multispectral optoacoustic tomography. *J. Investig. Dermatol.* **2016**, *136*, 753–761. [CrossRef] [PubMed]

41. Attia, A.B.E.; Chuah, S.Y.; Razansky, D.; Ho, C.J.H.; Malempati, P.; Dinish, U.; Bi, R.; Fu, C.Y.; Ford, S.J.; Lee, J.S.-S.; et al. Noninvasive real-time characterization of non-melanoma skin cancers with handheld optoacoustic probes. *Photoacoustics* **2017**, *7*, 20–26. [CrossRef]
42. Deán-Ben, X.L.; Razansky, D. Functional optoacoustic human angiography with handheld video rate three dimensional scanner. *Photoacoustics* **2013**, *1*, 68–73. [CrossRef] [PubMed]
43. Ivankovic, I.; Merčep, E.; Schmedt, C.-G.; Deán-Ben, X.L.; Razansky, D. Real-time volumetric assessment of the human carotid artery: Handheld multispectral optoacoustic tomography. *Radiology* **2019**, *291*, 45–50. [CrossRef] [PubMed]
44. Deán-Ben, X.; Fehm, T.F.; Razansky, D. Universal Hand-held Three-dimensional Optoacoustic Imaging Probe for Deep Tissue Human Angiography and Functional Preclinical Studies in Real Time. *J. Vis. Exp.* **2014**, *93*, e51864. [CrossRef]
45. Fehm, T.F.; Deán-Ben, X.L.; Razansky, D. Four dimensional hybrid ultrasound and optoacoustic imaging via passive element optical excitation in a hand-held probe. *Appl. Phys. Lett.* **2014**, *105*, 173505. [CrossRef]
46. Ozsoy, C.; Cossettini, A.; Ozbek, A.; Vostrikov, S.; Hager, P.; Dean-Ben, X.L.; Benini, L.; Razansky, D. LightSpeed: A Compact, High-Speed Optical-Link-Based 3D Optoacoustic Imager. *IEEE Trans. Med. Imaging* **2021**, *40*, 2023–2029. [CrossRef] [PubMed]
47. Deán-Ben, X.L.; Özbek, A.; Razansky, D. Accounting for speed of sound variations in volumetric hand-held optoacoustic imaging. *Front. Optoelectron.* **2017**, *10*, 280–286. [CrossRef]
48. Ron, A.; Deán-Ben, X.L.; Reber, J.; Ntziachristos, V.; Razansky, D. Characterization of Brown Adipose Tissue in a Diabetic Mouse Model with Spiral Volumetric Optoacoustic Tomography. *Mol. Imaging Biol.* **2018**, *21*, 620–625. [CrossRef]
49. Ozsoy, C.; Cossettini, A.; Hager, P.; Vostrikov, S.; Dean-Ben, X.L.; Benini, L.; Razansky, D. Towards a compact, high-speed optical linkbased 3D optoacoustic imager. In Proceedings of the 2020 IEEE SENSORS, Rotterdam, The Netherlands, 25–28 October 2020; pp. 1–4. [CrossRef]
50. Knauer, N.; Dean-Ben, X.L.; Razansky, D. Spatial Compounding of Volumetric Data Enables Freehand Optoacoustic Angiography of Large-Scale Vascular Networks. *IEEE Trans. Med. Imaging* **2019**, *39*, 1160–1169. [CrossRef] [PubMed]
51. Chen, Z.; Deán-Ben, X.L.; Gottschalk, S.; Razansky, D. Hybrid system for in vivo epifluorescence and 4D optoacoustic imaging. *Opt. Lett.* **2017**, *42*, 4577–4580. [CrossRef]
52. Chuah, S.Y.; Attia, A.B.E.; Long, V.; Ho, C.J.H.; Malempati, P.; Fu, C.Y.; Ford, S.J.; Lee, J.S.S.; Tan, W.P.; Razansky, D.; et al. Structural and functional 3D mapping of skin tumours with non-invasive multispectral optoacoustic tomography. *Ski. Res. Technol.* **2016**, *23*, 221–226. [CrossRef] [PubMed]
53. Fehm, T.F.; Deán-Ben, X.L.; Schaur, P.; Sroka, R.; Razansky, D. Volumetric optoacoustic imaging feedback during endovenous laser therapy—An ex vivo investigation. *J. Biophotonics* **2015**, *9*, 934–941. [CrossRef] [PubMed]
54. Liu, S.; Tang, K.; Jin, H.; Zhang, R.; Kim, T.T.H.; Zheng, Y. Continuous wave laser excitation based portable optoacoustic imaging system for melanoma detection. In Proceedings of the 2019 IEEE Biomedical Circuits and Systems Conference (BioCAS), Nara, Japan, 17–19 October 2019; pp. 1–4. [CrossRef]
55. Liu, S.; Tang, K.; Feng, X.; Jin, H.; Gao, F.; Zheng, Y. Toward Wearable Healthcare: A Miniaturized 3D Imager with Coherent Frequency-Domain Photoacoustics. *IEEE Trans. Biomed. Circuits Syst.* **2019**, *13*, 1417–1424. [CrossRef] [PubMed]
56. Liu, S.; Feng, X.; Jin, H.; Zhang, R.; Luo, Y.; Zheng, Z.; Gao, F.; Zhenga, Y. Handheld Photoacoustic Imager for Theranostics in 3D. *IEEE Trans. Med. Imaging* **2019**, *38*, 2037–2046. [CrossRef] [PubMed]
57. Liu, S.; Song, W.; Liao, X.; Kim, T.T.-H.; Zheng, Y. Development of a Handheld Volumetric Photoacoustic Imaging System with a Central-Holed 2D Matrix Aperture. *IEEE Trans. Biomed. Eng.* **2020**, *67*, 2482–2489. [CrossRef] [PubMed]
58. Lee, C.; Choi, W.; Kim, J.; Kim, C. Three-dimensional clinical handheld photoacoustic/ultrasound scanner. *Photoacoustics* **2020**, *18*, 100173. [CrossRef] [PubMed]
59. Park, B.; Bang, C.H.; Lee, C.; Han, J.H.; Choi, W.; Kim, J.; Park, G.S.; Rhie, J.W.; Lee, J.H.; Kim, C. 3D wide-field multispectral photoacoustic imaging of human melanomas in vivo: A pilot study. *J. Eur. Acad. Dermatol. Venereol.* **2020**, *35*, 669–676. [CrossRef]
60. Yoon, C.; Lee, C.; Shin, K.; Kim, C. Motion Compensation for 3D Multispectral Handheld Photoacoustic Imaging. *Biosensors* **2022**, *12*, 1092. [CrossRef] [PubMed]
61. Aguirre, J.; Schwarz, M.; Garzorz, N.; Omar, M.; Buehler, A.; Eyerich, K.; Ntziachristos, V. Precision assessment of label-free psoriasis biomarkers with ultra-broadband optoacoustic mesoscopy. *Nat. Biomed. Eng.* **2017**, *1*, 0068. [CrossRef]
62. Bost, W.; Lemor, R.; Fournelle, M. Optoacoustic Imaging of Subcutaneous Microvasculature with a Class one Laser. *IEEE Trans. Med. Imaging* **2014**, *33*, 1900–1904. [CrossRef]
63. Hajireza, P.; Shi, W.; Zemp, R.J. Real-time handheld optical-resolution photoacoustic microscopy. *Opt. Express* **2011**, *19*, 20097–20102. [CrossRef]
64. Lin, L.; Zhang, P.; Xu, S.; Shi, J.; Li, L.; Yao, J.; Wang, L.; Zou, J.; Wang, L.V. Handheld optical-resolution photoacoustic microscopy. *J. Biomed. Opt.* **2016**, *22*, 041002. [CrossRef]
65. Qin, W.; Chen, Q.; Xi, L. A handheld microscope integrating photoacoustic microscopy and optical coherence tomography. *Biomed. Opt. Express* **2018**, *9*, 2205–2213. [CrossRef] [PubMed]
66. Chen, Q.; Guo, H.; Jin, T.; Qi, W.; Xie, H.; Xi, L. Ultracompact high-resolution photoacoustic microscopy. *Opt. Lett.* **2018**, *43*, 1615–1618. [CrossRef] [PubMed]
67. Chen, J.; Zhang, Y.; Zhu, J.; Tang, X.; Wang, L. Freehand scanning photoacoustic microscopy with simultaneous localization and mapping. *Photoacoustics* **2022**, *28*, 100411. [CrossRef] [PubMed]

68. Fournelle, M.; Hewener, H.; Gunther, C.; Fonfara, H.; Welsch, H.-J.; Lemor, R. Free-hand 3d optoacoustic imaging of vasculature. In Proceedings of the 2009 IEEE International Ultrasonics Symposium, Rome, Italy, 20–23 September 2009; pp. 116–119. [CrossRef]
69. Jiang, D.; Chen, H.; Zheng, R.; Gao, F. Hand-held free-scan 3D photoacoustic tomography with global positioning system. *J. Appl. Phys.* **2022**, *132*, 074904. [CrossRef]
70. Holzwarth, N.; Schellenberg, M.; Gröhl, J.; Dreher, K.; Nölke, J.-H.; Seitel, A.; Tizabi, M.D.; Müller-Stich, B.P.; Maier-Hein, L. Tattoo tomography: Freehand 3D photoacoustic image reconstruction with an optical pattern. *Int. J. Comput. Assist. Radiol. Surg.* **2021**, *16*, 1101–1110. [CrossRef] [PubMed]
71. Deán-Ben, X.L.; Razansky, D. On the link between the speckle free nature of optoacoustics and visibility of structures in limited-view tomography. *Photoacoustics* **2016**, *4*, 133–140. [CrossRef] [PubMed]
72. Jeon, S.; Park, J.; Managuli, R.; Kim, C. A novel 2-D synthetic aperture focusing technique for acoustic-resolution photoacoustic microscopy. *IEEE Trans. Med. Imaging* **2018**, *38*, 250–260. [CrossRef] [PubMed]
73. Yao, J.; Wang, L.V. Sensitivity of photoacoustic microscopy. *Photoacoustics* **2014**, *2*, 87–101. [CrossRef]
74. Cho, S.-W.; Park, S.M.; Park, B.; Kim, D.Y.; Lee, T.G.; Kim, B.-M.; Kim, C.; Kim, J.; Lee, S.-W.; Kim, C.-S. High-speed photoacoustic microscopy: A review dedicated on light sources. *Photoacoustics* **2021**, *24*, 100291. [CrossRef] [PubMed]
75. Ozsoy, C.; Cossettini, A.; Ozbek, A.; Vostrikov, S.; Hager, P.; Dean-Ben, X.L.; Benini, L.; Razansky, D. *Compact Optical Link Acquisition for High-Speed Optoacoustic Imaging (SPIE BiOS)*; SPIE: Bellingham, WA, USA, 2022.
76. Hager, P.A.; Jush, F.K.; Biele, M.; Düppenbecker, P.M.; Schmidt, O.; Benini, L. LightABVS: A digital ultrasound transducer for multi-modality automated breast volume scanning. In Proceedings of the IEEE International Ultrasonics Symposium (IUS), Glasgow, Scotland, UK, 6–9 October 2019.
77. Ozbek, A.; Deán-Ben, X.L.; Razansky, D. Realtime parallel back-projection algorithm for three-dimensional optoacoustic imaging devices. In *Opto-Acoustic Methods and Applications*; Ntziachristos, V., Lin, C., Eds.; Optica Publishing Group: Washington, DC, USA, 2013; Volume 8800, p. 88000I. Available online: https://opg.optica.org/abstract.cfm?URI=ECBO-2013-88000I (accessed on 12 May 2013).
78. Wang, Y.; Erpelding, T.N.; Jankovic, L.; Guo, Z.; Robert, J.-L.; David, G.; Wang, L.V. In vivo three-dimensional photoacoustic imaging based on a clinical matrix array ultrasound probe. *J. Biomed. Opt.* **2012**, *17*, 0612081–0612085. [CrossRef]
79. Xu, M.; Wang, L.V. Universal back-projection algorithm for photoacoustic computed tomography. *Phys. Rev. E* **2005**, *71*, 016706. [CrossRef]
80. Rosenthal, A.; Ntziachristos, V.; Razansky, D. Acoustic Inversion in Optoacoustic Tomography: A Review. *Curr. Med. Imaging Former. Curr. Med. Imaging Rev.* **2013**, *9*, 318–336. [CrossRef] [PubMed]
81. Wang, L.; Maslov, K.; Yao, J.; Rao, B.; Wang, L.V. Fast voice-coil scanning optical-resolution photoacoustic microscopy. *Opt. Lett.* **2011**, *36*, 139–141. [CrossRef]
82. Wang, L.; Maslov, K.; Wang, L.V. Single-cell label-free photoacoustic flowoxigraphy in vivo. *Proc. Natl. Acad. Sci. USA* **2013**, *110*, 5759–5764. [CrossRef] [PubMed]
83. Rohling, R.; Gee, A.; Berman, L. A comparison of freehand three-dimensional ultrasound reconstruction techniques. *Med. Image Anal.* **1999**, *3*, 339–359. [CrossRef] [PubMed]
84. Chen, H.-B.; Zheng, R.; Qian, L.-Y.; Liu, F.-Y.; Song, S.; Zeng, H.-Y. Improvement of 3-D Ultrasound Spine Imaging Technique Using Fast Reconstruction Algorithm. *IEEE Trans. Ultrason. Ferroelectr. Freq. Control.* **2021**, *68*, 3104–3113. [CrossRef]
85. Lasso, A.; Heffter, T.; Rankin, A.; Pinter, C.; Ungi, T.; Fichtinger, G. PLUS: Open-Source Toolkit for Ultrasound-Guided Intervention Systems. *IEEE Trans. Biomed. Eng.* **2014**, *61*, 2527–2537. [CrossRef] [PubMed]
86. Xu, S.; Li, S.; Zou, J. A micromachined water-immersible scanning mirror using BoPET hinges. *Sensors Actuators A Phys.* **2019**, *298*, 111564. [CrossRef]
87. Xu, S.; Huang, C.-H.; Zou, J. Microfabricated water-immersible scanning mirror with a small form factor for handheld ultrasound and photoacoustic microscopy. *J. Micro/Nanolithography MEMS MOEMS* **2015**, *14*, 035004. [CrossRef]
88. Li, Y.; Wang, Y.; Huang, W.; Zhang, Z. Automatic image stitching using SIFT. In Proceedings of the 2008 International Conference on Audio, Language and Image Processing, Shanghai, China, 7–9 July 2008; pp. 568–571. [CrossRef]
89. Domokos, C.; Kato, Z. Parametric estimation of affine deformations of planar shapes. *Pattern Recognit.* **2010**, *43*, 569–578. [CrossRef]
90. Szeliski, R. *Image Alignment and Stitching: A Tutorial*; Foundations and Trends®in Computer Graphics and Vision: Hanover, MA, USA, 2007; Volume 2, pp. 1–104. [CrossRef]
91. AA, T.; Arfanakis, K. Motion correction in PROPELLER and turboprop-MRI. *Magn. Reson Med.* **2009**, *62*, 174–182.
92. Mozaffarzadeh, M.; Moore, C.; Golmoghani, E.B.; Mantri, Y.; Hariri, A.; Jorns, A.; Fu, L.; Verweij, M.D.; Orooji, M.; de Jong, N.; et al. Motion-compensated noninvasive periodontal health monitoring using handheld and motor-based photoacoustic-ultrasound imaging systems. *Biomed. Opt. Express* **2021**, *12*, 1543–1558. [CrossRef] [PubMed]
93. Heinrich, M.P.; Jenkinson, M.; Bhushan, M.; Matin, T.; Gleeson, F.V.; Brady, S.M.; Schnabel, J.A. MIND: Modality independent neighbourhood descriptor for multi-modal deformable registration. *Med. Image Anal.* **2012**, *16*, 1423–1435. [CrossRef]
94. Wang, Y.; Wang, D.; Hubbell, R.; Xia, J. Second generation slit-based photoacoustic tomography system for vascular imaging in human. *J. Biophotonics* **2017**, *10*, 799–804. [CrossRef] [PubMed]
95. Wang, Y.; Wang, D.; Zhang, Y.; Geng, J.; Lovell, J.F.; Xia, J. Slit-enabled linear-array photoacoustic tomography with near isotropic spatial resolution in three dimensions. *Opt. Lett.* **2015**, *41*, 127–130. [CrossRef] [PubMed]

96. Li, G.; Xia, J.; Wang, K.; Maslov, K.; Anastasio, M.A.; Wang, L.V. Tripling the detection view of high-frequency linear-array-based photoacoustic computed tomography by using two planar acoustic reflectors. *Quant. Imaging Med. Surg.* **2015**, *5*, 57–62. [CrossRef]
97. Westerveld, W.J.; Hasan, M.U.; Shnaiderman, R.; Ntziachristos, V.; Rottenberg, X.; Severi, S.; Rochus, V. Sensitive, small, broadband and scalable optomechanical ultrasound sensor in silicon photonics. *Nat. Photon.* **2021**, *15*, 341–345. [CrossRef]
98. Hazan, Y.; Levi, A.; Nagli, M.; Rosenthal, A. Silicon-photonics acoustic detector for optoacoustic micro-tomography. *Nat. Commun.* **2022**, *13*, 1488. [CrossRef]
99. Pan, J.; Li, Q.; Feng, Y.; Zhong, R.; Fu, Z.; Yang, S.; Sun, W.; Zhang, B.; Sui, Q.; Chen, J.; et al. Parallel interrogation of the chalcogenide-based micro-ring sensor array for photoacoustic tomography. *Nat. Commun.* **2023**, *14*, 3250. [CrossRef] [PubMed]
100. Park, J.; Park, B.; Kim, T.Y.; Jung, S.; Choi, W.J.; Ahn, J.; Yoon, D.H.; Kim, J.; Jeon, S.; Lee, D.; et al. Quadruple ultrasound, photoacoustic, optical coherence, and fluorescence fusion imaging with a trans-parent ultrasound transducer. *Proc. Natl. Acad. Sci. USA* **2021**, *118*, e1920879118. [CrossRef] [PubMed]
101. Yang, J.; Choi, S.; Kim, J.; Park, B.; Kim, C. Recent advances in deep-learning-enhanced photoacoustic imaging. *Adv. Photonics Nexus* **2023**, *2*, 054001. [CrossRef]
102. Park, B.; Park, S.; Kim, J.; Kim, C. Listening to drug delivery and responses via photoacoustic imaging. *Adv. Drug Deliv. Rev.* **2022**, *184*, 114235. [CrossRef] [PubMed]

sensors

MDPI

Review

The Convenience of Polydopamine in Designing SERS Biosensors with a Sustainable Prospect for Medical Application

Lulu Tian [1], Cong Chen [1], Jing Gong [1], Qi Han [1], Yujia Shi [1], Meiqi Li [1], Liang Cheng [1,*], Lin Wang [1,*] and Biao Dong [2,*]

[1] Department of Oral Implantology, School and Hospital of Stomatology, Jilin University, Changchun 130021, China; tianll22@mails.jlu.edu.cn (L.T.); gongjing22@mails.jlu.edu.cn (J.G.); hanqi22@mails.jlu.edu.cn (Q.H.)

[2] State Key Laboratory on Integrated Optoelectronics, College of Electronic Science and Engineering, Jilin University, Changchun 130021, China

* Correspondence: chengliang@jlu.edu.cn (L.C.); wanglin1982@jlu.edu.cn (L.W.); dongb@jlu.edu.cn (B.D.)

Abstract: Polydopamine (PDA) is a multifunctional biomimetic material that is friendly to biological organisms and the environment, and surface-enhanced Raman scattering (SERS) sensors have the potential to be reused. Inspired by these two factors, this review summarizes examples of PDA-modified materials at the micron or nanoscale to provide suggestions for designing intelligent and sustainable SERS biosensors that can quickly and accurately monitor disease progression. Undoubtedly, PDA is a kind of double-sided adhesive, introducing various desired metals, Raman signal molecules, recognition components, and diverse sensing platforms to enhance the sensitivity, specificity, repeatability, and practicality of SERS sensors. Particularly, core-shell and chain-like structures could be constructed by PDA facilely, and then combined with microfluidic chips, microarrays, and lateral flow assays to provide excellent references. In addition, PDA membranes with special patterns, and hydrophobic and strong mechanical properties can be used as independent platforms to carry SERS substances. As an organic semiconductor material capable of facilitating charge transfer, PDA may possess the potential for chemical enhancement in SERS. In-depth research on the properties of PDA will be helpful for the development of multi-mode sensing and the integration of diagnosis and treatment.

Keywords: polydopamine; surface-enhanced Raman scattering; hybrid materials; biosensors; SERS labels; sustainability

Citation: Tian, L.; Chen, C.; Gong, J.; Han, Q.; Shi, Y.; Li, M.; Cheng, L.; Wang, L.; Dong, B. The Convenience of Polydopamine in Designing SERS Biosensors with a Sustainable Prospect for Medical Application. *Sensors* **2023**, *23*, 4641. https://doi.org/10.3390/s23104641

Academic Editor: Evgeny Katz

Received: 11 April 2023
Revised: 1 May 2023
Accepted: 4 May 2023
Published: 10 May 2023

1. Introduction

Engineered nanomaterials need to be designed and synthesized in safe and sustainable ways to unlock their potential global economic, social, and environmental benefits. Therefore, while striving to invent new materials with better properties, an inescapable issue is to mitigate the risks for human health and the environment at all stages of their life cycle, including material extraction and processing, manufacturing, use, and the end of their life. More emphasis should be placed on the following aspects: (1) utilizing ingredients that can be quickly and harmlessly recycled as much as possible; (2) selecting a reaction pathway that consumes less energy and maximizes raw materials utilization; (3) elucidating the detailed mechanism of material synthesis; (4) finely characterizing physicochemical properties of end products; and (5) accurately predicting their short- and long-term biological activity. Among these, using renewable feedstocks is the focus of the beginning of the life cycle and an essential element of the green chemistry principles [1,2].

In the surface-enhanced Raman scattering (SERS) technique, a lot of innovative nanomaterials are invented to enhance the weak Raman scattering intensity of adjacent molecules by electromagnetic enhancement (EE) and chemical enhancement (CE). With the narrow and sharp fingerprint peaks and unique spectral profiles, the information and number of those molecules are clearly presented. Highly sensitive and selective results of qualitative and quantitative detection can be obtained rapidly and minimally invasively, with an acquisition time of only a few seconds and with proper laser power [3–5]. Some new SERS sensors achieved the purpose of reuse through cleaning [6,7], heating [8], light decomposition [9], hydrophobic modification [10] of the substrate, and good use of reversible intermolecular force [11]. Recently, it has been popular to design SERS tags, which can stably capture target analytes and output a Raman signal that can be distinguished clearly [12]. Polymers as functional interfaces have excellent structures and properties which can be used to fabricate robust, sensitive, and selective SERS sensors. Smarter sensors are available thanks to the properties of the polymers themselves and the convenience of introducing other functional ingredients in large quantities. For example, being target-matched, stimuli-responsive, non-fouling, electro-conductive, and biocompatible are all key points in designing sensors. Typical structures include (1) hair-like open polymer brushes of repeat units attaching to the core material; (2) surface molecularly imprinted polymer (MIP) coatings with mechanically and thermally robust interconnected structures; (3) layer-by-layer assemblies [13].

Cellulose, starch, silk fibroin, collagen, chitosan, alginate, lignin, and other materials from animals or plants not only have tailorable chemical components and mechanical properties, but also have rich sources, biocompatibility, and biodegradability [14]. They are ideal materials for the preparation of environmentally friendly biosensors. Inspired by the catechol and amino groups of Mytilus edulis foot protein 5 with strong adhesion, polydopamine (PDA) was first discovered as a biomimicry that could coat diverse materials [15]. It is also the main pigment in natural eumelanin, possessing biocompatibility and biodegradability [16]. Typically, the solid substrate is immersed in Tris buffer (pH 8.5) containing a low concentration of dopamine hydrochloride at ambient temperature and pressure for 5–6 h to generate the PDA coating. Compared with other surface modification methods, it requires less and low-cost raw materials, facile operation, and a mild reaction condition. Most importantly, PDA can adhere to almost all organic and inorganic materials, without special surface preparation and aggressive cleaning, and then directly utilize its rich functional groups for secondary modification to obtain the desired properties easily [17].

However, considering the simplified synthesis step and multiple performance for practical and sustainable utilization, this review focused on the advantages of PDA as a polymer in designing SERS sensors and its application in the biomedical field (Figure 1) in recent years. Firstly, the mechanism of SERS, and the molecular structure and synthesis techniques of PDA will be briefly described. Secondly, guidance for the design of SERS sensors will be proposed: (1) the introduction of noble metal or semiconductor materials; (2) the strategies for loading abundant signal molecules; (3) the methods for facilitating the identification of target objects; (4) the routes for enhancing the practicality of the biosensor. Thirdly, recent cases of SERS biosensors in diagnosis will be summarized. Finally, it is hoped that this review will inspire us to spare no efforts to explore more properties and a more detailed theory of PDA to enhance the intensity of Raman scattering, and to stimulate its potential in combination with other sensing and therapeutic modalities through flexible conception and simple synthesis.

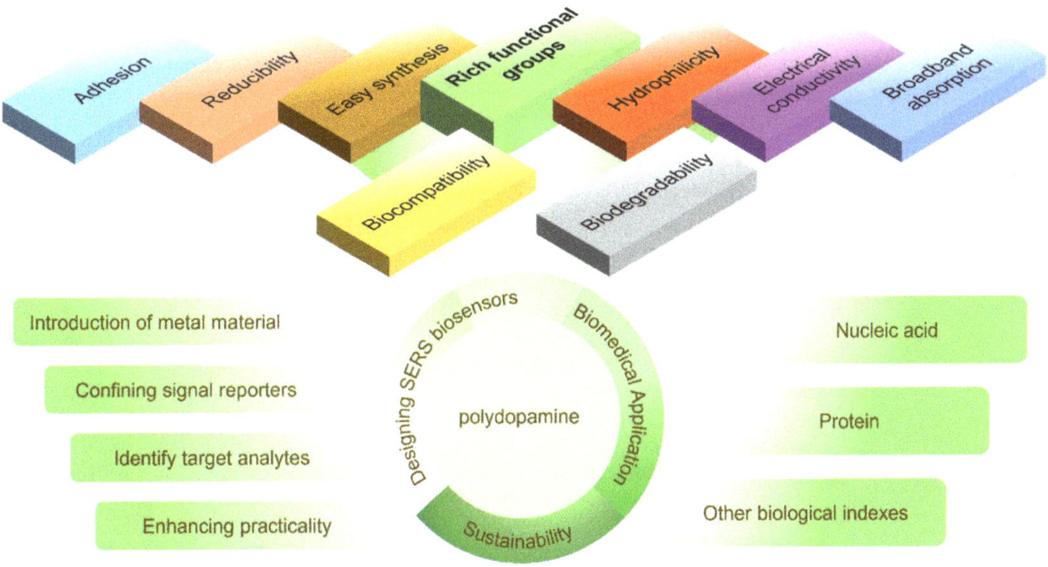

Figure 1. Overview of the various properties of PDA that can be used to design SERS biosensors with sustainable prospects for biomedical applications.

2. Mechanism of SERS

The Raman scattering effect was discovered by Indian scientist C. V. Raman in 1928 [18]. Under the excitation of light, there is little scattered light, mainly changing the vibrational energy. In the scattering process, the molecule absorbs energy and jumps from the initial state to a virtual energy level, and then falls back to a lower energy level and emits scattered photons. This phenomenon is also called inelastic scattering and includes Stokes and anti-Stokes scattering. The Stokes scattering leads to a smaller frequency of the emitted photons than that of the excitation source and red-shifted inelastic scattering. Anti-Stokes Raman scattering is the opposite [19]. The shift of inelastic scattered light with respect to the applied excitation wavelength appears together on the Raman spectrum. The fingerprint-like Raman spectrum reflects the intrinsic molecular vibrational information for determining the chemical components, the molecular conformation, and the interaction between molecules. Additionally, commendably, the interference signal from water in the sample can be negligible under the excitation of the visible and near-infrared light [5]. Unfortunately, only 10^{-8} of the incident photons are scattered inelastically [19], and the Raman scattering cross-section is merely about 10^{-30}–10^{-25} cm^2 per molecule [20], resulting in poor sensitivity.

In 1974, Fleischmann et al. found that a Ag electrode with a rough surface enhanced the Raman signal of pyridine adsorbed on it [21]. In 1977, Van Duyne et al. theoretically calculated that the Raman signal of each pyridine molecule was enhanced by 10^5–10^6 and initially called this phenomenon surface Raman scattering [22]. Considering the low and indistinguishable Raman scattering, noble metals (e.g., Au/Ag/Cu) and semiconductor materials (e.g., graphene and its analogues, transition metal sulfides or oxides, conjugated organic compounds) [23] are used to adsorb target molecules on their surfaces, resulting in bright Raman signals via EE or CE. Compared to the fluorescence (FL) signal, SERS has higher sensitivity (even down to the single molecule level) and better selectivity (obtaining multiple peaks with narrow bandwidth for multiplex detection) under single laser irradiation. Besides, SERS sensors are more suitable for long-term utilization due to the ability of SERS signal molecules to resist photobleaching and photodegradation [5]. Meaningfully, damage to SERS-active substances and molecules from the laser should be paid more attention to [24,25]. It is necessary to reduce the power of the laser, shorten the irradiation time, and use the laser with longer wavelengths. The metal–semiconductor complexes can be affected by these parameters minimally, according to some examples of reusable SERS sensors. A label-free SERS tag, Fe_3O_4@TiO_2@Ag, could be reused for nine cycles to detect prostate specific antigen under the laser with a power of 20 mW and an integration time of 10 s. The UV lamp (365 nm, 6 W cm^{-2}) used for the photocatalytic degradation of targets caused little damage to it [9]. Moreover, Chi et al. observed that monolayer graphene could retain the original shape of Au triangular nanoarrays for 16 annealing cycles [26]. However, the nanoporous structures were worn out after 22 reuse cycles, shown by the SEM photo [27]. Actually, the SERS activity of most materials decreases significantly after a few cycles [7,8]. The source of this destruction can be further investigated and solved for the development of sustainable SERS sensors. Apart from that, the protection of molecules from photo-induced invasion is crucial for medical application [28]. SERS substrates can be optimized for the excitation of near-infrared light to reduce the light damage to biological tissues [29] and export the detection result of less autofluorescence interference [30].

The SERS effect is regarded as the result of a combination of the enhanced electric field around the molecule and the increased molecular polarizability, and can be explained by two classical theoretical models, EE and CE [31]. In the theory of EE (Figure 2a,b), this electric field is usually composed of two steps. Firstly, the nanoparticle (NP) can be excited by a field at the incident wavelength to create a local field. Secondly, the molecule at the vicinity of the NPs can be polarized by the local field to produce a scattered field at the Raman wavelength. Then, the scattered field can interact with the NP to create a re-radiated field [5,32]. The intensity of the surface electric field is negatively correlated with the distance of molecules from the metal surface [33,34]. The enhancement of the local electric field mainly comes from the localized surface plasmon resonance (LSPR) of the noble metal. When the frequency of the incident light matches the natural oscillation frequency of the free electrons in the metal, the electromagnetic wave can drive the electrons of the conduction band (CB) at the metal–dielectric interface to produce collective oscillation, and LSPR will occur when it is highly localized to a specific site [4,5,35]. Moreover, when the diameter of semiconductor NPs without significant LSPR is close to the wavelength of the incident laser, Mie scattering theory cannot be ignored. This phenomenon is related to the resonance of charges with a three-dimensional distribution within the NP [36,37]. Mie resonances depend on the refractive index contrast between the dielectric sphere and the surrounding dielectric to affect the spatial distribution of the electromagnetic field. Particles with cavity structures or larger particle sizes will exhibit better performance in this resonance [38–41]. The CE is relatively weak compared to the EE, generally induced by the formation of chemical bonds, the resonance, and photo-induced charge transfer (PICT) between the adsorbed molecule and the substrate [42–44]. Notably, PICT (Figure 2c) is related to the charge transfer between the lowest unoccupied molecular orbital (LUMO) or the highest unoccupied molecular orbital (HUMO) of the molecule, and the Fermi level of a metal or the CB and valence band (VB) of a semiconductor [3,23]. Moreover, the coupled resonance strategy should be researched while designing novel SERS substances containing semiconductors. In this theory, it is suggested that two or more of the three factors—molecule-semiconductor CT resonance, molecular or exciton resonance, and plasmon or Mie scattering resonance—are at or near the laser excitation wavelength simultaneously.

The activity of the SERS substrate was related to its size, thickness, shape, structure (such as gap, tip, and edge) and refraction coefficient. Regrettably, the irregular morphology of NPs is incompatible with their uniform dispersion, both of which have a positive effect on SERS activity. Meanwhile, colloidal plasmonic metal NPs tend to aggregate and have a short service life [3], which makes the repeatability of the SERS spectrum poor while the quantitative values of multiple sites must be recorded in a specific time window to be calculated to obtain an average. The formation of dimers contributes to a high electric field intensity, namely one kind of hot spot, due to the gap of 2–10 nm [4]. If the sample content is too small [5] and the gap between two particles is too small [4], the unreliability of the detection structure will be increased (Figure 2d). Therefore, designing a SERS substrate with a more elaborate structure is considered to be a reliable means to improve the sensitivity, specificity, and reproducibility of SERS. The Raman signal of the final detection result can come from the target detection molecule itself or from the signal molecule pre-incorporated in SERS tags that have captured the target molecule. To form relatively complete SERS labels, the following components should be included: metal or semiconductor materials, Raman reporter molecules, protective shells, and identification elements [12].

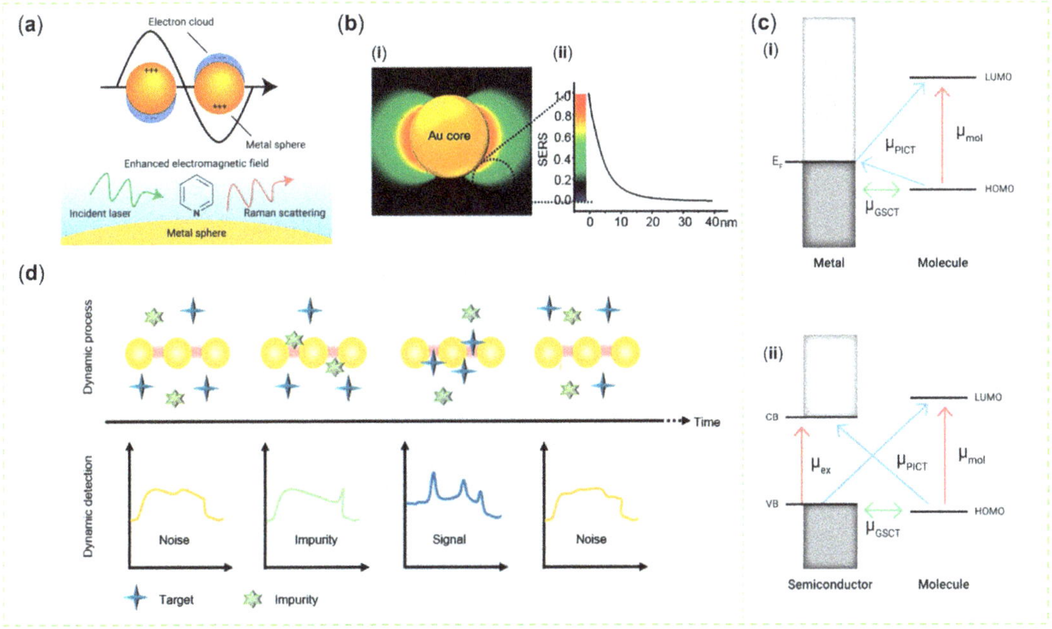

Figure 2. Mechanism of SERS about EE and CE. (**a**) Electromagnetic enhancement in SERS based on plasmonic nanospheres [23]. (**b**) Dependence of the electric field (**i**) and SERS (**ii**) enhancement on the distance from Au surfaces [5]. (**c**) Charge transfer transitions in metal–molecule (**i**) and semiconductor–molecule systems (**ii**). (Some notes for the abbreviations in the picture. E_F: the Fermi level in the metal; μ_{GSCT}: the interfacial ground-state charge transfer; μ_{PICT}: the photoinduced charge transfer resonance; μ_{ex}: exciton resonance in the semiconductor; μ_{mol}: molecule resonance.) [23] (**d**) A Challenge from the dynamic process of targets and impurities moving into the hot spot for qualitative and quantitative analysis of SERS substances [5]. Copyrights: all images have been adapted and reproduced with permission from: (**a**,**c**) © 2020 Shan Cong et al. published by Elsevier; (**b**,**d**) © 2018, American Chemical Society.

3. Synthesis and Molecular Structure of PDA

PDA stands out among the many polymers for surface functionalization because of its easy polymerization and many other excellent properties. Electropolymerization, enzymatic oxidation, and solution oxidation are simple methods for preparing PDA without the need for sophisticated equipment. We can select different buffers and solvents, regulate pH value and temperature, add various oxidants, change dopamine (DA) concentration and supporting substrates, and introduce external stimuli to obtain PDA film with ideal thickness and roughness [45].

The detailed molecular structure of PDA is not clear from current research. It is generally believed that PDA could be obtained by covalent polymerization and non-covalent self-assembly [46]. In the process of DA oxidation polymerization, various small-molecule intermediates are formed, e.g., DA, dopamine quinone, leucodopachrome, aminochrome, 5,6-dihydroxyindole (DHI), and 5,6-indolequinone (Figure 3a) [47]. Including these small molecules, the final products may also consist of oligomeric components and high-molecular-weight polymers [45]. By C-C bonding, hydrogen bonding, π-π stacking, and cation–π interactions, each monomer is linked [48]. Trimers of the $(DA)_2/DHI$ complex, via non-covalent stacking [46,49] and flake and cyclic tetramers formed by the C-C bond, are two distinct oligomers [50,51]. Under basic conditions with hydrogen peroxide, the catechol or o-quinone ring can be transformed to pyrrole carboxylic acids by oxidative degradation. Each part is the breakthrough point for probing the properties of PDA, including amino, imine, catechol, quinone, carboxy functional groups, and benzene rings. PDA is a versatile material that can be used to design SERS sensors. Adhesion is one of its most prominent and well-studied properties. With diverse functional moieties, it can be coated on precious metals, semiconductors, oxides, polymers, and ceramics with complex morphology through metal coordination or chelation, hydrogen bonding, π-π stacking, and cation–π interactions [15–17,52]. The most commonly used form of covalent adhesion is the introduction of sulfhydryl by Michael addition and of amino by Michael addition or Schiff base reaction (Figure 3b) [53]. This kind of chemical reaction and the reducibility of PDA are attributed to the quinone moiety. As a zwitterion, PDA can possess a positive charge at a lower pH for the protonated amino groups and a negative charge at a higher pH for the deprotonated catechol groups. The electrical conductivity and paramagnetic properties of PDA can be regulated by the π-electron. PDA is usually hydrophilic with low surface energy and can absorb the energy of light ranging from UV to infrared to protect molecules from damage [16,48]. Due to the satisfying biocompatibility and biodegradability, PDA is a burgeoning material in the biomedical field [16].

The molecular structure, thickness, and roughness of PDA depend strongly on the mode of synthesis [48]. The following cases can serve as references for synthesizing of PDA-modified nanomaterials to improve the SERS properties. The thickness of PDA is positively correlated with the concentration of DA and the pH value of the solution. The use of a low concentration of DA and multiple coating steps by shortening each immersion time contributes to inhibiting the formation and aggregation of PDA particles to reduce the roughness of the PDA film [17]. For example, Badillo-Ramirez et al. used SERS characterization technology to observe the changes in peak value on different bands. They also proved that when PDA was covered with Ag colloidal particles reduced by varied reducing agents, laser wavelength, irradiation time, and pH would produce diverse molecular structures in PDA [54]. Alfieri et al. found that the pre-existing PDA membrane was able to adsorb DA and other intermediate monomers and oligomers in solution, even at low concentrations. And in the polymerization system with periodate oxidizer, the conversion rate of catechol to quinone groups was found to be higher compared to that in the autoxidation system. It was also suggested that oxidation fission into carboxyl groups may occur. Long chain aliphatic amines provided hydrophobic moieties, inhibited intramolecular cyclization and the formation of large insoluble aggregates to promote the growth of PDA films, and thus improved underwater adhesion [55]. Intriguingly, biomolecules such as proteins, peptide chains, and nucleotides can also control DA oxidation and aggregation and enhance the stability and biocompatibility of PDA. The products of oxidative polymerization are diverse due to the interaction between oligomeric species and the bases of DNA, electrostatic attraction, and spatial complementarity [47]. Moreover, Raman spectra reflect the existence of PDA, especially the two obvious wideband peaks, near 1365 cm^{-1} and 1575 cm^{-1} [56], covering the signals of target molecules [57]. The Raman peaks from the undesired materials are difficult to circumvent. Amine-free organic (e.g., bicine) or inorganic (e.g., phosphate) buffers can be used as alternatives to avoid mixing Tris into the end product. Microwave irradiation can speed up the reaction rate [17].

According to the requirements of green synthesis, the amount and type of ingredients fed into the reaction system need to be minimized and simplified not to affect the purity of the final product, and must also be environmentally friendly. Given the prospect of mass production, the utilization of external power for the reaction system can improve the reaction efficiency. However, unreasonable energy supply goes against the concept of sustainable development. Notably, an increasing number of researchers are anticipated to concentrate on investigating the molecular structure and functionality of PDA. At present, PDA mainly played an auxiliary role in designing SERS substrates and failed to play a direct role in Raman signal enhancement, although PDA as an organic semiconductor material has such potential.

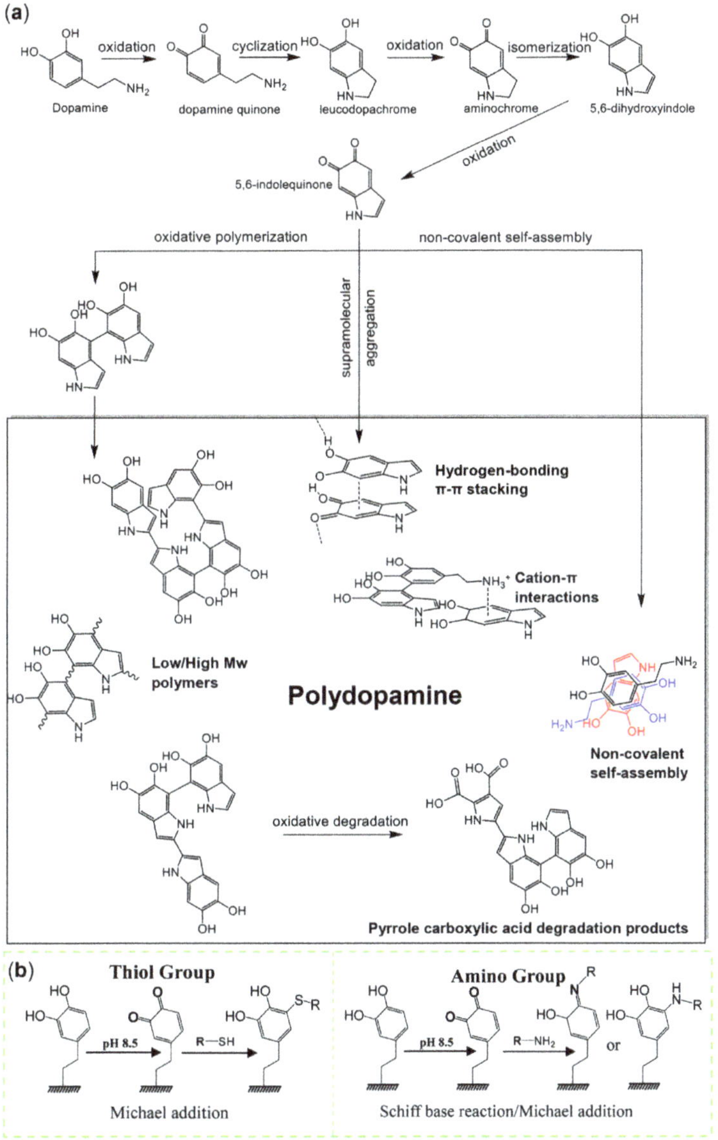

Figure 3. (**a**) Diagram of the synthesis mechanism of PDA [47]. (**b**) Schematic illustration of the Michael addition or Schiff base reaction [53]. Copyrights: all images have been adapted and reproduced with permission from: (**a**) © 2022 Elsevier B.V; (**b**) © 2010, American Chemical Society.

4. Guideline for Designing SERS Substrates Using PDA

Nanomaterials for SERS biosensing can be divided into three types in modality. Firstly, the label-free strategy can directly bond biological components to the material surface, but it is very difficult to select the characteristic peaks of target analytes for qualitative and quantitative analysis. Secondly, in the reaction-based strategy, with the occurrence and variation of the surrounding chemical composition, the functional groups of the probe molecules attached to the interface of the nanostructure will change, reflected in the Raman spectrum. Thirdly, the fabrication of SERS tags is currently a prevalent practice in the field [58]. The most important point is that the Raman characteristic peaks enhanced by SERS substrates for identifying and combining analytes can come from the analytes intrinsically, or from the reporters extrinsically in SERS labels [3]. Enlightened by the review of Gong et al. [12], this paper will firstly discuss the convenience brought by PDA in designing SERS tags containing metal or semiconductor materials, Raman reporters, and recognition elements. Then, the advantages of PDA will be revealed from the standpoint of enhancing the practicality of the SERS sensor.

4.1. Introduction of Metal or Semiconductor Materials

At present, metal and semiconductor materials are an essential part of each SERS substrate. PDA can self-polymerize on pre-synthesized materials to form coatings of appropriate thickness and rough morphology. Amino and catechol groups play an important role in fixing and reducing metal ions to form crystal seeds without other reducing agents, and a uniformly distributed metal matrix is eventually obtained. Just like glue, PDA has great convenience and charm in synthesizing a variety of metal or semiconductor materials into beaded, core-shell, or other peculiar configurations that help to heighten SERS activity. In addition, noble metals are expensive and in short supply, so the exploration of alternative materials in the SERS field is highly anticipated. For example, some semiconductor substrates with lower enhancement factors than noble metals can be improved by oxygen incorporation and extraction [59]. In this case, the reducibility of PDA can strengthen the stability of the material. It has also been found that PDA can assist in charge transfer and redistribution. There are a few relevant pieces of literature on this aspect, which are worthy of further study.

4.1.1. Adhesive Property

Adhesion is one of the most special properties of PDA. PDA can be a main material, providing ample surface area for other components to adhere to. Alternatively, it can be an auxiliary material by adhering to the core component for complex configurations. In virtue of abundant catechol and amino groups on the PDA nanosphere, several small Au seeds could be deposited for subsequent growth of a Au shell layer with a rough surface topography and numerous voids [60]. Additionally, PDA could adhere to linear [61], spherical [62], cuboid [63], rod-shaped [64], and other forms of materials with tips and sharp edges.

A moderate amount of PDA could connect 4–5 plasmonic nanospheres into a stable worm-like structure with a large surface area and adjustable LSPR responses. Choi et al. demonstrated that protonated DA of a certain concentration bound to Cit-Au NPs (Figure 4a) or Ag NPs (Figure 4b) in a mildly alkaline environment and subsequently polymerized to form Au/Ag@PDA NWs by facile sonication. Hydrogen bonds between N atoms of DA [shorted as N(DA)] and O atoms of Cit [O(Cit)], O(DA), and O(Cit), as well as N(DA) and N(DA) were essential for this assembly, as proven by the molecular dynamics simulation of intermolecular interactions. In addition, the electrostatic attraction between protonated primary amine groups of DA and carboxyl groups of Cit anions was strong. In their experiments, well-dispersed PDA-coated Au NPs could be synthesized by capping Au cores with anionic ligands, such as tannic acid, carboxylate-terminated or methoxy-terminated poly (ethylene glycol) (PEG). However, compared with Cit-Au NWs, these nanomaterials did not have a special worm-like structure and a second pronounced plasmonic peak. Beyond that, coating Cit-Au cores with some structural analogues of DA, e.g., norepinephrine, epinephrine, catechol, and 3,4-dimethoxyphenethylamine, without the primary amine or catechol group, even led to products without complete shells. As a result, Cit as a capping ligand and DA with primary amine and catechol groups were necessary compositions to form such a characteristic nanohybrid. More importantly, as opposed to granular and rod-like structures, more one-dimensional nanochains protected by PDA with good biocompatibility could enter cells (Figure 4c) via micropinocytosis and not be damaged for up to 24 h. They also scattered red light for label-free, dark-field scattering cell imaging, carried water-insoluble fluorescent dyes for FL imaging, and absorbed near-infrared light for photothermal therapy [65]. What makes their research worth learning from is that they explored the detailed intermolecular forces that ensured the stability of a particular structure and demonstrated that the selected materials were not substitutable. The worm-like nanomaterials had the advantage of entering cells for detection, and had the potential for SERS applications. Later, inspired by this configuration, some researchers adjusted the thickness of PDA and added 2D MoS_2 nanosheets (NSs) to construct Au NWs@PDA@MoS_2 nanohybrids (Figure 4d) for SERS sensing. RhB and MB were used to detect its SERS activity, which showed that this combination had a better effect than either component used alone (Figure 4e) [66].

In addition to the unique structure of the chain, PDA also shows great involvement in the construction of petal-like, porous, hollow nanomaterials with large surface areas. Park et al. incubated $CuSO_4$ and dopamine hydrochloride in PBS for 3 days at room temperature. Cu^{2+} catalyzed the polymerization of DA to PDA. At the same time, Cu^{2+} bound to the free amino group of PDA, which made $Cu_3(PO_4)_2$ seeds adhere to the surface of PDA for further anisotropic growth. A large number of Ag^+ were reduced by the catechol part of PDA to attach to $Cu_3(PO_4)_2$ nanoflowers (NFs) formed in the previous step. The SERS ship prepared by the NFs coated on a silicon wafer could be reused three times to detect the content of thiocholine, the hydrolyzed product of acetylcholine, in the extracellular fluid of patients with pesticide poisoning, and its limit of detection (LOD) value was 60.0 pM [67]. $Mo_7O_{24}{}^{6-}$ anions were chelated with catechol groups to form hydrophobic units, while amino groups formed hydrophilic units. After the addition of ammonia hydroxide, these Mo–DA with nanobubbles, dispersed in an ethanol solution, could polymerize into a porous Mo–PDA complex after the addition of ammonia hydroxide. After that, Fe ions could also be chelated to PDA and Au NPs could be captured by amino groups [68]. PDA was attached to the template material of different sizes, shapes, and compositions, and the template was etched off by a hydrothermal method to obtain a spherical, elliptical hollow or yolk-shell structure. During the above process, Wang et al. used the SiO_2 nanosphere loaded with Au NPs as the template. After etching, Au NPs would remain on the inner wall of the hollow PDA nanospheres, and the outer wall for secondary modification could continue to load plasmonic metals, which was conducive to achieving stronger SERS activity. It was also a way to introduce other components for drug delivery, catalysis, bioimaging, and other functions [69].

The materials with fancy configurations for highly active SERS and strong adaptability for biological applications still based on PDA can be adhered to. Their species need to be expanded and their mechanisms of adhesion should be explored. PDA coated on the Ti_3C_2 surface with hydroxyl groups after etching could further absorb Cd^{2+} by its catechol and amine groups [70]. Mn:ZnCdS quantum dots relying on carboxylation beforehand were connected to the amino group in the PDA [71]. The hydrophilic PDA layer could improve the dispersion of Fe_3O_4 NPs in solution [72]. PDA could self-polymerize on natural and cheap reduced graphene oxide including the many functional groups of epoxy, carbonyl, and carboxylic acid that could replace expensive noble metals in order to devise novel SERS substrates [73]. Metal organic frameworks (MOFs) with a porous structure and special functional groups are considered as a variety of ideal capture for analytes, and have a remarkable application prospect in the SERS field. The rich amino and phenolic hydroxyl groups in PDA could trap the metal ions of MOFs [72]. UIO-66(NH_2) was built by Zr^{4+} and 2-aminoterephthalic acid, which could be wrapped by PDA with excellent adhesiveness. Ag NPs and MIP could be modified on PDA. The PDA layer needed to be thin to maintain the original octahedral form of the MOFs, not to reduce "hot spots" and ensure a conspicuous SERS signal [74]. Special phenolic hydroxyl groups of PDA could capture Zn^{2+} for clusters of ZnO nanorods (NRs) [75].

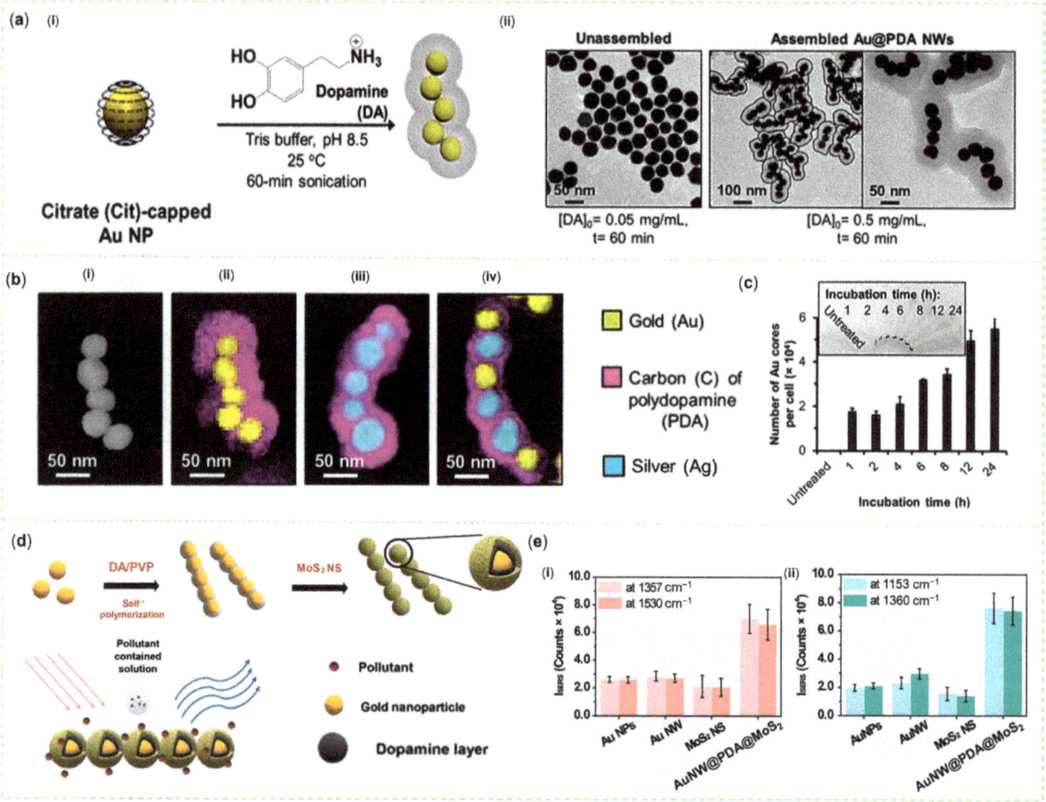

Figure 4. Worm-like SERS tags constructed by PDA. (**a**) (**i**) Conventional diagram of DA-mediated assembly of 40 nm citrate-capped Au (Cit-Au NPs) into Au@PDA nanoworms (NWs). (**ii**) TEM images of unassembled Au NPs and assembled Au@PDA NWs under different concentrations of DA. Black lines outline the morphology of individual NW. (**b**) (**i**) HAADF-STEM image of an Au@PDA NW and (**ii**) EDX elemental map of an Au@PDA NW, (**iii**) an Ag@PDA NW, and (**iv**) a bimetallic Au/Ag@PDA NW. (**c**) Kinetics of the association of Au@PDA NWs to HeLa cells through collecting cell pellets for ICP-MS after treatment with different incubation times [65]. (**d**) Schematic representation of the synthesis of Au NWs@PDA@Molybdenum disulfide (MoS_2) nanohybrid material and its application for SERS biosensing. (**e**) SERS intensity of two critical Raman peaks with (**i**) Rhodamine B (RhB) at 1357 and 1530 cm^{-1} and (**ii**) Methylene blue (MB) at 1153, 1360 cm^{-1} [66]. Copyrights: all images have been adapted and reproduced with permission from: (**a–c**) © 2019, American Chemical Society, (**d**,**e**) © 2022 Elsevier B.V.

As mentioned above, noble metals, semiconductors, quantum dots, MOFs, and other components can be introduced with the help of PDA to reveal their SERS capability. SERS substrates often require a unique structural design. On the one hand, PDA with a hollow or porous structure as the main supporting component has a large surface area for late modification. On the other hand, PDA as an auxiliary material can be attached to different forms of the main material for remodification. Most notably, the PDA-coated NPs also form stable chains that allow cells to swallow them. When we design the complex SERS substrates and study the Raman signal enhancement ability of new materials, PDA is like a nano-level double-sided adhesive with simple synthesis and varied performance.

4.1.2. Reductive Property

The brilliant reductive property of PDA for metallic cations has been attributed to the electrons released during the oxidation of its catechol groups to the quinone groups [16]. The strong reducibility enabled PDA to introduce Au [76], Ag [61,77,78], and other common noble metals [79] into SERS substrates. PDA nanospheres have the ability to provide a large number of attachment sites for metal particles. Ag particles with a diameter of 10 nm were reduced and then distributed evenly on a PDA nanosphere of about 400 nm. The gap between each Ag particle was close to its radius. Compared with the Ag particles of the same size freely distributed, this design obtained a greater electromagnetic field intensity, which was conducive to SERS signal amplification. Furthermore, in the results from X-ray photoelectron spectroscopy (XPS), after the material was stored at 4 °C for 45 days, the characteristic peaks of Ag ions were not obviously different from those peaks observed in the freshly prepared ones. This demonstrated that PDA could help to maintain the stability of this SERS sensor for long-time utilization [80]. Wang et al. proved that PDA coated on the hydrophobic plastic well plates could also offer electrons to Ag ions. Then, small Ag seeds synthesized in advance as surface-bound catalysts were imperative for the electroless deposition of Ag with uniform size distribution [81]. In addition to the enhanced SERS performance, non-aggregated Ag NPs showed a stronger FL quenching ability to realize the dual-mode sensing combined by SERS and FL [82].

Notably, core-shell SERS tags were popular because the gap-enhanced Raman effect could be stably obtained and the signal molecules could be placed between gaps without being affected by the harsh and changeable environment of the detected samples [83]. PDA, like double-sided tape, could easily introduce several metal shells and adjust the gap thickness [84]. Intriguingly, PDA could use its reductive property to help add plasmonic materials through layer by layer to form multi-shell NPs with nanogaps of controllable size such as a rocking snowball for ultrahigh SERS activity (Figure 5a) [85].

Uniquely, Jiao et al., designed a yolk-shell $Fe/Fe_4N@Pd/C$ magnetic nanocomposite as a recyclable SERS sensor. Through annealing at a high temperature under a H_2/Ar atmosphere, the PDA layer with Pd^{2+} coated on the Fe_3O_4 nanospheres with citrate groups became a porous N-doped defective carbon shell. Owing to the reductive property of PDA and H_2, Pd NPs were synthesized successfully and dispersed uniformly in the carbon shell. Meanwhile, it has good electrical conductivity for electrocatalytic application due to interfacial charge polarization formed between the defective carbon shell and Pd NPs. The carbon layer accumulated electrons while Pd NPs accumulated holes. Eventually, the following three factors, positively charged Pd NPs attracting more rhodamine 6G (R6G) molecules, the electromagnetic field between individual Pd NPs, and the magnetic field from iron cores contributed to the excellent SERS performance synergistically. Moreover, the PDA shell could effectively shield the iron cores from oxidation and maintain material stability, even if its performance was slightly compromised due to damage to its integrity in this work [79].

More oxygen vacancies for metal oxides had been proved to be able to help semiconductor materials to obtain a stronger SERS effect [87]. However, the reaction conditions were rather difficult to obtain, such as calcination in a hydrogen atmosphere, calcination in argon by grinding and mixing with $NaBH_4$, and pulse laser irradiation [88]. Worse still, the painstakingly synthesized materials were easily oxidized, causing them to lose their valuable properties. PDA was found to be a green material that could facilely introduce and protect oxygen vacancies to synthesize MoO_{3-x} with a tunable color and LSPR. In a certain range, the higher the pH value in the environment, the more reducing DA became. However, it was still necessary to consider and find semiconductor materials with an appropriate oxygen vacancy for the suitable CB and VB, and proper signal molecules with the LUMO and HOMO. One of the two possible PICT transitions, from the VB to the LUMO or from the CB to the HOMO, should match the excitation light energy to realize charge transfer and molecular resonance, so as to yield the most striking SERS signal [89].

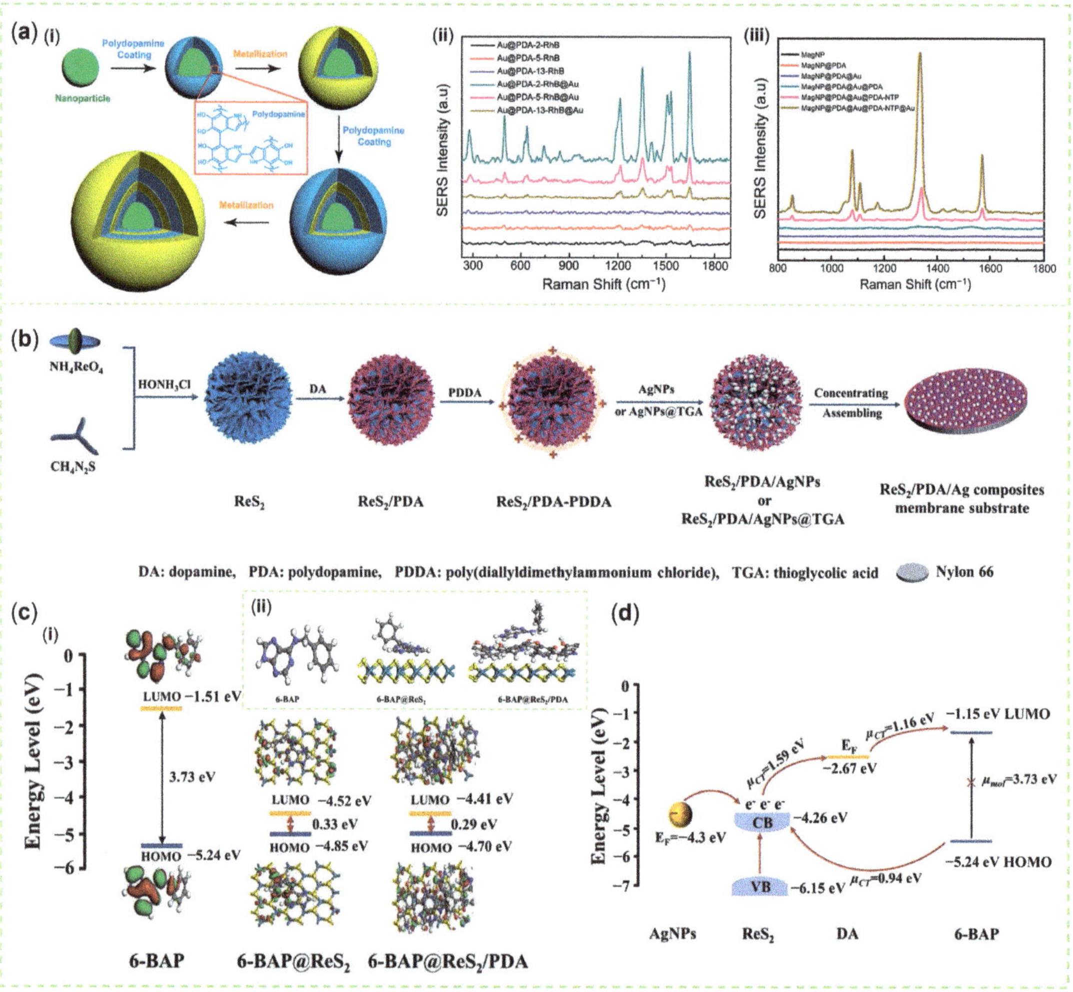

Figure 5. The function of PDA in building a nice structure with nano gaps and a case accounting for the process of charge transfer by PDA. (**a**) (**i**) Schematic illustration of the synthesis of versatile plasmonic nanogapped NPs based on PDA coating. (**ii**) SERS spectra of different Au nanostructures with RhB tags positioned on the PDA layer. (**iii**) SERS spectra of 4-nitrothiophenol (NTP)-encoded magnetic nanogapped NPs and the control NPs [85]. (**b**) A synthetic flow diagram of rhenium disulfide (ReS_2)/PDA/Ag. (**c**) (**i**) Effect of ReS_2/PDA for the band gap of 6-benzylaminopurine (6-BAP). (**ii**) The interaction models between ReS_2/PDA and 6-BAP of density functional theory calculation for HOMO and LUMO. (**d**) The route map of charge transfer in the composites of ReS_2/PDA/Ag and 6-BAP [86]. Copyrights: all images have been adapted and reproduced with permission from: (**a**) © 2019, American Chemical Society; (**b**–**d**) © 2023 Elsevier B.V.

The reducing and antioxidant properties of PDA have been demonstrated in the process of synthesis and protection of noble metals and metal oxides, even in the case where PDA was calcined to form a nitrogen-containing carbon layer. Noble metals are still the ideal materials to achieve strong SERS activity. The reduction of metal ions on the PDA layer is a convenient way to obtain a stable metal layer. Additionally, in the search for alternatives to expensive precious metals, PDA can obviously solve the problem of the easy oxidation of materials.

4.1.3. Charge Transfer

CE is a short-range effect, so there are two critical conditions for the acquisition of efficient charge transfer. One is that the chemical bonding needs to be formed between molecules and the substance for the flow of charge. The other is that the energy level of the substrate needs to match that of the molecule. Importantly, the energy of the incident light should be greater than the minimum required for charge transfer [23]. While conducting a popular multicomponent SERS substance with the couple resonance effect, the inclusion of materials with mismatched energy levels may inhibit this transfer [90,91]. In addition, research on tunning morphology [87], doping elements [92], and creating crystal defects [93] in metal and semiconductor components has proliferated. However, the development of SERS platforms based on π-conjugated organic semiconductors is still in its infancy. A PDA coating that can not only tightly wrap the semiconductor material part but also promote the proximity between the substrate and the analyte is a prerequisite [86]. Many past experiments have proven that for overcoming the shortcomings of wide band-gap semiconductor materials that do not respond to visible and near-infrared light, PDA with strong light harvesting and broadband-absorption ability can be a charge transfer mediator [94–97]. Selecting this kind of nano-glue that can transfer electrons through a conjugated π structure [94], catechol, or quinone groups [98] may be a shortcut.

As previously mentioned, the utilization of PDA in the research of Yuan et al. played a pivotal role in constructing a distinctive and logical plasmon-coupled 2D nanohybrid system consisting of Au NPs and MoS_2 NSs. On the interfaces of AuNPs prepared by citrate reduction, the polymerization of DA was controlled by polyvinyl pyrrolidone (PVP) to reduce the PDA thickness. The TEM images showed that worm-like structures were formed with a PDA thickness of 2 nm and an interspace within 1 nm, as well as a high dispersion due to the steric hindrance effect of PVP. Using finite-difference time-domain calculations, multiple hot spots could be observed between each NP. The SERS effect could be enhanced by charge transfer after further loading of MoS_2 NSs on the Au NWs. The ultrathin PDA between Au NWs and MoS_2 NSs is an electroactive molecule [66]. Additionally, Chin et al. affirmed that a conductive PDA acted as a charge transport bridge by donating plentiful electrons for charge redistribution between ZnO and Ag [77].

There is no critical evidence in their work to support this hypothesis, whereas the following examples may provide clues for later verification. Lin et al. evidenced that the Cu $2p_{3/2}$ peak in the XPS spectrum shifted to a lower energy level gradually with the sequential addition of PDA and Ag to CuO. This phenomenon revealed that the middle layer of PDA provided a large number of π electrons, which could affect the charge redistribution between Ag and CuO. Raman spectroscopy also showed that this PDA interlayer between the noble metal and semiconductor material could indeed achieve additional Raman scattering enhancement [99].

The thin-layer PDA was an organic semiconductor material that could transfer electrons to Ag NPs to enhance the surface plasmon resonance [100]. In this case, by electrostatic interaction, the positively charged PDA could be coated on the negatively charged ReS_2 nanoflake, and then the negatively charged Ag NPs could be carried by ReS_2/PDA that had been converted to a positive charge by PDDA (Figure 5b). Hu et al. found that the intermolecular electron transition from HOMO to LUMO of 6-BAP was not achieved under the excitation of light at 785 nm with low energy. As an energy level coupling platform, composites composed of ReS_2 and PDA (Figure 5c) reduced the band gap and stimulated efficient PICT. Simultaneously, the proposed process that the electrons were injected into the CB of ReS_2 from Ag NPs, and that the charges were transferred from the HOMO of 6-BAP to the CB of ReS_2 to the Fermi level of DA and finally to the LUMO of 6-BAP (Figure 5d) provided a reasonable explanation for the SERS phenomenon.

Nevertheless, this property of PDA cannot only be applied to the design of nanocomposite materials with the combination of precious metals and metal semiconductors, and its potential for small molecules and polymers needs to be further investigated. Doping PDA into a semiconducting conjugated polymer would increase electron transfer and reduce the optical bandgap energy to amplify Raman scattering. For example, DA and 5,6-dihydroxyinode, the intermediate in the assembly of PDA, have a similar five-member nitrogen-containing heterocyclic ring with a pyrrole monomer. This phenomenon led to the formation of π-π stacking between PDA and polypyrrole (PPy). In fact, the DA monomers were seen as the better dopant to break the chain of PPy, but they tended to polymerize under the reaction condition with oxidative reagents. It was reasonable to decrease the initial pH and the amount of DA molecules to inhibit the conformation of non-hybridized polymers. By replacing PPy with another conjugated polymer, polyaniline, the same effect could be achieved. Then, Liu et al. added chondroitin sulfate, DA, and pyrrole on the outer layer of SiO_2 nanospheres to synthesize a novel material for Raman and photoacoustic imaging. Beyond that, plenty of reactive moieties in the molecular construction of PDA provided sites for other functional units to achieve magnetic resonance imaging and targeted therapy [101]. Yan et al. constructed an elaborate photoelectrochemical sensor and proved that PDA could assist charge transfer by electrochemical impedance spectroscopy. The photocurrent was increased after the PDA film was coated on the electrode. This phenomenon further supports the theory that PDA acts as an electron donor, reinforcing the tendency for photogenerated electron–hole pairs to separate [71].

Therefore, in this section, it is found that the sensor with PDA may have the opportunity to realize the goal of the integration of diagnosis and therapeutics, produce a dual-mode sensor with SERS and photoelectric activity, and be employed for bioimaging.

4.2. *Confining Signal Reporters*

The unique functional groups and rough morphology allow PDA to combine quantities of signal molecules easily, which is favorable for sensors that require SERS tags to obtain an external signal for analysis. The dynamic cycle signal amplifying system can also be constructed on the PDA. However, it is worth noticing that the amount of PDA needs to be controlled because the two Raman peaks of PDA are generally considered as background noise and can interfere with the recognition of target peaks [57].

Zhou et al. found that the distribution and quantity of Au NPs on the PDA nanospheres can be modulated by adjusting the mass ratio of $HAuCl_4$ and PDA and the reaction temperature, thereby preserving more binding sites for signal molecules [76]. The signal molecules modified with sulfhydryl groups could be chemically bonded to the Au NPs, and the signal stability could be longer lasting after being wrapped with PDA of strong adhesion [102]. Small aromatic thiols are popular candidates for optimal Raman signal reporters, having a single and distinct peak [4]. Directly wrapping the reporter molecules containing thiols on a plasmonic particle with a thin layer is called a self-assembled monolayer (SAM). Li et al. discovered that compared with SAM, the introduction of a thicker PDA film could better utilize the electromagnetic field from a single particle to enhance the intensity of the SERS signal. Thiol groups can covalently bind to DA molecules via the Michael addition reaction. Compared to bare Ag NPs, PDA-coated Ag NPs exhibited an almost 20-fold increase in the optimal amount of 4-nitrobenzenethiol that could be carried. They also tested six other signal molecules, and the results all showed that this variety of three-dimensional volume-active SERS probes deadly confined more reporters and achieved more significant SERS signals than SAM-based SERS tags. Finally, regarding the strength of this design, 4-(phenylethynyl)benzenethiol including alkynyl, having a distinct peak in the signal silence region of biological tissue, was selected to construct probes for tumor imaging [103].

MB was a common, cheap, and low-toxicity dye that could be absorbed on PDA via electrostatic attraction and π-π stacking interaction for both FL and Raman sensing [76]. Other fluorescent substances and their derivatives with strong absorption are also worth examining to generate distinguishable signals for dual-mode sensing [28,104]. Both FL and SERS intensities depend on the distance between molecules and metal particles with enhanced electric and magnetic fields. Unlike SERS, the FL intensity tends to first increase and then decrease as the molecules move away from the metal surface. To avoid the instability of dynamic analysis, a thin dielectric material shell, being chemically and electrically inert, can be inserted between molecules and metal particles to adjust the distance between them [105–107]. Compared with Ag nanocubes combined with SiO_2, Ag nanocubes modified with PDA could trap more Raman reporters through π-π stacking interaction and hydrogen bonding between PDA and R6G, resulting in a more prominent SERS signal [63]. Wu et al. indicated that hydrogen bonding and π-π stacking forces between PDA and the aptamer nucleobases locked R6G tightly in the pores of mesoporous SiO_2 NPs. Through the collaboration of the acidic environment and strong hydrogen bonding between the target analyte and the aptamer under PDA, PDA was degraded and the signal molecules were released to combine with additional SERS substates to achieve quantitative analysis [108]. PDA-functionalized Au bipyramids can be utilized in conjunction with a dual-mode probe that exhibits both SERS and fluorescent. This probe was a phenylboronic acid-substituted distyryl boron dipyrromethene [109], which enabled the boric acid to interact with the catechol groups of PDA.

After all, a single substrate has a fixed surface area, and the number of reporter molecules it can carry is limited. Therefore, the SERS biosensor modified by PDA may be able to cooperate with some automatic cyclic signal amplification systems. For example, when an aptamer with one signal molecule bound to the target analyte and then was away from the sensing substrate, the complementary nucleotide chain formed a hairpin structure, and another signal molecule at its tail approached the sensing substrate and output the corresponding signal. Usually, better statistical results would be reached by the ratio of two signals rather than a single signal [102,110]. The aptamer that recognized the target analyte was then digested by the exonuclease, allowing the analyte to be released to continue reacting with the sensing substrate [111].

To sum up, modifying the thickness, roughness, and other spatial structures of the PDA layer leads to an increase in the number of binding sites available for signal molecules. Of course, it is also desirable to use the adhesive properties of PDA to bond more components that can be directly connected to signal molecules. Moreover, PDA can isolate signal molecules from the complex environment as an internal standard to output more stable and accurate results. Importantly, the signal molecules could be close to or far from the surface of SERS substrates as expected to enable the Raman signal to be turned on or off, which may be a more promising design. Since the molecular structure of the Raman reporter is not damaged in the process of Raman signal detection, it is advantageous to realize multi-mode sensing with the assistance of multifunctional molecules. Finally, a dynamic cycle signal amplification system is mentioned, which can not only help the sensor to obtain obvious target signals but can also save consumables.

4.3. *Identify Target Analytes*

Most proteins, peptide chains, nucleotides, small molecules, and polymers can bind to PDA through electrostatic attraction, hydrogen bonding, π-π stacking, and cation–π interactions [17]. On the one hand, it was convenient for PDA-smeared materials to conjugate with thiol or amino group modified biomolecules, such as antibody (Ab) [112], nucleic acid [110], and aptamer [47], for specific recognition. Without PDA, the introduction of amino-containing biomolecules required the additional carboxyl groups to be activated by 1-ethyl-3-(3-dimethylaminopropyl) carbodiimide and N-hydroxy succinimide [113]. This synthetic process was tedious. Additionally, several ingredients might not be completely cleaned, which affects the presentation of the Raman peak. The amino group on the Ab could be assembled to the PDA shell by the Michael addition reaction [102]. Through Schiff base reactions, anti-human epidermal growth factor receptor monoclonal Ab was added on the PDA surface to obtain the selectivity of SERS probes [103]. Single-strand nucleotides and proteins could be attached to PDA with numerous functional groups by π-π stacking to shorten the incubation time of the immunoassay [114]. The PDA layer with high density and roughness could hold more Abs and keep them active for a long time. However, it was necessary to find a proper concentration of DA for self-assembly because the active sites of the Ab would be buried by PDA of too high density [115]. On the other hand, a PDA film also makes a contribution to non-specific sensors. 4-carboxyphenylboric acid could be introduced by PDA, and boronic acid could interact with the bacterial diol group of the saccharide [116]. The PDA layer could also use π-π and hydrogen bonds to attach some target molecules directly [63]. Intriguingly, the combination of PDA and a phospholipid bilayer [47] may have promising applications in targeted sensing.

In the process of oxidative polymerization of DA for MIP, the target molecule can be used as a template doped in the molecular structure of PDA. After the template is removed, the remaining cavities corresponding to the size, shape, and functional groups of the target can be specifically matched to the target to be recognized. For example, amino and carboxyl groups in peptide chains could link to hydroxyl and amino groups in PDA [117]. Yang et al. synthesized a PDA-MIP-coated SERS sensor with moderate Au NPs that did not mask the recognition site for analyzing three varieties of phthalate plasticizers [118]. To increase the amount of binding sites, this PDA-based MIP could be coated on the SiO_2 NPs, wherein SiO_2 NPs were removed by hydrofluoric acid to produce a stable, homogeneously-dispersed, hollow configuration [119]. Nevertheless, in real samples with a lot of interferences, the selectivity of this artificial Ab would be weakened. Morphological and molecular weight analogs of target analytes may occupy sites. Liu et al. used the target biomarker, C-reactive protein (CPR), as the template to mediate the structural rearrangement of $(DA)_2$/DHI trimers in the PDA film of MIP-PDA microfiber sensors. The hydrophobic area of CRP was combined with the hydrophobic area of the benzene ring in DA and the negatively charged carboxyl groups on CRPs were attached to the positively charged -NH_2 or -NH groups on DA and DHI (Figure 6a). Even if the template was finally eluted, this hydrophilic/hydrophobic and charged distribution compatible with the substance to be recognized could be retained to enhance the specificity of the sensor. Eventually, MIP-PDA showed superior recognition ability when compared with an antigen-Ab binding assay (Figure 6b) and under the interference of glutathione (GSH), NaCl, and immunoglobulin G (IgG) (Figure 6c) [49].

The enantiomers of a chiral molecule may exert completely different effects on the health of the body, but they are hard to tell apart. Arabi et al. selected a linear-shaped aminothiol molecule as an inspector for the PDA chiral imprinted cavities coated on SERS nanotags (Figure 6d). The nanotags were composed of gold nanostars (Au NSs) and 3.3′-diethylthiatricarbocyanine iodide (DTTC). The adjustable density and thickness of the PDA layer was a determining factor in changing the permeability of the inspector. Passing through cavities unsuccessfully combined with good enantiomers, qualified inspectors could irreversibly and quickly degrade DTTC to obtain a decreased SERS signal [120]. Moreover, Kong et al. found that the molecular structure of PDA synthesized on the surface of SiO_2 with a certain chiral characteristic would change accordingly. PDA could discern the corresponding tyrosine and phenylalanine enantiomers because of the identical chirality from a phenethylamine molecule-like unit and π-π stacking interactions from benzene rings [121].

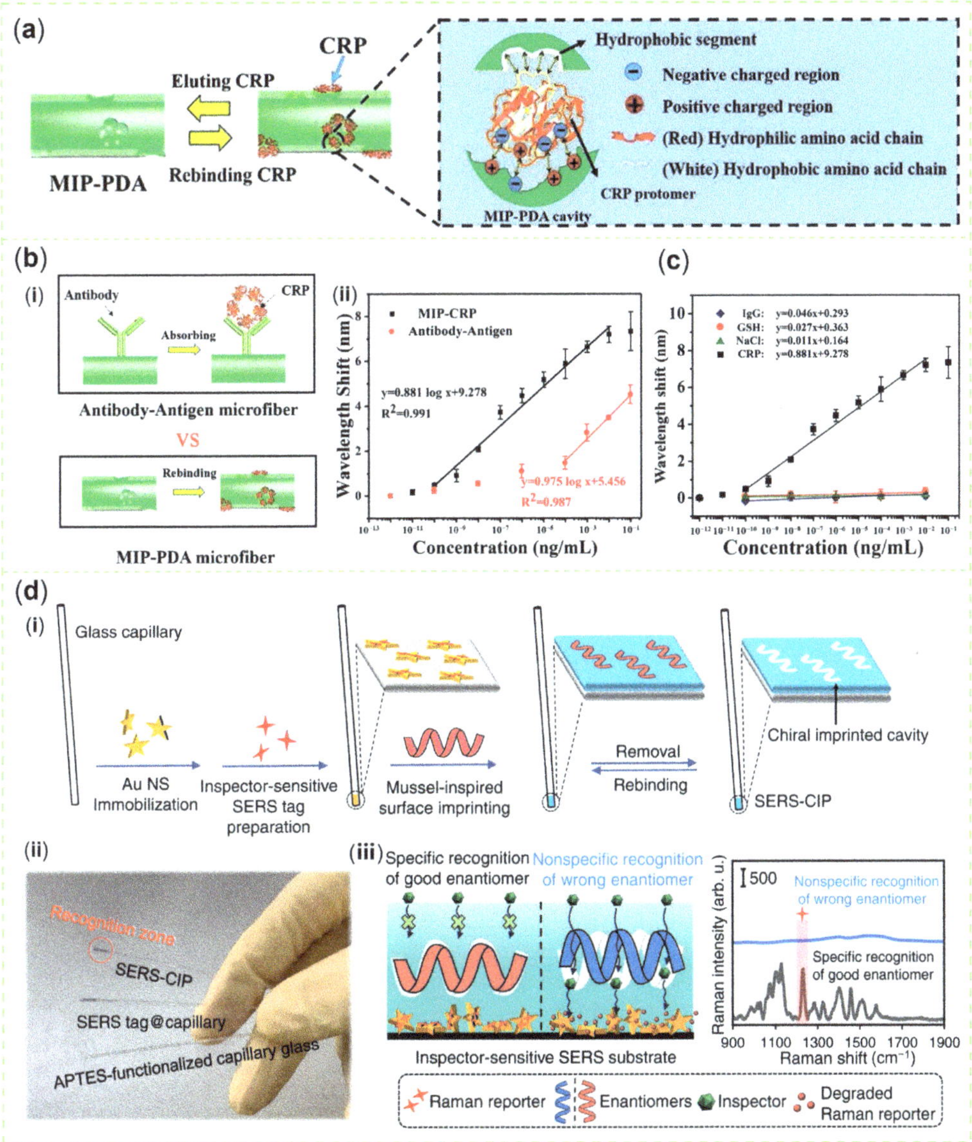

Figure 6. Design and characteristics of SERS sensor related to PDA MIP. (**a**) Diagram of CRP-induced polymerization of PDA. (**b**) (**i**) Schematic of the Ab-antigen microfiber probe functionalization. (**ii**) Linear fitting of Ab-antigen microfiber response to CRP injections. (**c**) Spectral wavelength shifts of MIP-PDA microfiber chemosensor responding to CRP, NaCl, GSH, and IgG at gradient concentrations [49]. (**d**) (**i**) Schematic illustration of the SERS-chiral imprinted platform (CIP) construction. (**ii**) Photo images of SERS-CIP. The recognition zone is illustrated by a red circle. (**iii**) Principle of an "inspector" recognition mechanism implemented on SERS-CIP [120]. Copyrights: all images have been adapted and reproduced with permission from: (**a**–**c**) © 2022 Elsevier B.V.; (**d**) © 2022, Maryam Arabi et al. published by Springer Nature.

Electrokinetic pre-separation enhanced the concentration of target charge molecules on the AuNP@PDA-MIP surface, and improved the selectivity and sensitivity of SERS sensors. One thing to note is the thickness of the PDA-MIP layer. For the layer, being too thick would reduce the electromagnetic field intensity around AuNPs, negatively affecting SERS performance; meanwhile, if it was too thin, it would be destroyed under a high electric field and lose its specificity [122]. In the meantime, PDA was pH responsive due to amine and phenolic hydroxyl groups [123], and could be negative in the alkaline environment, attracting identified molecules with a positive charge [51].

To enable rapid point-of-care testing in vitro, interfering factors from blood, urine, saliva, and other various biological samples need to be eliminated without pretreating the sample [124]. Undoubtedly, with regard to in vivo testing, this issue necessitates further attention [125]. Raman peaks, like fingerprints, bring convenience to the analysis of results, but the SERS sensors capturing substances non-specifically make the follow-up work extremely tedious. Therefore, SERS biosensors can specifically capture target analytes and cooperate with signal molecules to construct signal enhancement or attenuation mode, especially in the case of no sample pretreatment. Fortunately, PDA can directly chelate metal ions, directly bind small molecules and biomacromolecules, select targets with different charge and chirality, or use recognition elements. At present, PDA is also expected to construct advanced and environmentally friendly MIP.

Essentially, the improvement in the above three functions is far from sufficient for the commercialization and reuse of SERS sensors. Therefore, the fourth part of this review discusses the enhancement of their practicality.

4.4. Enhancing the Practicality of Sensors

Here, as can be seen from some cases concerning powder materials, enhanced stability, uniform dispersion, and resistance to contamination can ensure the reliability of test results and can facilitate long-term preservation. To overcome the challenges of uneven distribution, difficult collection, and reuse of powder materials, it is recommended to incorporate medical consumables, such as test paper, cotton swabs, and glass rods, and smart platforms, such as microfluidic devices, microarray, and lateral flow assays, for rapid sample collection and detection. In most experiments, the method of dripping liquid samples on the silicon wafer and then drying them has been adopted to conduct multi-site detection. However, in the actual testing process, the distribution of analytes to be tested is quite uneven, which reduces the repeatability. The hydrophobic treatment of platforms carrying liquid samples can improve this situation. Additionally, the research on two-dimensional PDA membranes with special textures and strong mechanical capacity needs to be accelerated, which can reduce the consumption of raw materials. Fortunately, the biocompatibility and biodegradability of PDA have been extensively researched in therapeutic and sensing applications, providing strong support for its use. However, its problem of long-term preservation after being administrated into the organism remains unsolved. The practicality of SERS sensors can be improved through the implementation of various approaches, as listed in Table 1.

Table 1. Various approaches were used to enhance the practicality of SERS-sensors.

Properties	Materials	Ref.
biocompatibility	Ag/PDA/ZnO-filter paper	[77]
	Au-HA *-PDA-PLGA *-microneedles	[126]
inoxidizability	Ag nanocubes@PDA	[63]
	Fe/Fe_4N@Pd/C	[79]
	PDA@Ag-anti-cTn I *	[80]
	Al nanocrystals@PDA	[127]
	Al NPs@PDA on cellulose paper	[128]

Table 1. *Cont.*

Properties	Materials	Ref.
antifouling property	Au@EB *@PDA@Ab@BSA	[102]
	Magchains *@PDA@PEG@Ab	[113]
uniformity	Capillary glass@Au NSs@MIP	[117]
flexibility and uniformity	PET/PDA/ZnO/Ag	[75]
	filter paper@PDA@Ag NPs	[78]
	cotton swab@PDA@Ag NPs	[129]
	$W_{18}O_{49}$@Ag/PDA@PVDF-MIP membranes	[130]
	Ag@PDA@SiO_2 nanofibrous membranes	[131]
	non-woven fabrics@PDA@Ag NPs	[132]
	polyurethane sponge@PDA@Ag NPs	[133]
mechanical stability and uniformity	Ag@DNA/PDA-CNF	[51]
	Au-HA-PDA-PLGA-microneedles	[126]
hydrophobic or superhydrophobic	PDA-Ag microbowl array	[134]
	PDA film patterned by microcup or breath figure arrays	[135]
	cellulose filter paper@PDA@Ag NPs	[136]

* HA: hydroxyapatite; PLGA: poly (lactic-co-glycolic acid); cTn I: cardiac troponin I; EB: ethynylbenzene; Magchains: magnetic nanochians.

A qualified sensing element must first ensure stable performance in complex practical application environments. Additionally, it cannot have a destructive effect on the targets. Especially in the biological samples, good biocompatibility is highly important. PDA could decrease the biotoxicity of various nanomaterials, which is conducive to the production of biosensors without damaging normal tissues and cells [77,137].

The PDA layer can protect the highly oxidizable metal matrix [128]. Renard et al. synthesized octahedral aluminum nanocrystals coated with a 5–7 nm PDA layer binding to Al_2O_3 by catechol groups. PDA functionalization provided them with structural stability in Milli-Q H_2O for up to two weeks. Since aluminum nanocrystals are extremely unstable in aqueous environments, the common Tris buffer could be replaced by isopropanol and NH_4OH in the synthesis of PDA [127]. This feature could also extend the shelf life of materials and save the cost of preservation [63].

Bovine serum albumin (BSA) could be attached by the formation of covalent bonding between its amino terminus and the quinone groups in the PDA to block non-specific binding sites [71]. PEG with the thiol or amino group could be loaded on the multifunctional nanomaterials by PDA to obtain the antifouling property [113] and inhibit cell or protein adhesion [15]. Liu et al. found that by remodeling DNA, PEG, and PVP on the outer layer of PDA, the NPs could remain stable, based on electrostatic and steric repulsion, under repeated freeze-thaw cycles and in the harsh environment of high ionic strength, serum, and cell lysate [138].

It was difficult to avoid the problem that the SERS matrix in powder form was unstable, easily aggregated, and not suitable for recycling. The smooth interface of a glass capillary made the test result repeatable due to the uniform coating. In addition, the LOD was low, and the mass production was easy. PDA could be attached to it by covalent and non-covalent bonding, and then loaded with Au NSs [117]. Chang et al. placed the PDA-attached matrix in water, spread silicon or polystyrene (PS) nanospheres with different diameters over the air–water interface, and then discharged the water to obtain a colloidal crystal monolayer, stuck stably and uniformly on the matrix. The versatile application of PDA also enabled the formation of bilayers or trilayers in colloidal crystal. The metal grid, polydimethylsiloxane, polyethylene terephthalate (PET), glass slide, silicon wafer, single mode fiber, and glass capillary could be functionalized by this simple method for a promising SERS application [139].

Fortunately, flexible platforms, such as PET film [75], filter paper [78], or cotton swabs [129], can uniformly disperse the powder over its surface and conform well to the shape of the test object. Li et al. used PDA to assist with the deposition of $W_{18}O_{49}/Ag$ composites on the hydrophilic modified polyvinylidene difluoride (PVDF) membrane [130]. The selection of an ideal paper with simple Raman spectra as a carrier for the SERS substrate should be considered [128]. Through chemical bonding, PDA could be coated on SiO_2 nanofibrous membranes prepared by electrostatic spinning and calcination with strong thermal stability, no fracture after folding, and no SERS peaks [131]. PDA could be firmly attached to the conical poly (lactic-*co*-glycolic acid) microneedles. The catechol moieties in PDA were combined with Ca^{2+} and then electrostatically attracted PO_4^{3-} to obtain randomly-directed, petal-like hydroxyapatite which provided an interface for massive Au nucleation and growth. This device has high mechanical stability, biocompatibility, and detection capability to be inserted into the epidermis and dermis for biosensing [126]. The amino and catechol moieties of PDA facilitated its adhesion to the non-woven fabric with high permeability and a large surface area, followed by reduction of Ag ions resulting in a strong SERS signal [132]. PDA-decorated polyurethane sponges retained high porosity, allowing rapid swab sampling, and could internally generate photon diffusion due to the flexible, hierarchical, and three-dimensional interconnected framework [133].

Xu et al. discovered that the presence of hydrogen bonds, van der Waals forces, and electrostatic attraction helped the flake PDA tetramers adhere to the rod-like CNF. The PDA NSs were rearranged by the introduction of DNA molecules with double helix structures through π-π conjugate bindings between DNA bases and PDA oligomers (Figure 7a). This ordered configuration made the nanofibers possess charge selectivity (Figure 7b), and stronger mechanical properties (Figure 7c), and Ag NPs attached to them were evenly distributed, resulting in a wide range of hot spots. Using R6G as the Raman probe, smaller relative standard deviations at 607 and 1503 cm^{-1} (Figure 7d) were attributed to a more uniform assignment of hot spots of the Ag@DNA/PDA-CNF compared to the Ag@PDA-CNF [51]. This result indicated that the participation of DNA in the polymerization process of PDA could improve the repeatability of SERS sensors. The use of bio-derived materials to produce wearable flexible sensors adapted to human tissue curves has been extensively investigated [14]. Hence, PDA could be combined with CNF, cellulose nanocrystals, SF, carboxymethyl cellulose, and other sustainable materials to prepare intelligent SERS sensors in the form of controllably degradable biomedical implants or air-permeable, easily-detachable, adhesive artificial skin for in situ diagnosis and on-demand treatment.

As the volatile solvent evaporates from sessile droplets, capillary flows caused by contact line pinning transport non-volatile solutes to the contact line, resulting in the coffee-ring effect (CRE). Suppressing or exploiting the CRE so that the solute is concentrated and uniformly distributed at a central point or on the contact line at the outer edge [140,141] is also a matter of concern in constructing a SERS detection device with good reproducibility. The sample preparation platform with hydrophobic properties could condense the solution droplet and distribute the effective analytes relatively uniformly in a small area after drying (Figure 8a). For SERS sensors, which conducted detection at multiple points on the sample and then obtained the average value, more valuable repeatability and smaller error could be acquired [142,143]. Particularly, PDA films with a variety of patterns also showed better hydrophobicity with a positive effect on SERS activity [135]. It was generally considered that the introduction of low surface energy chemicals and the construction of a rough surface with micro-, nano-, and hierarchical structures were two ways to obtain hydrophobicity [144]. Some hydrophobic alkanes with sulfhydryl groups, such as n-dodecyl mercaptan and hexadecanethiol, could be easily imported into biosensing materials through the Michael addition reaction with PDA [145]. Dong et al. modified PDA-decorated cellulose filter paper with PFDT to gain more sensitive SERS results than with non-hydrophobic substrates [136]. With respect to the rough surface with special structures formed by PDA, Shang et al. used trichloromethane to remove monodisperse PS photonic crystal nanospheres to prepare a superhydrophobic PDA microbowl array (Figure 8b). The array not only had hydrophobicity due to PFDT and roughness from Ag

NPs (Figure 8c(i–iii)) but also had high water adhesion (Figure 8c(iv–vi)) both to constrain the sample well and to obtain a lower LOD (Figure 8d). The PDA-Ag microbowl array maintained good hydrophobicity and SERS activity even under a harsh environment [134]. In developing this kind of sensor, two contact states, the Wenzel state and Cassie state, between the substrate with a special texture and the droplet need to be explored [146]. Clearly, Yu et al. found that although honeycomb porous and pincushion-like PDA films had high hydrophobicity, they had completely different water adhesion [147]. It is necessary to point out that a PDA film could be merely fabricated at the air/water interface in the Tris buffer containing DA. Making good use of this feature, Kozma et al. placed different sacrificed templates to float on the liquid surface in order to pattern the PDA film. Microcup arrays, an inverse opal structure with 500 nm voids, were obtained by closely packed PS microspheres (Figure 8e(i)), and honeycomb structures were made by breath figure arrays (Figure 8e(ii),f). Furthermore, the resulting film, transferred to a piece of glass, was annealed at 100 °C for 10 min to enhance its mechanical properties. Subsequently, Au NPs were attached by reduction to confer SERS ability on the PDA film. Such SERS substrates with a large surface area, hydrophobicity, and new photonic features showed brilliant SERS performance (Figure 8g). They also reported that bacterial cellulose and silk fibroin, some natural polymers, could support them for promising applications [135].

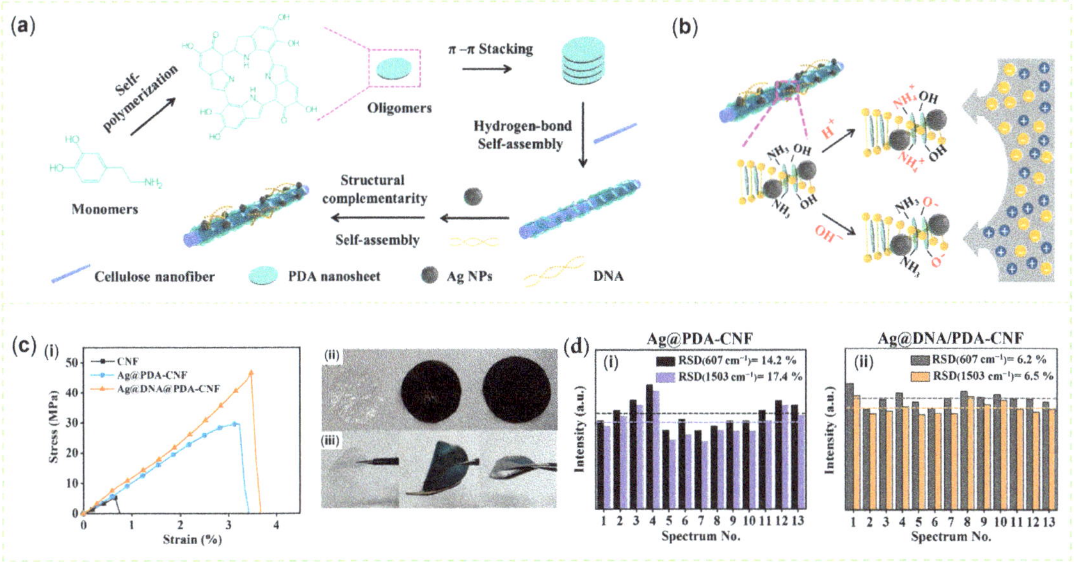

Figure 7. The advantages from the DNA-modified molecular structure of PDA for the SERS sensor. (**a**) Formation mechanism and structure diagram of the Ag@DNA/PDA-cellulose nanofibers (CNF). (**b**) Schematic showing the charge selectivity of the Ag@DNA/PDA-CNF film to solution molecules in acidic and alkaline media. (**c**) (**i**) Stress–strain curves of the samples under a dry condition. Photograph of each membrane (**ii**) in the natural state and (**iii**) when it was folded by tweezers. (**d**) SERS spectra of the R6G solution of 10^{-5} M were obtained by randomly collecting 13 points on Ag@PDA-CNF and Ag@DNA/PDA-CNF. The corresponding SERS intensities distributed at 607 and 1503 cm^{-1}: (**i**) Ag@PDA-CNF and (**ii**) Ag@DNA/PDA-CNF [51]. Copyrights: all images have been adapted and reproduced with permission from: (**a**–**d**) © 2021, American Chemical Society.

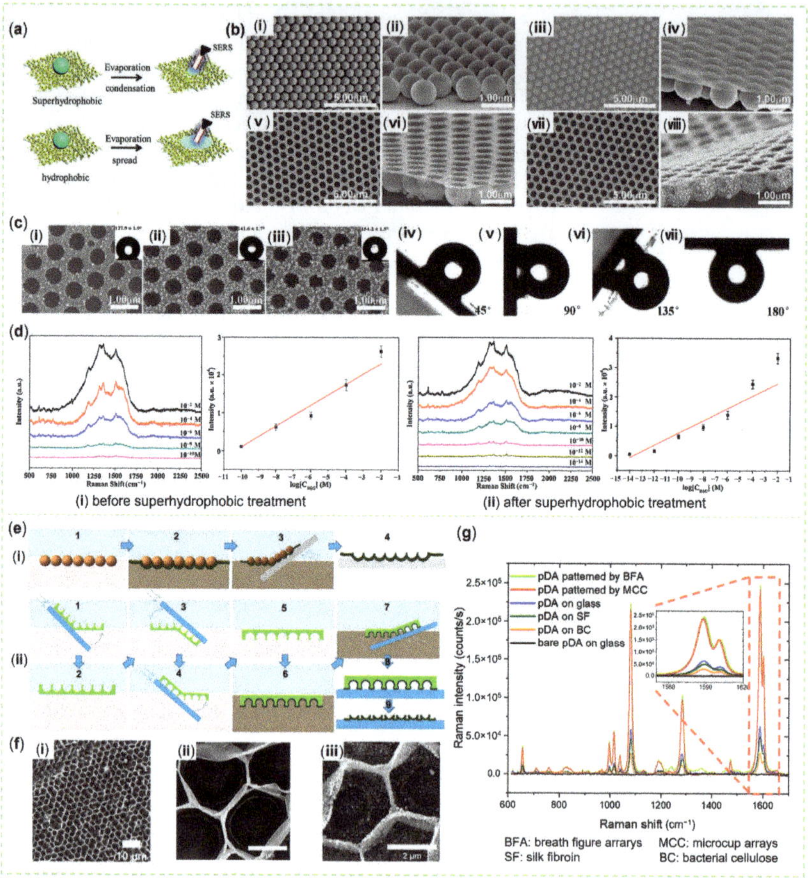

Figure 8. The hydrophobic PDA films for SERS sensors. (**a**) Schematic illustration for the evaporating and spreading process of the droplet [142]. (**b**) Scanning electron microscopy (SEM) images of (**i,ii**) monolayer PS colloidal crystals; (**iii,iv**) PS-PDA colloidal monolayer film; (**v,vi**) PDA microbowl array film and (**vii,viii**) PDA-Ag microbowl array substrate. SEM images were observed from different directions: (**i,iii,v,vii**) from the top view; (**ii,iv,vi,viii**) from the side view. (**c**) SEM images of PDA-Ag array substrates prepared using different concentrations of $[Ag(NH_3)_2]^+$ ions: (**i**) 1.2×10^{-2} M, (**ii**) 2.4×10^{-2} M, and (**iii**) 4.8×10^{-2} M and modified with *1H, 1H, 2H, 2H*- perfluorodecanethiol (PFDT). Insets: the corresponding water contact angle measurement; (**iv–vii**) photographs showing a water drop on the substrate (**iii**) with tilted angles from 45° to 180°. (**d**) Comparison of the detection range of the PDA-Ag array between before (**i**) and after (**ii**) the superhydrophobic treatment [134]. (**e**) Fabrication scheme of patterned PDA surfaces by using closely packed colloidal spheres (**i**) and breath figure arrays (**ii**) as templates. (**f**) SEM images of PDA film patterned with breath figure arrays (**i,ii**) and further decorated with Ag NPs (**iii**). (**g**) SERS activity of PDA films with or without special patterns [135]. Copyrights: all images have been adapted and reproduced with permission from: (**a**) © The Royal Society of Chemistry 2017; (**b–d**) © 2017 Elsevier B.V.; (**e–g**) © 2022 Elsevier B.V.

The mechanical properties of PDA, such as compressive strength, adhesive strength, and elastic modulus, will be the hot spots for practical applications in the future [17]. Coy et al. prepared free-standing and transferable PDA thin films with high hardness and elasticity [148]. The novel semi-permeable PDA film, similar to the cell membrane, could incorporate molecules with specific biological functions for the separation of target analytes [149]. However, the long-term accumulation of SERS tags functionalized by PDA in the livers of mice may lead to side effects in the body, which needs to be carefully considered [84].

5. Biomedical Application

In the past, multiple studies have proved that the realization of sensing and imaging through innovative SERS substances was conducive to monitoring diverse diseases and evaluating post-treatment conditions [30,150,151]. Nonetheless, there is still a long way to go to produce biosensors that can meet the budget and are suitable for a variety of site applications to achieve portable, field-deployable, accurate, rapid, and affordable point-of-care diagnostics [30,124,152].

PDA-modified materials cooperating with proteins, enzymes, peptides, nucleotides, polysaccharides, lipids, and living cells have attracted much attention in the field of biomedicine. PDA is usually able to retain the biological activity of the biomolecules. Particularly, after in vitro and in vivo tests, valuable biocompatibility can be demonstrated at the level of the tissue, cell, protein, and gene. They are beneficial for cell adhesion and growth, avoid damage to mitochondria, reduce the production of reactive oxygen species, and resist inflammation and senescence [47]. As the basic unit of the human body, cells can produce proteins, nucleic acids, lipids, reactive oxygen species, and other biomolecules. The specific abnormal expression of certain biomolecules can be a sign of serious disease [153]. Biosensors are often prepared with the goal of distinguishing and monitoring the type and level of this substance for prevention, diagnosis, and follow-up. The following section discusses the medical diagnostic applications of related materials in SERS biosensing as reported in recent research.

5.1. Detection of Nucleic Acid

Nucleic acids, containing DNAs and RNAs, as promising biomarkers of several diseases have complex molecular structures that can be adhered to PDA film easily to endow SERS materials with the ability of biorecognition. There were some excellent instances, as follows, offering instructive advice for designing such biosensors.

Obtaining a good balance between constructing a lot of hot spots in different dimensions [154] and improving the quantitative capacity is not an easy task, especially for actual sample testing [155]. Molecules entering the region of SERS-active materials with hot spots show an anomalously significant signal [156]. Therefore, it is a top priority to obtain all the target analytes in the sample into the signal enhancement, and to eliminate the disruptors. This purpose can be attained by porous enclosures that restrict the proximity of distractors [155] and materials that achieve anti-fouling effects through hydration and steric hindrance [157]. To control analyte–particle and particle–particle distances in the detection matrix and improve the stability and repeatability of the sensing, Zhou et al., designed host–guest nanosensor particles (MPDA@Au-SAM) (Figure 9a) that were uniformly assembled into a coffee ring pattern on the surface of a porous PVDF film. In their work, mesoporous PDA (MPDA) NPs with sizes of 150 ± 20 nm and an average pore size of 35 nm were synthesized for the in situ reduction of chloroauric acid. At the optimized $HAuCl_4$/MPDA weight ratio of 2.4, largely enriched and uniformly distributed Au NPs could be installed on the outer surface of MPDA. Additionally, then, a SAM of 11-Mercapto-1-undecanol was covered on AuNPs by Au-S bonds. As a result, through the strong π-π stacking interaction of PDA and the low affinity of SAM to aromatic molecules, this nanomaterial could select and enrich SERS active analytes into the pores of MPDA. Moreover, porous materials with a larger surface area could immobilize plentiful plasmonic particles and analytes. In particular, the powerful FL quenching effect of PDA was also used to highlight the Raman signal when they employed a Cy5-labeled DNA probe to recognize miRNA-21 of a lower concentration. The permeable porous platform was utilized to enhance the CRE (Figure 9b) to obtain a lower relative standard deviation (Figure 9c) [158].

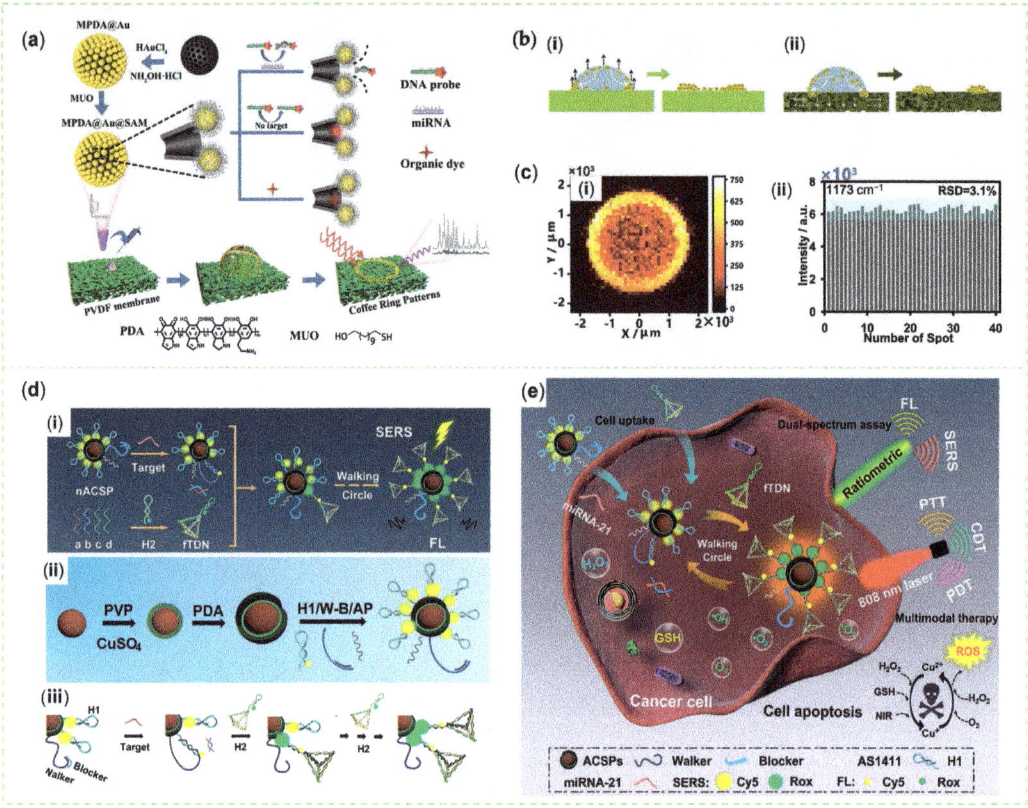

Figure 9. SERS biosensors for nucleic acid detection facilitated by PDA. (**a**) Schematic representation of the construction of SERS sensing devices based on two-tier host–guest architectures and regioselective enrichment effect. (**b**) Schematic illustrations for the process of forming a coffee ring pattern on the nonporous support-glass sheet (**i**) and porous support-PVDF membrane (**ii**). (**c**) SERS mapping results when using PVDF membrane (**i**); the bar chart for the peak intensity on PVDF membrane (**ii**) [158]. (**d**) (**i–iii**) Design and detailed working principle of the fTDN (fuel DNA-conjugated tetrahedral DNA nanostructures)-assisted DNA walking nanomachine for simultaneous ratiometric SERS-FL assay of microRNA (miRNA)-21. (**e**) Illustration of the developed nanodevice for miRNA detection and imaging in living cells and Au@$Cu_{2-x}S$ NPs (ACSPs)-mediated multimodal synergistic therapy [110]. Copyrights: all images have been adapted and reproduced with permission from: (**a–c**) © 2022 Elsevier B.V.; (**d**,**e**) © 2021 American Chemical Society.

The invention of sensors that meet the needs of multi-mode sensing and the integration of diagnosis and treatment is the future trend, which improves the reliability of diagnostic results and is in line with the concept of sustainable development. Zhai et al. produced a SERS/EC sensor, embracing the sensitivity of SERS sensing and the reproducibility of electrochemical sensing. When gastric cancer-related miRNA-106a was present in the sample solution, a sandwich structure was constructed by combining MPA@MB-P2 and the Ag NRs array electrode through the linkage between miR-106a and ssDNAs P1 and P2. Consequently, MB molecules were loaded on the MPA localized in the gap and a featured peak at 1622 cm^{-1} with enhanced intensity could be chosen for quantitative analysis [159]. In their previous research, the MPA could kill *Staphylococcus aureus*, consisting of MoS_2 NSs, PDA, and Ag NPs. The PDA film coated on the MoS_2 could be synthesized via a microwave in ten minutes. Ag^+ ions were absorbed by various nitrogen-containing and phenolic groups of PDA, and then reduced by glucose to form 5 nm particles uniformly distributed on the NSs [160].

Unfortunately, the one-to-one binding of the biometric element to the target nucleic acid strand compromised the sensitivity of the biosensor. However, the implementation of dynamic DNA nanomechanics results in a reduction of the LOD and an expansion of the linear range. For example, using a PDA layer covered on the ACSPs, He et al. were inspired by the dynamic DNA walking nanomachine and constructed a stable signal amplification system by quinone groups to link the sulfhydryl group modified on the nucleic acid molecule. When the target miRNA was present, a DNA walker (Figure 9d) binding the blocker was released, unfolding a hairpin-shaped Cy5-labeled DNA strand H1 on the particle, catalyzing its interaction with the corresponding ROX-labeled H2 strand of a DNA tetramer outside the particle. After the completion of this automatic cycle, the NP was filled with H1–H2 complexes. Additionally, Cy5, initially located close to the surface of the particle and exhibiting strong SERS and weak FL emission, ultimately migrated away from the surface resulting in enhanced FL and diminished SERS response. Conversely, ROX showed the opposite trend. This phenomenon was attributed to the surface plasmon resonance of the metal substrate and the FL quenching ability of PDA for calculating the SERS and FL signal ratio. Moreover, the AS1411 aptamer could be added by PDA to make it easier for ACSPs to enter the diseased cells. The photothermal conversion ability of PDA could also work in cancer therapy (Figure 9e) [110]. Jiang et al. demonstrated that the cooperation of Au@PDA@Ag NPs with 4-mercaptobenzoic acid (4-MBA), magnetic NPs (MNPs), and a re-circulating enzyme amplification system could detect tumor genes. A kind of hairpin DNA with three regions could coexist with the bridging DNA and nicking endonuclease in the absence of target RNA. The 5′ end of the bridging DNA hybridized with aptamer S_1 on the MNPs. Additionally, the remaining oligonucleotide chain at the 3′ end was related to PDA on the Au@PDA@Ag SERS tag by π-π bonding. Finally, they were all separated magnetically from the supernatant. However, when miRNA-31 was present, the hairpin DNA unfolded, and the target could be fixed on Region I. The bridging DNA bound to Region II and Region III of the hairpin DNA could be cleaved by the nicking enzyme to obtain the S_1 complementary strand. Importantly, the resulting duplex on the MNPs would not react with PDA to make the SERS tags stay in the supernatant. Next, the released Region II and Region III were further hybridized with the newly introduced bridging DNA in a cyclic manner to achieve signal enlargement. Consequently, quantitative results were obtained according to the amount of bridging DNA and the SERS intensity of the supernatant [114]. Wu et al. designed a recyclable biosensor to identify miRNA 155, including a magnetic core-shell material constructed with the assistance of PDA, which was biocompatible and highly dispersed, and a signal amplification system that focused on enzymatic DNA digestion [161].

In summary, PDA have been mainly used in the preparation of biosensors for monitoring nucleic acids to reflect the health status by loading probes. Based on the above cases, nucleotide chains with bases and signal molecules can be linked by PDA to enter or leave the electromagnetic field.

Based on various designs, including colloidal solutions and three intelligent platforms, Table 2 summarizes the data of PDA-based SERS biosensors applied in the biomedical field.

Table 2. Data on SERS biosensors with participation of PDA.

Different Design	Material	Method	Targets	Linear Detection Range	LOD	Ref.
Colloidal solutions	ACE2 *-mag-MoO_3-PDA@Au-4-MBA	SERS	SARS-CoV-2 spike protein	10 fg mL^{-1} –1 ng mL^{-1}	4.5 fg mL^{-1} (in PBS) 9.7 fg mL^{-1} (in whole-cell lysate media)	[68]
	PDA@Ag-anti-cTn I *	SERS	cTn I	- -	0.01 ng mL^{-1} (in PBS) 0.025 ng mL^{-1} (in human serum)	[80]
	Au@$Cu_{2-x}S$@PDA	FL	miRNA-21	1 pM–10 nM	0.11 pM	[110]
		SERS		10 aM–1 nM (in vitro)	4.95 aM	
				0.29 fM–9.30 pM (in living cells)	0.11 fM	
	Au@PDA@Ag	SERS	miRNA-31	0.6–1.8 fM	0.2 fM	[114]
	Fe_3O_4@PDA/Pt-TB/S1 and AuNFs-modified S2	SERS	miRNA 155	1 fM–10 μM	0.28 fM	[161]
	PDR*@Au NPs and Au nanocages@4-MBA	SERS	SCCA	10 pg mL^{-1} –1 μg mL^{-1}	7.16 pg mL^{-1} (in PBS) 8.03 pg mL^{-1} (in peripheral blood)	[162]
Modified chips	Au@EB *@PDA@Ab NPs and GCMS *@PDA-Au@MB@Ag@4-ATP */MPB * NPs-MIP circular array	SERS	CEA *	0.1 pg mL^{-1} –10 μg mL^{-1}	0.064 pg mL^{-1}	[102]
	S-agCDs@PDA-MNPs-Ag NCs and a single-layer graphene substrate	SERS	NoV *	1 fg mL^{-1} –10 ng mL^{-1}	0.1 fg mL^{-1} (in PBS) 0.95 fg mL^{-1} (in 10% human serum)	[112]
				10–10^6 RNA copies mL^{-1} (clinical NoV detection)	10 RNA copies mL^{-1}	
		FL		10 fg mL^{-1} –10 ng mL^{-1}	5.8 fg mL^{-1} (in PBS) 6.5 fg mL^{-1} (in 10% human serum)	
				10^2–10^6 RNA copies mL^{-1} (clinical NoV detection)	80 RNA copies mL^{-1}	
	Au@Ag/4-ATP@PDA@Ab and PDA-modified glass ship	SERS	migration inhibitory factor on exosome	5.44×10^2–2.72×10^4 particles/mL	one exosome in a 2 μL (9×10^{-19} mol L^{-1})	[115]
			glypican-1	5.44×10^2–2.72×10^4 particles/mL	9×10^{-19} mol L^{-1}	
			epidermal growth factor receptor	5.44×10^2–2.72×10^4 particles/mL	9×10^{-19} mol L^{-1}	
			CD63	2.72×10^3–2.72×10^4 particles/mL	4.5×10^{-18} mol L^{-1}	
			EpCAM	5.44×10^2–2.72×10^4 particles/mL	9×10^{-19} mol L^{-1}	
	Capillary glass@Au NSs@MIP	SERS	trypsin enzyme	0.01–1000 μg L^{-1}	4.1×10^{-3} μg L^{-1}	[117]
			pepsin BSA hemoglobin	1×10^{-3} –1000 μg L^{-1}	0.6×10^{-3} μg L^{-1} 0.4×10^{-3} μg L^{-1} 0.4×10^{-3} μg L^{-1}	
	MPDA@Au-SAM and PVDF	SERS	miRNA-21	1 pM–10 μM	308.5 fM	[158]

Table 2. *Cont.*

Different Design	Material	Method	Targets	Linear Detection Range	LOD	Ref.
	MPA@MB-P2 And AgNRs array electrode	SERS EC	miRNA-106a	100 fM–100 nM	67.44 fM 433.34 fM	[159]
	Au-PS-PDA-Si chip	SERS	cTn I creatine kinase isoenzyme MB	0.01–100 ng mL^{-1}	3.16 pg mL^{-1} 4.27 pg mL^{-1}	[163]
	Au@Ag/4-MPY*@BSA@PDA@Ab and PDA-modified glass chip	SERS	albumin	10–300 mg/L	0.2mg/L	[164]
	Au@MPB PDA-MIPs glass slide	SERS	acid phosphatase	18.2–1.82 × 10^6 pM (1 ng mL^{-1} –100 μg mL^{-1})	1.82 pM (0.1 ng mL^{-1})	[165]
			horseradish peroxidase	1 ng mL^{-1} –100 μg mL^{-1}	-	
			transferrin	0.1 ng mL^{-1} –10 μg mL^{-1}	-	
Microfluidic devices	Au NRs and MiChip *	SERS	prostate-specific antigen CEA α-fetoprotein	0.1–100 ng mL^{-1}	10 pg mL^{-1}	[113]
			Escherichia coli O157:H7 *Staphylococcus aureus*	10^0–10^4 CFU μL^{-1}	-	
	Au NPs@NTP@Ag and MiChip		prostate-specific antigen	0–1 pg mL^{-1}	0.2 pg mL^{-1}	
Lateral Flow devices	Fe_3O_4@PDA@AuNPs	colorimetry	HCG *	0–500 mIU mL^{-1}	10 mIU mL^{-1}	[166]
		magnetic signal		0–500 mIU mL^{-1}	1.2 mIU mL^{-1}	
		SERS		0–50 mIU mL^{-1}	0.2 mIU mL^{-1}	
	PDA@Ag-NPs	SERS	SCCA *	10 pg mL^{-1} –10 μg mL^{-1}	7.156 pg mL^{-1} (in PBS) 8.093 pg mL^{-1} (in human serum)	[167]
			cancer antigen 125		7.182 pg mL^{-1} (in PBS) 7.370 pg mL^{-1} (in human serum)	

* ACE2: angiotensin-converting enzyme 2; cTn I: cardiac troponin I; PDR: PDA resin microspheres; EB: ethynylbenzene; GCMS: Au coated microarray substrate; 4-ATP: 4-aminothiophenol; MPB: 4-mercaptophenylboronic acid; CEA: carcinoembryonic antigen; NoV: norovirus; 4-MPY: 4-Mercaptopyridine; MiChip: microfluidic chip; HCG: human chorionic gonadotropin; SCCA: squamous cell carcinoma antigen.

5.2. Detection of Protein

Abs, peptide sequences, and aptamers are commonly used to recognize protein and are easily captured by PDA. Additionally, interestingly, MIP synthesized by PDA further simplifies recognition components and can be reused for sustainability. In this part, relevant research cases are classified into four sections, powder reagents, LIFA, microarray, and microfluidic platform.

Lu et al., synthesized PDR with a loose surface and functional groups for loading Au NPs. The PDR@Au NPs with SCCA monoclonal Ab cooperated with the hollow Au nanocages with 4-MBA and SCCA polyclonal Ab to recognize SCCA. The material still had good sensitivity and reproducibility when applied to clinical blood samples from patients with cervical cancer and cervical intraepithelial neoplasia, and healthy subjects [162]. Magnetic ingredients can facilitate the separation of powered particles dispersed in the sample, enhance the sensitivity of detection, and recycle the active sensors. Achadu et al. inserted semiconducting oxides, plasmonic metals, and magnetic materials into a hollow and porous PDA nanoskeleton to synthesize mag-MoO_3-PDA@Au nanospheres with the tunable charge-transfer and LSPR effect via magnetic actuation. 4-MBA with a higher enhancement factor and PICT potential was designated as the Raman reporter attached to Au NPs. Angiotensin-converting enzyme 2 was the capturing agent for the SARS-CoV-2 spike protein. In the detection process, the immunocomplexes with a sandwich type could be enriched and extracted simply by an external magnet from PBS and whole-cell lysate media [68].

A variety of flexible and intelligent devices are needed to lift the usefulness of biosensors. Xia et al. utilized PDA nanospheres to carry Ag NPs and be loaded with 4-ATP and SCCA, or 5,5′-dithiobis-(2-nitrobenzoic acid) and cancer antigen 125. Combined with LFIA, the high precision of SERS detection results in clinical samples showed the potential for the diagnosis of cancer [167]. Exploring the potential of various properties of nanomaterials for multimode detection is conducive to strengthening the reliability of detection results. Liu et al. designed CSSNS combined with LFIA to detect the level of HCG through three signals (Figure 10a), namely color, magnetic, and Raman, for the rapid, accurate, and convenient diagnosis of malignant trophoblastic carcinoma. The dark brown Fe_3O_4 nanobeads were hydrophilic and were used as the core for PDA coating. PDA could bind Au^{3+} by amino groups and reduce Au^{3+} to Au seeds by catechol groups, providing an interface for 4-MB. 4-MB also showed reducibility in the formation of Au NPs. The intensity of the 4-MB Raman peak was related to its amount, but it should not be superfluous to occupy the sites for Ab in order to ultimately weaken the SERS effect. After layer-upon-layer modification, polycrystalline Fe_3O_4 nanobeads still had high magnetic strength. An ideal diagnostic effect could be obtained by such a sophisticated formulation [166].

The adhesion of PDA has shown an extraordinary effect on constructing arrays for catching target analysis. Wang et al. used the adhesiveness of PDA to arrange PS microspheres on the silicon wafer one by one. Optical confinement from the microsphere's cavity and the electromagnetic field from Au NPs were beneficial for improving SERS activity. Ab-antigen-Ab sandwich immunocomplexes brought SERS reporters near the chip for a quantitative analysis of the two signature proteins, cTn I and creatine kinase isoenzyme MB, that could monitor cardiovascular disease [163]. Huang et al. combined a PDA self-polymerized glass chip possessing albumin Ab with Ab-modified bimetallic SERS tags doped with 4-MPY to detect albumin in urine. The intensity of the peak at 1096 cm^{-1} was of great significance for monitoring the prognosis of cardiovascular disease and diabetic nephropathy [164]. Abundant exosomes derived from cancer cells are continuously released into body fluids, including serum, plasma, and urine. These exosomes play a crucial role in promoting immunosuppression, angiogenesis, cell migration, and invasion during the occurrence and progression of cancer. Carrying information about the tumor microenvironment, they are reliable biomarkers for non-invasive early cancer diagnosis, and the evaluation of therapeutic efficacy. A large number of characteristic proteins on the surface of exosomes are ideal sites for biometrics [168]. Li et al. chose the optimum Ab to capture exosomes by a specific protein expressed on their membranes for the early diagnosis, clinical staging, and tumor metastasis monitoring of pancreatic cancer. A PDA-coated glass chip could easily bind to a variety of Abs. The chip modified by PDA and an Ab was homogeneous and hydrophilic. As a result, the uniform distribution of SERS labels could be achieved without CRE, after the solvent evaporated from the sample on the chip [115].

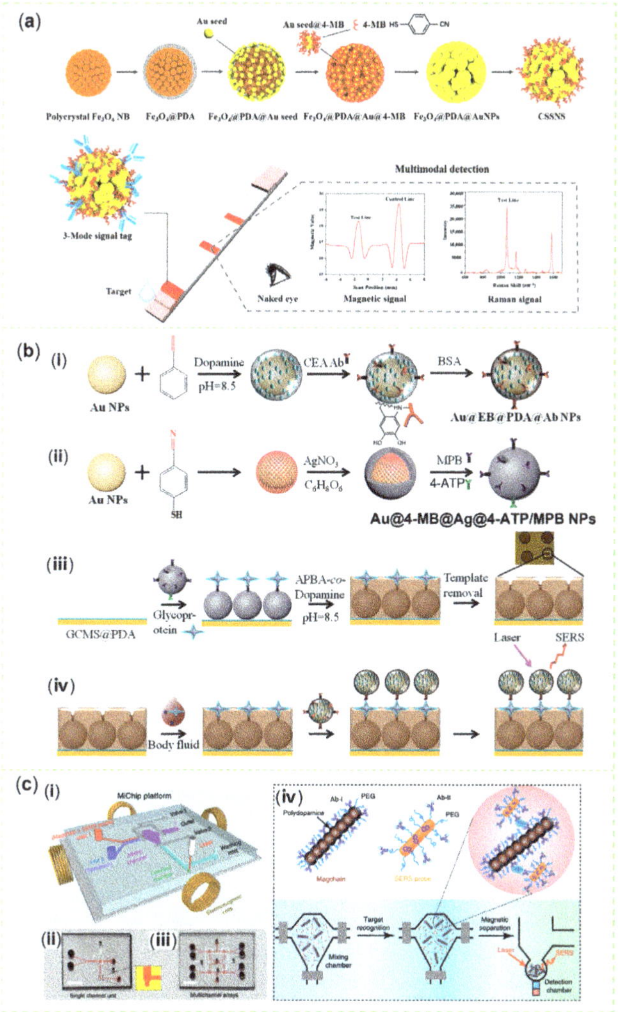

Figure 10. PDA-containing materials cooperated with diverse smart platforms for protein monitoring. (**a**) Schematic illustration of core-shell-shell nanosunflowers (CSSNS) preparation and the multimodal detection of HCG using CSSNS-based lateral flow immunoassay (LFIA) [166]. (**b**) The fabrication process of PDA-encapsulated SERS probes. Components of the (i) Au@EB@PDA@Ab NPs and (ii) Au@4-mercaptobenzonitrile (4-MB) @Ag@4-ATP/MPB NPs. (**iii**) Synthetic route of preparing boronated affinity-based glycoprotein imprinted arrays. (**iv**) The MIP-based SERS approach for the detection of glycoprotein [102]. (**c**) Design of the magnetic nanochain integrated MiChip. (**i**) Schematic diagram of the MiChip assay platform. (**ii**,**iii**) Photographs of the MiChip: single channel unit (**ii**) and multichannel arrays (**iii**). The microchannel was filled with a red dye for better visualization. Scale bar: 0.5 cm. (**iv**) The MiChip assay for the detection of biomarkers. The specimen, Ab-conjugated magnetic nanochains and SERS probes are mixed in the mixing chamber. The targets of interest in the sample are recognized by the Abs on the Magchains and the SERS probes to form sandwich immune complexes. The immune complexes are then isolated into the detection chamber and subjected to Raman spectroscopic detection [113]. Copyrights: all images have been adapted and reproduced with permission from: (**a**) © 2021 American Chemical Society; (**b**) © 2019 Elsevier B.V.; (**c**) © 2018, Qirong Xiong et al. published by Springer Nature.

The selectivity of the PDA-MIP SERS substrate has been fully described above. The following research work will reveal its great potential in biomedical detection. Arabi et al. produced a cheap sensor that could specifically identify four kinds of proteins for quantitative analysis with a SERS signal-reduction mode. A glass capillary coated with Au NSs and a PDA-MIP was immersed in the sample without pretreatment. After the target protein occupied the cavities, a Coomassie brilliant blue G-250 staining step blocked empty pockets between peptide chains so that DTTC could not approach Au NSs [117]. Li et al. immobilized a glycoprotein template in a polymer skeleton containing boric acid and then produced PDA-MIP on it to identify a variety of glycoproteins playing an important regulatory role in metabolism [165]. Significantly, the combination of MIP and Ab may elevate the sensitivity of biosensors. Lin et al. incorporated 4-MB as an internal standard within the gap of Au and Ag bimetal SERS labels to fabricate core-molecule-shell-molecule (CMSM) NPs. A circular array was formed on the modified GCMS through the Michael addition reaction between 4-ATP and PDA (Figure 10b). Then, on the array, CEA glycoproteins were used as the template to form a PDA-MIP layer with appropriate thickness and were eluted cleanly. There were also MPB ligands on CMSM NPs which could covalently and reversibly capture the cis-dihydroxyl of glycoprotein from CEA by boric acid. CEA could still bind to the Ab on PDA-coated Au NPs with EB. The EB molecule had a sharp band at 2024 cm^{-1} for the alkynyl group, and the 4-MB molecule had a narrow band at 2224 cm^{-1} for the cyan croup. Both peaks were located in a Raman silent region, avoiding the interference of fingerprint peaks from endogenous molecules. The ratio of featured SERS peaks of 4-MB and EB had a stronger linear relationship with the content of CEA [102].

Xiong et al. observed the formation of chain-like structures of PDA and Fe_3O_4 NPs in an alkaline environment. Notably, its length was increased by lengthening the magnetic stirring time. The shorter chain could be obtained by ultrasonic treatment. By increasing the initial DA concentration, a wider chain was obtained. Stirring bars with the right length and width should be synthesized to achieve the highest rotating speed and the most obvious convection to accelerate mixing in a corresponding rotating magnetic field. The combination of the bioconjugated magnetic nanochains with the MiChip enabled full contact and identification of the target analytes by stirring. Most importantly, it was able to facilitate the reuse of this system without blocking its channels. Finally, Au NRs with different Raman reporters were added to form a sandwich immune recognition suit with nanochains to detect various tumor markers and bacteria in saliva (Figure 10c). This new sensor gained excellent results rapidly, easily, sensitively, and in an environmentally friendly way compared to ELISA and bacterial culture methods commonly applied in the clinic [113].

Obviously, for high sensitivity, selectivity, repeatability, and recyclability, the functions of PDA in the invention of novel and smart biosensors cannot be limited to introducing elements for biological recognition, but all aspects of the performance mentioned in the guideline for designing SERS substrate should be incorporated into our train of thought.

5.3. Detection of Other Biological Indexes

Here are some examples of the non-specific recognition of bacteria. Wang et al. captured and magnetically separated bacteria bound to boronic acid-modified and PDA-stabilized SERS labels by IgG@Fe_3O_4. Then, through principal component analysis and hierarchical cluster analysis, five pathogenic bacteria can be classified and quantified with a LOD value of 10 CFU mL^{-1} [77]. Wan et al. determined the optimal amount of Ag NPs deposited on SiO_2 nanofibers for effective differentiation and eradication of diverse bacteria, while the amino and hydroxyl groups in PDA immobilized Ag^+. The components such as tyrosine, adenine, guanine, saturated lipid, and amide II of proteins in bacteria had Raman peaks, which facilitated the label-free SERS discrimination of Gram-negative and positive bacteria [96]. Hong et al. utilized PDA to facilitate the anisotropic growth of Au on Fe_3O_4 clusters, resulting in the formation of short microvilli and ultimately achieving a three-dimensional leukocyte-shaped structure. The large surface area of Fe_3O_4 microvilli could capture and isolate bacteria for label-free SERS analysis. Moreover, the material to achieve high photothermal conversion under near-infrared light could inhibit Gram-negative and positive, anaerobic, drug-resistant pathogens [169].

Reactive oxygen species consist of the hydroxyl radical (•OH), superoxide (O_2^-), and hydrogen peroxide (H_2O_2). They can cause a higher level of oxidative stress at organismal, cellular, and molecular levels and have a close relationship with abnormal metabolic activity and mitochondrial function. Thus, they can act as diagnostic markers for various diseases [170–173]. A kind of chemistry-based SERS nanoprobe with a core-satellite structure assembled by PDA could differentiate the cancerous cells and monitor the autophagy process in living cells. Fe(IV)=O could be obtained with an evident change in SERS intensity at 1386 cm^{-1} through the reaction between reactive oxygen species and the six-coordinated Fe(III)-OH_2 of myoglobin. The amine groups of myoglobin were proven to be carried by the Michael addition [174].

6. Conclusions and Perspectives

This review mainly discussed the application potential of PDA, a biomimetic polymer with rich functions, in the design of new, sensitive, and specific SERS sensors. PDA is considered an eco-friendly material due to its reliable biocompatibility and biodegradability, as well as its simple synthesis process that does not require energy-intensive equipment. Furthermore, the PDA-based SERS sensor finally obtained also had the possibility of reuse. Therefore, the combination of the two was selected to increase the possibility of preparing a biosensor that could be sustainably used. Using the various advantages of PDA to prepare SERS sensors for food safety and environmental protection has received a lot of attention and achieved good research results, such as for the detection of pesticide residues [175], and toxins [108] on the surface of fruits and vegetables, harmful substances in water, etc. However, there is a paucity of research studies conducted on PDA-based SERS sensors within medical field.

The SERS assay has some disadvantages that can be solved by combination with other analytic methods to increase confidence. For instance, the laser excitation region is small so multiple spots should be randomly selected on the sample to be tested for SERS characterization to calculate the average value. However, the SERS signal results obtained at each site can vary greatly, which leads to worse reproducibility. Thus, the electrochemical sensor with relatively stable signals from the entire electrode can be introduced to compensate for this shortcoming [159]. PDA also demonstrated the ability to amplify photoacoustic signals [137]. Integrated diagnostic and therapeutic materials can provide precise location and dosage guidance, minimizing drug waste while optimizing treatment outcomes. PDA was not only a material with a good photothermal treatment effect, but also could carry multiple chemical drugs for chemodynamical treatment [137]. In the field of tissue regeneration engineering, amyloid fibers were employed as catalysts and peptide nanostructures were utilized to direct the polymerization of DA. Subsequently, boric acid was bound to catechol groups in PDA for pH-responsive properties [176]. In the molecule of DA, there are four sites on amino, alkyl, and aromatic rings that can be chemically derived. DA derivatives and their analogues with better surface modification abilities can learn from the cases of PDA in order to obtain more practical SERS sensors. For example, plant-derived polyphenols exhibited swift metal–ligand coordination kinetics for the formation of metal-polyphenol network coatings, in addition to possessing antibacterial and antioxidant biological properties [17].

Through this review, we not only demonstrate the convenience of PDA in fabricating SERS biosensors, but also provide insights into further developments of multimodal sensors and theranostic applications. We believe that this study will be of interest to diverse researchers in the interdisciplinary fields of nanomedicine, biosensing, organic chemistry, and materials engineering, and we hope that it will rapidly stimulate more applied research in this area.

Author Contributions: L.T. writing—original draft; L.T., L.W. and B.D. conceptualization; C.C. data curation; C.C. and J.G. validation and investigation; Q.H. and Y.S. resources and formal analysis; M.L. software; L.C., L.W. and B.D. supervision; L.W. funding acquisition; B.D. project administration. All authors have read and agreed to the published version of the manuscript.

Funding: This work was supported by the National Science Foundation of China (82170998).

Institutional Review Board Statement: Not applicable.

Informed Consent Statement: Not applicable.

Data Availability Statement: Data sharing not applicable.

Conflicts of Interest: The authors declare no conflict of interest.

References

1. Gilbertson, L.M.; Zimmerman, J.B.; Plata, D.L.; Hutchison, J.E.; Anastas, P.T. Designing nanomaterials to maximize performance and minimize undesirable implications guided by the Principles of Green Chemistry. *Chem. Soc. Rev.* **2015**, *44*, 5758–5777. [CrossRef] [PubMed]
2. Bartolucci, C.; Scognamiglio, V.; Antonacci, A.; Fraceto, L.F. What makes nanotechnologies applied to agriculture green? *Nano Today* **2022**, *43*, 101389. [CrossRef]
3. Han, X.X.; Rodriguez, R.S.; Haynes, C.L.; Ozaki, Y.; Zhao, B. Surface-enhanced Raman spectroscopy. *Nat. Rev. Methods Prim.* **2022**, *1*, 87. [CrossRef]
4. Langer, J.; Jimenez de Aberasturi, D.; Aizpurua, J.; Alvarez-Puebla, R.A.; Auguié, B.; Baumberg, J.J.; Bazan, G.C.; Bell, S.E.J.; Boisen, A.; Brolo, A.G.; et al. Present and Future of Surface-Enhanced Raman Scattering. *ACS Nano* **2020**, *14*, 28–117. [CrossRef] [PubMed]
5. Zong, C.; Xu, M.; Xu, L.J.; Wei, T.; Ma, X.; Zheng, X.S.; Hu, R.; Ren, B. Surface-Enhanced Raman Spectroscopy for Bioanalysis: Reliability and Challenges. *Chem. Rev.* **2018**, *118*, 4946–4980. [CrossRef]
6. Ouyang, L.; Hu, Y.; Zhu, L.; Cheng, G.J.; Irudayaraj, J. A reusable laser wrapped graphene-Ag array based SERS sensor for trace detection of genomic DNA methylation. *Biosens. Bioelectron.* **2017**, *92*, 755–762. [CrossRef]
7. Zhou, B.; Ou, W.; Zhao, C.; Shen, J.; Zhang, G.; Tang, X.; Deng, Z.; Zhu, G.; Li, Y.Y.; Lu, J. Insertable and reusable SERS sensors for rapid on-site quality control of fish and meat products. *Chem. Eng. J.* **2021**, *426*, 130733. [CrossRef]
8. He, Q.; Han, Y.; Huang, Y.; Gao, J.; Gao, Y.; Han, L.; Zhang, Y. Reusable dual-enhancement SERS sensor based on graphene and hybrid nanostructures for ultrasensitive lead (I I) detection. *Sens. Actuators B* **2021**, *341*, 130031. [CrossRef]
9. Du, Y.; Liu, H.; Chen, Y.; Tian, Y.; Zhang, X.; Gu, C.; Jiang, T.; Zhou, J. Recyclable label-free SERS-based immunoassay of PSA in human serum mediated by enhanced photocatalysis arising from Ag nanoparticles and external magnetic field. *Appl. Surf. Sci.* **2020**, *528*, 146953. [CrossRef]
10. Min, K.; Choi, K.S.; Jeon, W.J.; Lee, D.K.; Oh, S.; Lee, J.; Choi, J.-Y.; Yu, H.K. Hierarchical Ag nanostructures on Sn-doped indium oxide nano-branches: Super-hydrophobic surface for surface-enhanced Raman scattering. *RSC Adv.* **2018**, *8*, 12927–12932. [CrossRef]
11. Li, C.; Huang, Y.; Li, X.; Zhang, Y.; Chen, Q.; Ye, Z.; Alqarni, Z.; Bell, S.E.J.; Xu, Y. Towards practical and sustainable SERS: A review of recent developments in the construction of multifunctional enhancing substrates. *J. Mater. Chem. C* **2021**, *9*, 11517–11552. [CrossRef]
12. Gong, T.; Das, C.M.; Yin, M.J.; Lv, T.R.; Singh, N.M.; Soehartono, A.M.; Singh, G.; An, Q.F.; Yong, K.T. Development of SERS tags for human diseases screening and detection. *Coord. Chem. Rev.* **2022**, *470*, 214711. [CrossRef]
13. Sun, J.; Ma, Q.; Xue, D.; Shan, W.; Liu, R.; Dong, B.; Zhang, J.; Wang, Z.; Shao, B. Polymer/inorganic nanohybrids: An attractive materials for analysis and sensing. *TrAC Trends Anal. Chem.* **2021**, *140*, 116273. [CrossRef]
14. Lan, L.; Ping, J.; Xiong, J.; Ying, Y. Sustainable Natural Bio-Origin Materials for Future Flexible Devices. *Adv. Sci.* **2022**, *9*, 2200560. [CrossRef]
15. Lee, H.; Dellatore, S.M.; Miller, W.M.; Messersmith, P.B. Mussel-inspired surface chemistry for multifunctional coatings. *Science* **2007**, *318*, 426–430. [CrossRef]
16. Liu, Y.; Ai, K.; Lu, L. Polydopamine and Its Derivative Materials: Synthesis and Promising Applications in Energy, Environmental, and Biomedical Fields. *Chem. Rev.* **2014**, *114*, 5057–5115. [CrossRef]
17. Ryu, J.H.; Messersmith, P.B.; Lee, H. Polydopamine Surface Chemistry: A Decade of Discovery. *Acs Appl. Mater. Interfaces* **2018**, *10*, 7523–7540. [CrossRef]
18. Raman, C.V.; Krishnan, K.S. A New Type of Secondary Radiation. *Nature* **1928**, *121*, 501–502. [CrossRef]
19. Cialla-May, D.; Schmitt, M.; Popp, J. Theoretical principles of Raman spectroscopy. *Phys. Sci. Rev.* **2019**, *4*, 6. [CrossRef]
20. Kneipp, J.; Kneipp, H.; Kneipp, K. SERS—A single-molecule and nanoscale tool for bioanalytics. *Chem. Soc. Rev.* **2008**, *37*, 1052–1060. [CrossRef]
21. Fleischmann, M.; Hendra, P.J.; McQuillan, A.J. Raman spectra of pyridine adsorbed at a silver electrode. *Chem. Phys. Lett.* **1974**, *26*, 163–166. [CrossRef]

22. Jeanmaire, D.L.; Van Duyne, R.P. Surface raman spectroelectrochemistry: Part I. Heterocyclic, aromatic, and aliphatic amines adsorbed on the anodized silver electrode. *J. Electroanal. Chem. Interfacial Electrochem.* **1977**, *84*, 1–20. [CrossRef]
23. Cong, S.; Liu, X.; Jiang, Y.; Zhang, W.; Zhao, Z. Surface Enhanced Raman Scattering Revealed by Interfacial Charge-Transfer Transitions. *Innovation* **2020**, *1*, 100051. [CrossRef] [PubMed]
24. Viets, C.; Hill, W. Laser power effects in SERS spectroscopy at thin metal films. *J. Phys. Chem. B* **2001**, *105*, 6330–6336. [CrossRef]
25. Xia, M. 2D Materials-Coated Plasmonic Structures for SERS Applications. *Coatings* **2018**, *8*, 137. [CrossRef]
26. Zhang, X.; Si, S.; Zhang, X.; Wu, W.; Xiao, X.; Jiang, C. Improved thermal stability of graphene-veiled noble metal nanoarrays as recyclable SERS substrates. *Acs Appl. Mater. Interfaces* **2017**, *9*, 40726–40733. [CrossRef]
27. Chi, H.; Wang, C.; Wang, Z.; Zhu, H.; Mesias, V.S.D.; Dai, X.; Chen, Q.; Liu, W.; Huang, J. Highly reusable nanoporous silver sheet for sensitive SERS detection of pesticides. *Analyst* **2020**, *145*, 5158–5165. [CrossRef]
28. Wang, C.; Mu, X.; Huo, J.; Zhang, B.; Zhang, K. Highly-efficient SERS detection for E. coli using a microfluidic chip with integrated NaYF4: Yb, Er@ SiO2@ Au under near-infrared laser excitation. *Microsyst. Technol.* **2021**, *27*, 3285–3291. [CrossRef]
29. Li, L.; Jiang, R.; Shan, B.; Lu, Y.; Zheng, C.; Li, M. Near-infrared II plasmonic porous cubic nanoshells for in vivo noninvasive SERS visualization of sub-millimeter microtumors. *Nat. Commun.* **2022**, *13*, 5249. [CrossRef]
30. Hang, Y.; Boryczka, J.; Wu, N. Visible-light and near-infrared fluorescence and surface-enhanced Raman scattering point-of-care sensing and bio-imaging: A review. *Chem. Soc. Rev.* **2022**, *51*, 329–375. [CrossRef]
31. Cao, Y.; Zhang, J.; Yang, Y.; Huang, Z.; Long, N.V.; Fu, C. Engineering of SERS Substrates Based on Noble Metal Nanomaterials for Chemical and Biomedical Applications. *Appl. Spectrosc. Rev.* **2015**, *50*, 499–525. [CrossRef]
32. Guillot, N.; de la Chapelle, M.L. The electromagnetic effect in surface enhanced Raman scattering: Enhancement optimization using precisely controlled nanostructures. *J. Quant. Spectrosc. Radiat. Transf.* **2012**, *113*, 2321–2333. [CrossRef]
33. Nitzan, A.; Brus, L. Theoretical model for enhanced photochemistry on rough surfaces. *J. Chem. Phys.* **1981**, *75*, 2205–2214. [CrossRef]
34. Stiles, P.L.; Dieringer, J.A.; Shah, N.C.; Van Duyne, R.P. Surface-enhanced Raman spectroscopy. *Annu. Rev. Anal. Chem.* **2008**, *1*, 601–626. [CrossRef] [PubMed]
35. Willets, K.A.; Van Duyne, R.P. Localized surface plasmon resonance spectroscopy and sensing. *Annu. Rev. Phys. Chem.* **2007**, *58*, 267–297. [CrossRef] [PubMed]
36. Hergert, W.; Wriedt, T. Mie theory: A review. In *The Mie Theory: Basics and Applications*, 2nd ed.; Springer: Berlin/Heidelberg, Germany, 2012; Volume 169, pp. 53–71.
37. Alessandri, I.; Lombardi, J.R. Enhanced Raman Scattering with Dielectrics. *Chem. Rev.* **2016**, *116*, 14921–14981. [CrossRef]
38. Alessandri, I. Enhancing Raman scattering without plasmons: Unprecedented sensitivity achieved by TiO_2 shell-based resonators. *J. Am. Chem. Soc.* **2013**, *135*, 5541–5544. [CrossRef]
39. Shin, H.Y.; Shim, E.L.; Choi, Y.J.; Park, J.H.; Yoon, S. Giant enhancement of the RMaman response due to one-dimensional ZnO nanostructures. *Nanoscale* **2014**, *6*, 14622–14626. [CrossRef]
40. Cao, L.; Laim, L.; Valenzuela, P.D.; Nabet, B.; Spanier, J.E. On the Raman scattering from semiconducting nanowires. *J. Raman Spectrosc.* **2007**, *38*, 697–703. [CrossRef]
41. Cao, L.; Nabet, B.; Spanier, J.E. Enhanced Raman scattering from individual semiconductor nanocones and nanowires. *Phys. Rev. Lett.* **2006**, *96*, 157402. [CrossRef]
42. Saikin, S.K.; Olivares-Amaya, R.; Rappoport, D.; Stopa, M.; Aspuru-Guzik, A. On the chemical bonding effects in the Raman response: Benzenethiol adsorbed on silver clusters. *Phys. Chem. Chem. Phys.* **2009**, *11*, 9401–9411. [CrossRef]
43. Liang, X.; Liang, B.; Pan, Z.; Lang, X.; Zhang, Y.; Wang, G.; Yin, P.; Guo, L. Tuning plasmonic and chemical enhancement for SERS detection on graphene-based Au hybrids. *Nanoscale* **2015**, *7*, 20188–20196. [CrossRef] [PubMed]
44. Jensen, L.; Aikens, C.M.; Schatz, G.C. Electronic structure methods for studying surface-enhanced Raman scattering. *Chem. Soc. Rev.* **2008**, *37*, 1061–1073. [CrossRef] [PubMed]
45. Cheng, W.; Zeng, X.; Chen, H.; Li, Z.; Zeng, W.; Mei, L.; Zhao, Y. Versatile Polydopamine Platforms: Synthesis and Promising Applications for Surface Modification and Advanced Nanomedicine. *ACS Nano* **2019**, *13*, 8537–8565. [CrossRef] [PubMed]
46. Hong, S.; Na, Y.S.; Choi, S.; Song, I.T.; Kim, W.Y.; Lee, H. Non-Covalent Self-Assembly and Covalent Polymerization Co-Contribute to Polydopamine Formation. *Adv. Funct. Mater.* **2012**, *22*, 4711–4717. [CrossRef]
47. Alfieri, M.L.; Weil, T.; Ng, D.Y.W.; Ball, V. Polydopamine at biological interfaces. *Adv. Colloid Interface Sci.* **2022**, *305*, 102689. [CrossRef] [PubMed]
48. Liebscher, J. Chemistry of Polydopamine—Scope, Variation, and Limitation. *Eur. J. Org. Chem.* **2019**, *31–32*, 4976–4994. [CrossRef]
49. Liu, X.; Lin, W.; Xiao, P.; Yang, M.; Sun, L.P.; Zhang, Y.; Xue, W.; Guan, B.O. Polydopamine-based molecular imprinted optic microfiber sensor enhanced by template-mediated molecular rearrangement for ultra-sensitive C-reactive protein detection. *Chem. Eng. J.* **2020**, *387*, 124074. [CrossRef]
50. Alfieri, M.L.; Micillo, R.; Panzella, L.; Crescenzi, O.; Oscurato, S.L.; Maddalena, P.; Napolitano, A.; Ball, V.; d'Ischia, M. Structural Basis of Polydopamine Film Formation: Probing 5,6-Dihydroxyindole-Based Eumelanin Type Units and the Porphyrin Issue. *ACS Appl. Mater. Interfaces* **2018**, *10*, 7670–7680. [CrossRef]
51. Xu, X.Y.; Hu, X.M.; Fu, F.Y.; Liu, L.; Liu, X.D. DNA-Induced Assembly of Silver Nanoparticle Decorated Cellulose Nanofiber: A Flexible Surface-Enhanced Raman Spectroscopy Substrate for the Selective Charge Molecular Detection and Wipe Test of Pesticide Residues in Fruits. *ACS Sustain. Chem. Eng.* **2021**, *9*, 5217–5229. [CrossRef]

52. Lee, H.A.; Park, E.; Lee, H. Polydopamine and Its Derivative Surface Chemistry in Material Science: A Focused Review for Studies at KAIST. *Adv. Mater.* **2020**, *32*, 1907505. [CrossRef] [PubMed]
53. Xu, L.Q.; Yang, W.J.; Neoh, K.G.; Kang, E.T.; Fu, G.D. Dopamine-Induced Reduction and Functionalization of Graphene Oxide Nanosheets. *Macromolecules* **2010**, *43*, 8336–8339. [CrossRef]
54. Badillo-Ramirez, I.; Saniger, J.M.; Popp, J.; Cialla-May, D. SERS characterization of dopamine and in situ dopamine polymerization on silver nanoparticles. *Phys. Chem. Chem. Phys.* **2021**, *23*, 12158–12170. [CrossRef] [PubMed]
55. Alfieri, M.L.; Panzella, L.; Oscurato, S.L.; Salvatore, M.; Avolio, R.; Errico, M.E.; Maddalena, P.; Napolitano, A.; D'Ischia, M. The Chemistry of Polydopamine Film Formation: The Amine-Quinone Interplay. *Biomimetics* **2018**, *3*, 26. [CrossRef] [PubMed]
56. Lee, K.; Park, M.; Malollari, K.G.; Shin, J.; Winkler, S.M.; Zheng, Y.; Park, J.H.; Grigoropoulos, C.P.; Messersmith, P.B. Laser-induced graphitization of polydopamine leads to enhanced mechanical performance while preserving multifunctionality. *Nat. Commun.* **2020**, *11*, 4848. [CrossRef] [PubMed]
57. Sun, C.; Zhang, L.; Zhang, R.; Gao, M.; Zhang, X. Facilely synthesized polydopamine encapsulated surface-enhanced Raman scattering (SERS) probes for multiplex tumor associated cell surface antigen detection using SERS imaging. *RSC Adv.* **2015**, *5*, 72369–72372. [CrossRef]
58. Lin, J.S.; Tian, X.D.; Li, G.; Zhang, F.L.; Wang, Y.; Li, J.F. Advanced plasmonic technologies for multi-scale biomedical imaging. *Chem. Soc. Rev.* **2022**, *51*, 9445–9468. [CrossRef]
59. Zheng, Z.; Cong, S.; Gong, W.; Xuan, J.; Li, G.; Lu, W.; Geng, F.; Zhao, Z. Semiconductor SERS enhancement enabled by oxygen incorporation. *Nat. Commun.* **2017**, *8*, 1993. [CrossRef]
60. Chen, D.; Zhu, X.; Huang, J.; Wang, G.; Zhao, Y.; Chen, F.; Wei, J.; Song, Z.; Zhao, Y. Polydopamine@Gold Nanowaxberry Enabling Improved SERS Sensing of Pesticides, Pollutants, and Explosives in Complex Samples. *Anal. Chem.* **2018**, *90*, 9048–9054. [CrossRef]
61. Zhang, Z.; Si, T.; Liu, J.; Han, K.; Zhou, G. Controllable synthesis of AgNWs@PDA@AgNPs core-shell nanocobs based on a mussel-inspired polydopamine for highly sensitive SERS detection. *RSC Adv.* **2018**, *8*, 27349–27358. [CrossRef]
62. Chen, M.; Zhang, L.; Yang, B.; Gao, M.; Zhang, X. Facile synthesis of terminal-alkyne bioorthogonal molecules for live -cell surface-enhanced Raman scattering imaging through Au-core and silver/dopamine-shell nanotags. *Anal. Bioanal. Chem.* **2018**, *410*, 2203–2210. [CrossRef] [PubMed]
63. Tegegne, W.A.; Mekonnen, M.L.; Beyene, A.B.; Su, W.N.; Hwang, B.J. Sensitive and reliable detection of deoxynivalenol mycotoxin in pig feed by surface enhanced Raman spectroscopy on silver nanocubes@polydopamine substrate. *Spectrochim. Acta Part A* **2020**, *229*, 117940. [CrossRef] [PubMed]
64. Yilmaz, M. Silver-Nanoparticle-Decorated Gold Nanorod Arrays via Bioinspired Polydopamine Coating as Surface-Enhanced Raman Spectroscopy (SERS) Platforms. *Coatings* **2019**, *9*, 198. [CrossRef]
65. Choi, C.K.K.; Chiu, Y.T.E.; Zhuo, X.; Liu, Y.; Pak, C.Y.; Liu, X.; Tse, Y.L.S.; Wang, J.; Choi, C.H.J. Dopamine-Mediated Assembly of Citrate-Capped Plasmonic Nanoparticles into Stable Core–Shell Nanoworms for Intracellular Applications. *ACS Nano* **2019**, *13*, 5864–5884. [CrossRef] [PubMed]
66. Yuan, H.; Yu, S.; Kim, M.; Lee, J.E.; Kang, H.; Jang, D.; Ramasamy, M.S.; Kim, D.H. Dopamine-mediated self-assembled anisotropic Au nanoworms conjugated with MoS_2 nanosheets for SERS-based sensing. *Sens. Actuators B* **2022**, *371*, 132453. [CrossRef]
67. Park, B.; Dang, T.V.; Yoo, J.; Tran, T.D.; Ghoreishian, S.M.; Lee, G.H.; Il Kim, M.; Huh, Y.S. Silver nanoparticle-coated polydopamine-copper hybrid nanoflowers as ultrasensitive surface-enhanced Raman spectroscopy probes for detecting thiol-containing molecules. *Sens. Actuators B* **2022**, *369*, 132246. [CrossRef]
68. Achadu, O.J.; Nwaji, N.; Lee, D.; Lee, J.; Akinoglu, E.M.; Giersig, M.; Park, E.Y. 3D hierarchically porous magnetic molybdenum trioxide@gold nanospheres as a nanogap-enhanced Raman scattering biosensor for SARS-CoV-2. *Nanoscale Adv.* **2022**, *4*, 871–883. [CrossRef] [PubMed]
69. Wang, Y.; Shang, B.; Liu, M.; Shi, F.; Peng, B.; Deng, Z. Hollow polydopamine colloidal composite particles: Structure tuning, functionalization and applications. *J. Colloid Interface Sci.* **2018**, *513*, 43–52. [CrossRef] [PubMed]
70. Xia, Y.; Liu, Y.; Hu, X.; Zhao, F.; Zeng, B. Dual-Mode Electrochemical Competitive Immunosensor Based on Cd^{2+}/Au/Polydopamine/Ti_3C_2 Composite and Copper-Based Metal–Organic Framework for 17β-Estradiol Detection. *ACS Sens.* **2022**, *7*, 3077–3084. [CrossRef] [PubMed]
71. Yan, T.; Wu, T.; Wei, S.; Wang, H.; Sun, M.; Yan, L.; Wei, Q.; Ju, H. Photoelectrochemical competitive immunosensor for 17β-estradiol detection based on $ZnIn_2S_4$@NH_2-MIL-125(Ti) amplified by PDA NS/Mn:ZnCdS. *Biosens. Bioelectron.* **2020**, *148*, 111739. [CrossRef]
72. Zhao, M.; Zhang, X.; Deng, C. Rational synthesis of novel recyclable Fe_3O_4@MOF nanocomposites for enzymatic digestion. *Chem. Commun.* **2015**, *51*, 8116–8119. [CrossRef] [PubMed]
73. Li, H.; Wang, X.; Wang, Z.; Wang, Y.; Dai, J.; Gao, L.; Wei, M.; Yan, Y.; Li, C. A polydopamine-based molecularly imprinted polymer on nanoparticles of type SiO_2@rGO@Ag for the detection of lambda-cyhalothrin via SERS. *Microchim. Acta* **2018**, *185*, 1–10.
74. Xue, Y.; Shao, J.; Sui, G.Q.; Ma, Y.Q.; Li, H.J. Rapid detection of orange II dyes in water with SERS imprinted sensor based on PDA-modified MOFs@Ag. *J. Environ. Chem. Eng.* **2021**, *9*, 106317. [CrossRef]
75. Cheng, D.S.; Zhang, Y.L.; Yan, C.W.; Deng, Z.M.; Tang, X.N.; Cai, G.M.; Wang, X. Polydopamine-assisted in situ growth of three-dimensional ZnO/Ag nanocomposites on PET films for SERS and catalytic properties. *J. Mol. Liq.* **2021**, *338*, 116639. [CrossRef]

76. Zhou, Y.; Zhou, J.; Wang, F.; Yang, H. Polydopamine-based functional composite particles for tumor cell targeting and dual-mode cellular imaging. *Talanta* **2018**, *181*, 248–257. [CrossRef]
77. Chin, H.K.; Lin, P.Y.; Chen, J.D.; Kirankumar, R.; Wen, Z.H.; Hsieh, S.C. Polydopamine-Mediated Ag and ZnO as an Active and Recyclable SERS Substrate for Rhodamine B with Significantly Improved Enhancement Factor and Efficient Photocatalytic Degradation. *Appl. Sci.* **2021**, *11*, 4914. [CrossRef]
78. Zhang, L.Z.; Liu, J.; Zhou, G.W.; Zhang, Z.L. Controllable In-Situ Growth of Silver Nanoparticles on Filter Paper for Flexible and Highly Sensitive SERS Sensors for Malachite Green Residue Detection. *Nanomaterials* **2020**, *10*, 826. [CrossRef]
79. Jiao, W.; Chen, C.; You, W.; Zhang, J.; Liu, J.; Che, R. Yolk–Shell Fe/Fe_4N@Pd/C Magnetic Nanocomposite as an Efficient Recyclable ORR Electrocatalyst and SERS Substrate. *Small* **2019**, *15*, 1805032. [CrossRef]
80. Wang, D.; Bao, L.P.; Li, H.J.; Guo, X.Y.; Liu, W.Z.; Wang, X.Y.; Hou, X.M.; He, B. Polydopamine stabilizes silver nanoparticles as a SERS substrate for efficient detection of myocardial infarction. *Nanoscale* **2022**, *14*, 6212–6219. [CrossRef]
81. Wang, C.D.; Wang, X.; Li, C.; Xu, X.H.; Ye, W.C.; Qiu, G.Y.; Wang, D.G. Silver mirror films deposited on well plates for SERS detection of multi-analytes: Aiming at 96-well technology. *Talanta* **2021**, *222*, 121544. [CrossRef]
82. Yu, W.; Lin, X.; Duan, N.; Wang, Z.; Wu, S. A fluorescence and surface-enhanced Raman scattering dual-mode aptasensor for sensitive detection of deoxynivalenol based on gold nanoclusters and silver nanoparticles modified metal-polydopamine framework. *Anal. Chim. Acta* **2023**, *1244*, 340846. [CrossRef] [PubMed]
83. Shen, W.; Lin, X.; Jiang, C.; Li, C.; Lin, H.; Huang, J.; Wang, S.; Liu, G.; Yan, X.; Zhong, Q.; et al. Reliable Quantitative SERS Analysis Facilitated by Core–Shell Nanoparticles with Embedded Internal Standards. *Angew. Chem. Int. Ed.* **2015**, *54*, 7308–7312. [CrossRef] [PubMed]
84. Yin, Y.; Mei, R.; Wang, Y.; Zhao, X.; Yu, Q.; Liu, W.; Chen, L. Silica-Coated, Waxberry-like Surface-Enhanced Raman Resonant Scattering Tag-Pair with Near-Infrared Raman Dye Encoding: Toward In Vivo Duplexing Detection. *Anal. Chem.* **2020**, *92*, 14814–14821. [CrossRef] [PubMed]
85. Zhou, J.; Xiong, Q.; Ma, J.; Ren, J.; Messersmith, P.B.; Chen, P.; Duan, H. Polydopamine-Enabled Approach toward Tailored Plasmonic Nanogapped Nanoparticles: From Nanogap Engineering to Multifunctionality. *ACS Nano* **2016**, *10*, 11066–11075. [CrossRef] [PubMed]
86. Hu, W.; Xia, L.; Hu, Y.; Li, G. Honeycomb-like ReS_2/polydopamine/Ag plasmonic composite assembled membrane substrate for rapid surface-enhanced Raman scattering analysis of 6-benzylaminopurine and alternariol in food. *Sens. Actuators B* **2023**, *380*, 133339. [CrossRef]
87. Yang, L.; Yin, D.; Shen, Y.; Yang, M.; Li, X.; Han, X.; Jiang, X.; Zhao, B. Mesoporous semiconducting TiO_2 with rich active sites as a remarkable substrate for surface-enhanced Raman scattering. *Phys. Chem. Chem. Phys.* **2017**, *19*, 18731–18738. [CrossRef]
88. Rajaraman, T.S.; Parikh, S.P.; Gandhi, V.G. Black TiO_2: A review of its properties and conflicting trends. *Chem. Eng. J.* **2020**, *389*, 123918. [CrossRef]
89. Wang, J.H.; Yang, Y.H.; Li, H.; Gao, J.; He, P.; Bian, L.; Dong, F.Q.; He, Y. Stable and tunable plasmon resonance of molybdenum oxide nanosheets from the ultraviolet to the near-infrared region for ultrasensitive surface-enhanced Raman analysis. *Chem. Sci.* **2019**, *10*, 6330–6335. [CrossRef]
90. Gu, L.J.; Ma, C.L.; Zhang, X.H.; Zhang, W.; Cong, S.; Zhao, Z.G. Populating surface-trapped electrons towards SERS enhancement of $W_{18}O_{49}$ nanowires. *Chem. Commun.* **2018**, *54*, 6332–6335. [CrossRef]
91. Wang, Y.; Liu, J.; Ozaki, Y.; Xu, Z.; Zhao, B. Effect of TiO_2 on Altering Direction of Interfacial Charge Transfer in a TiO_2-Ag-MPY-FePc System by SERS. *Angew. Chem. Int. Ed.* **2019**, *58*, 8172–8176. [CrossRef]
92. Yang, S.; Yao, J.; Quan, Y.; Hu, M.; Su, R.; Gao, M.; Han, D.; Yang, J. Monitoring the charge-transfer process in a Nd-doped semiconductor based on photoluminescence and SERS technology. *Light Sci. Appl.* **2020**, *9*, 117. [CrossRef] [PubMed]
93. Wang, X.; Shi, W.; Wang, S.; Zhao, H.; Lin, J.; Yang, Z.; Chen, M.; Guo, L. Two-Dimensional Amorphous TiO_2 Nanosheets Enabling High-Efficiency Photoinduced Charge Transfer for Excellent SERS Activity. *J. Am. Chem. Soc.* **2019**, *141*, 5856–5862. [CrossRef] [PubMed]
94. Guo, F.; Chen, J.; Zhao, J.; Chen, Z.; Xia, D.; Zhan, Z.; Wang, Q. Z-scheme heterojunction g-C_3N_4@PDA/BiOBr with biomimetic polydopamine as electron transfer mediators for enhanced visible-light driven degradation of sulfamethoxazole. *Chem. Eng. J.* **2020**, *386*, 124014. [CrossRef]
95. Xie, A.; Zhang, K.; Wu, F.; Wang, N.; Wang, Y.; Wang, M. Polydopamine nanofilms as visible light-harvesting interfaces for palladium nanocrystal catalyzed coupling reactions. *Catal. Sci. Technol.* **2016**, *6*, 1764–1771. [CrossRef]
96. Yu, Z.; Li, F.; Yang, Q.; Shi, H.; Chen, Q.; Xu, M. Nature-Mimic Method To Fabricate Polydopamine/Graphitic Carbon Nitride for Enhancing Photocatalytic Degradation Performance. *ACS Sustain. Chem. Eng.* **2017**, *5*, 7840–7850. [CrossRef]
97. Cheng, S.; Qi, M.; Li, W.; Sun, W.; Li, M.; Lin, J.; Bai, X.; Sun, Y.; Dong, B.; Wang, L. Dual-Responsive Nanocomposites for Synergistic Antibacterial Therapies Facilitating Bacteria-Infected Wound Healing. *Adv. Healthc. Mater.* **2022**, *12*, 2202652. [CrossRef]
98. Kim, J.H.; Lee, M.; Park, C.B. Polydopamine as a Biomimetic Electron Gate for Artificial Photosynthesis. *Angew. Chem. Int. Ed.* **2014**, *53*, 6364–6368. [CrossRef]
99. Lin, P.Y.; He, G.; Chen, J.; Dwivedi, A.K.; Hsieh, S. Monitoring the photoinduced surface catalytic coupling reaction and environmental exhaust fumes with an Ag/PDA/CuO modified 3D glass microfiber platform. *J. Ind. Eng. Chem.* **2020**, *82*, 424–432. [CrossRef]

100. Akin, M.S.; Yilmaz, M.; Babur, E.; Ozdemir, B.; Erdogan, H.; Tamer, U.; Demirel, G. Large area uniform deposition of silver nanoparticles through bio-inspired polydopamine coating on silicon nanowire arrays for practical SERS applications. *J. Mater. Chem. B* **2014**, *2*, 4894–4900. [CrossRef]
101. Lin, Q.; Yang, Y.; Ma, Y.; Zhang, R.; Wang, J.; Chen, X.; Shao, Z. Bandgap Engineered Polypyrrole–Polydopamine Hybrid with Intrinsic Raman and Photoacoustic Imaging Contrasts. *Nano Lett.* **2018**, *18*, 7485–7493. [CrossRef]
102. Lin, X.; Wang, Y.; Wang, L.; Lu, Y.; Li, J.; Lu, D.; Zhou, T.; Huang, Z.; Huang, J.; Huang, H.; et al. Interference-free and high precision biosensor based on surface enhanced Raman spectroscopy integrated with surface molecularly imprinted polymer technology for tumor biomarker detection in human blood. *Biosens. Bioelectron.* **2019**, *143*, 111599. [CrossRef] [PubMed]
103. Li, J.; Liu, F.G.; Ye, J. Boosting the Brightness of Thiolated Surface-Enhanced Raman Scattering Nanoprobes by Maximal Utilization of the Three- Dimensional Volume of Electromagnetic Fields. *J. Phys. Chem. Lett.* **2022**, *13*, 6496–6502. [CrossRef] [PubMed]
104. Yang, L.; Zhang, D.; Wang, M.; Yang, Y. Effects of solvent polarity on the novel excited-state intramolecular thiol proton transfer and photophysical property compared with the oxygen proton transfer. *Spectrochim. Acta Part A* **2023**, *293*, 122475. [CrossRef] [PubMed]
105. Fang, P.P.; Lu, X.; Liu, H.; Tong, Y. Applications of shell-isolated nanoparticles in surface-enhanced Raman spectroscopy and fluorescence. *TrAC Trends Anal. Chem.* **2015**, *66*, 103–117. [CrossRef]
106. Zhang, Y.; Qian, J.; Wang, D.; Wang, Y.; He, S. Multifunctional gold nanorods with ultrahigh stability and tunability for in vivo fluorescence imaging, SERS detection, and photodynamic therapy. *Angew. Chem. Int. Ed.* **2013**, *52*, 1148–1151. [CrossRef]
107. Li, Y.; Hao, Z.; Cao, H.; Wei, S.; Jiao, T.; Wang, M. Study on annealed graphene oxide nano-sheets for improving the surface enhanced fluorescence of silver nanoparticles. *Opt. Laser Technol.* **2023**, *160*, 109054. [CrossRef]
108. Wu, Z.; Sun, D.W.; Pu, H.; Wei, Q. A dual signal-on biosensor based on dual-gated locked mesoporous silica nanoparticles for the detection of Aflatoxin B1. *Talanta* **2023**, *253*, 124027. [CrossRef]
109. Cao, Y.; Han, S.; Zhang, H.; Wang, J.; Jiang, Q.Y.; Zhou, Y.; Yu, Y.J.; Wang, J.; Chen, F.; Ng, D.K. Detection of cell-surface sialic acids and photodynamic eradication of cancer cells using dye-modified polydopamine-coated gold nanobipyramids. *J. Mater. Chem. B* **2021**, *9*, 5780–5784. [CrossRef]
110. He, P.; Han, W.H.; Bi, C.; Song, W.L.; Niu, S.Y.; Zhou, H.; Zhang, X.R. Many Birds, One Stone: A Smart Nanodevice for Ratiometric Dual-Spectrum Assay of Intracellular MicroRNA and Multimodal Synergetic Cancer Therapy. *ACS Nano* **2021**, *15*, 6961–6976. [CrossRef]
111. Xu, G.; Hou, J.; Zhao, Y.; Bao, J.; Yang, M.; Fa, H.; Yang, Y.; Li, L.; Huo, D.; Hou, C. Dual-signal aptamer sensor based on polydopamine-gold nanoparticles and exonuclease I for ultrasensitive malathion detection. *Sens. Actuators B* **2019**, *287*, 428–436. [CrossRef]
112. Achadu, O.J.; Abe, F.; Hossain, F.; Nasrin, F.; Yamazaki, M.; Suzuki, T.; Park, E.Y. Sulfur-doped carbon dots@polydopamine-functionalized magnetic silver nanocubes for dual-modality detection of norovirus. *Biosens. Bioelectron.* **2021**, *193*, 113540. [CrossRef] [PubMed]
113. Xiong, Q.; Lim, C.Y.; Ren, J.; Zhou, J.; Pu, K.; Chan-Park, M.B.; Mao, H.; Lam, Y.C.; Duan, H. Magnetic nanochain integrated microfluidic biochips. *Nat. Commun.* **2018**, *9*, 1743. [CrossRef] [PubMed]
114. Jiang, N.; Hu, Y.; Wei, W.; Zhu, T.; Yang, K.; Zhu, G.; Yu, M. Detection of microRNA using a polydopamine mediated bimetallic SERS substrate and a re-circulated enzymatic amplification system. *Microchim. Acta* **2019**, *186*, 1–9. [CrossRef] [PubMed]
115. Li, T.D.; Zhang, R.; Chen, H.; Huang, Z.P.; Ye, X.; Wang, H.; Deng, A.M.; Kong, J.L. An ultrasensitive polydopamine bi-functionalized SERS immunoassay for exosome-based diagnosis and classification of pancreatic cancer. *Chem. Sci.* **2018**, *9*, 5372–5382. [CrossRef]
116. Wang, Y.L.; Li, Q.Y.; Zhang, R.; Tang, K.Q.; Ding, C.F.; Yu, S.N. SERS-based immunocapture and detection of pathogenic bacteria using a boronic acid-functionalized polydopamine-coated Au@Ag nanoprobe. *Microchim. Acta* **2020**, *187*, 290. [CrossRef]
117. Arabi, M.; Ostovan, A.; Zhang, Z.Y.; Wang, Y.Q.; Mei, R.C.; Fu, L.W.; Wang, X.Y.; Ma, J.P.; Chen, L.X. Label-free SERS detection of Raman-Inactive protein biomarkers by Raman reporter indicator: Toward ultrasensitivity and universality. *Biosens. Bioelectron.* **2021**, *174*, 112825. [CrossRef]
118. Yang, Y.Y.; Li, Y.T.; Li, X.J.; Zhang, L.; Kouadio Fodjo, E.; Han, S. Controllable in situ fabrication of portable AuNP/mussel-inspired polydopamine molecularly imprinted SERS substrate for selective enrichment and recognition of phthalate plasticizers. *Chem. Eng. J.* **2020**, *402*, 125179. [CrossRef]
119. Chen, W.; Fu, M.; Zhu, X.; Liu, Q. Protein recognition by polydopamine-based molecularly imprinted hollow spheres. *Biosens. Bioelectron.* **2019**, *142*, 111492. [CrossRef]
120. Arabi, M.; Ostovan, A.; Wang, Y.Q.; Mei, R.C.; Fu, L.W.; Li, J.H.; Wang, X.Y.; Chen, L.X. Chiral molecular imprinting-based SERS detection strategy for absolute enantiomeric discrimination. *Nat. Commun.* **2022**, *13*, 5757. [CrossRef]
121. Kong, H.J.; Sun, X.P.; Yang, L.; Liu, X.L.; Yang, H.F.; Jin, R.H. Polydopamine/Silver Substrates Stemmed from Chiral Silica for SERS Differentiation of Amino Acid Enantiomers. *Acs Appl. Mater. Interfaces* **2020**, *12*, 29868–29875. [CrossRef]
122. Yang, Y.Y.; Li, Y.T.; Zhai, W.L.; Li, X.J.; Li, D.; Lin, H.L.; Han, S. Electrokinetic Preseparation and Molecularly Imprinted Trapping for Highly Selective SERS Detection of Charged Phthalate Plasticizers. *Anal. Chem.* **2021**, *93*, 946–955. [CrossRef] [PubMed]
123. Yu, B.; Liu, J.; Liu, S.; Zhou, F. Pdop layer exhibiting zwitterionicity: A simple electrochemical interface for governing ion permeability. *Chem. Commun.* **2010**, *46*, 5900–5902. [CrossRef] [PubMed]

124. Nayak, S.; Blumenfeld, N.R.; Laksanasopin, T.; Sia, S.K. Point-of-Care Diagnostics: Recent Developments in a Connected Age. *Anal. Chem.* **2017**, *89*, 102–123. [CrossRef] [PubMed]
125. Liu, X.; Liu, X.; Rong, P.; Liu, D. Recent advances in background-free Raman scattering for bioanalysis. *TrAC Trends Anal. Chem.* **2020**, *123*, 115765. [CrossRef]
126. Linh, V.T.N.; Yim, S.G.; Mun, C.; Yang, J.Y.; Lee, S.; Yoo, Y.W.; Sung, D.K.; Lee, Y.I.; Kim, D.H.; Park, S.G.; et al. Bioinspired plasmonic nanoflower-decorated microneedle for label-free intradermal sensing. *Appl. Surf. Sci.* **2021**, *551*, 149411. [CrossRef]
127. Renard, D.; Tian, S.; Ahmadivand, A.; DeSantis, C.J.; Clark, B.D.; Nordlander, P.; Halas, N.J. Polydopamine-Stabilized Aluminum Nanocrystals: Aqueous Stability and Benzo[a]pyrene Detection. *ACS Nano* **2019**, *13*, 3117–3124. [CrossRef]
128. Chang, Y.L.; Su, C.J.; Lu, L.C.; Wan, D.H. Aluminum Plasmonic Nanoclusters for Paper-Based Surface- Enhanced Raman Spectroscopy. *Anal. Chem.* **2022**, *94*, 16319–16327. [CrossRef]
129. Liu, J.; Si, T.; Zhang, L.; Zhang, Z. Mussel-Inspired Fabrication of SERS Swabs for Highly Sensitive and Conformal Rapid Detection of Thiram Bactericides. *Nanomaterials* **2019**, *9*, 1331. [CrossRef]
130. Li, H.J.; Wang, J.F.; Fang, H.Q.; Xu, H.D.; Yu, H.C.; Zhou, T.Y.; Liu, C.B.; Che, G.B.; Wang, D.D. Hydrophilic modification of PVDF-based SERS imprinted membrane for the selective detection of L-tyrosine. *J. Environ. Manag.* **2022**, *304*, 114260. [CrossRef]
131. Wan, M.H.; Zhao, H.D.; Wang, Z.H.; Zhao, Y.B.; Sun, L. Preparation of Ag@PDA@SiO_2 electrospinning nanofibrous membranes for direct bacteria SERS detection and antimicrobial activities. *Mater. Res. Express* **2020**, *7*, 095012. [CrossRef]
132. Zhang, Z.L.; Si, T.T.; Liu, J.; Zhou, G.W. In-Situ Grown Silver Nanoparticles on Nonwoven Fabrics Based on Mussel-Inspired Polydopamine for Highly Sensitive SERS Carbaryl Pesticides Detection. *Nanomaterials* **2019**, *9*, 384. [CrossRef] [PubMed]
133. Liu, J.; Si, T.; Zhang, Z. Mussel-inspired immobilization of silver nanoparticles toward sponge for rapid swabbing extraction and SERS detection of trace inorganic explosives. *Talanta* **2019**, *204*, 189–197. [CrossRef] [PubMed]
134. Shang, B.; Wang, Y.B.; Yang, P.; Peng, B.; Deng, Z.W. Synthesis of superhydrophobic polydopamine-Ag microbowl/nanoparticle array substrates for highly sensitive, durable and reproducible surface-enhanced Raman scattering detection. *Sens. Actuators B* **2018**, *255*, 995–1005. [CrossRef]
135. Kozma, E.; Andicsova, A.E.; Siskova, A.O.; Tullii, G.; Galeotti, F. Biomimetic design of functional plasmonic surfaces based on polydopamine. *Appl. Surf. Sci.* **2022**, *591*, 153135. [CrossRef]
136. Dong, J.C.; Wang, T.C.; Xu, E.Z.; Bai, F.; Liu, J.; Zhang, Z.L. Flexible Hydrophobic CFP@PDA@AuNPs Stripes for Highly Sensitive SERS Detection of Methylene Blue Residue. *Nanomaterials* **2022**, *12*, 2163. [CrossRef]
137. Liu, Y.; Li, Z.; Yin, Z.; Zhang, H.; Gao, Y.; Huo, G.; Wu, A.; Zeng, L. Amplified Photoacoustic Signal and Enhanced Photothermal Conversion of Polydopamine-Coated Gold Nanobipyramids for Phototheranostics and Synergistic Chemotherapy. *ACS Appl. Mater. Interfaces* **2020**, *12*, 14866–14875. [CrossRef]
138. Liu, X.; Liao, G.; Zou, L.; Zheng, Y.; Yang, X.; Wang, Q.; Geng, X.; Li, S.; Liu, Y.; Wang, K. Construction of Bio/Nanointerfaces: Stable Gold Nanoparticle Bioconjugates in Complex Systems. *Acs Appl. Mater. Interfaces* **2019**, *11*, 40817–40825. [CrossRef]
139. Chang, N.; Wang, D.L.; Liu, B.; He, D.; Wu, H.; Zhao, X.W. Stable Plasmonic Coloration of Versatile Surfaces via Colloidal Monolayer Transfer Printing. *Adv. Eng. Mater.* **2019**, *21*, 1900313. [CrossRef]
140. Poulichet, V.; Morel, M.; Rudiuk, S.; Baigl, D. Liquid-liquid coffee-ring effect. *J. Colloid Interface Sci.* **2020**, *573*, 370–375. [CrossRef]
141. Mampallil, D.; Eral, H.B. A review on suppression and utilization of the coffee-ring effect. *Adv. Colloid Interface Sci.* **2018**, *252*, 38–54. [CrossRef]
142. Liang, X.; Zhang, H.; Xu, C.; Cao, D.; Gao, Q.; Cheng, S. Condensation effect-induced improved sensitivity for SERS trace detection on a superhydrophobic plasmonic nanofibrous mat. *RSC Adv.* **2017**, *7*, 44492–44498. [CrossRef]
143. Luo, X.; Pan, R.; Cai, M.; Liu, W.; Chen, C.; Jiang, G.; Hu, X.; Zhang, H.; Zhong, M. Atto-Molar Raman detection on patterned superhydrophilic-superhydrophobic platform via localizable evaporation enrichment. *Sens. Actuators B* **2021**, *326*, 128826. [CrossRef]
144. Bhushan, B.; Jung, Y.C. Natural and biomimetic artificial surfaces for superhydrophobicity, self-cleaning, low adhesion, and drag reduction. *Prog. Mater. Sci.* **2011**, *56*, 1–108. [CrossRef]
145. Wang, B.; Liang, W.; Guo, Z.; Liu, W. Biomimetic super-lyophobic and super-lyophilic materials applied for oil/water separation: A new strategy beyond nature. *Chem. Soc. Rev.* **2015**, *44*, 336–361. [CrossRef] [PubMed]
146. Kuang, M.; Wang, J.; Jiang, L. Bio-inspired photonic crystals with superwettability. *Chem. Soc. Rev.* **2016**, *45*, 6833–6854. [CrossRef] [PubMed]
147. Yu, X.; Zhong, Q.Z.; Yang, H.C.; Wan, L.S.; Xu, Z.K. Mussel-Inspired Modification of Honeycomb Structured Films for Superhydrophobic Surfaces with Tunable Water Adhesion. *J. Phys. Chem. C* **2015**, *119*, 3667–3673. [CrossRef]
148. Coy, E.; Iatsunskyi, I.; Colmenares, J.C.; Kim, Y.; Mrówczyński, R. Polydopamine Films with 2D-like Layered Structure and High Mechanical Resilience. *ACS Appl. Mater. Interfaces* **2021**, *13*, 23113–23120. [CrossRef]
149. Marchesi D'Alvise, T.; Harvey, S.; Hueske, L.; Szelwicka, J.; Veith, L.; Knowles, T.P.J.; Kubiczek, D.; Flaig, C.; Port, F.; Gottschalk, K.-E.; et al. Ultrathin Polydopamine Films with Phospholipid Nanodiscs Containing a Glycophorin A Domain. *Adv. Funct. Mater.* **2020**, *30*, 2000378. [CrossRef]
150. Lane, L.A.; Qian, X.; Nie, S. SERS Nanoparticles in Medicine: From Label-Free Detection to Spectroscopic Tagging. *Chem. Rev.* **2015**, *115*, 10489–10529. [CrossRef]
151. Song, L.; Chen, J.; Xu, B.B.; Huang, Y. Flexible plasmonic biosensors for healthcare monitoring: Progress and prospects. *ACS Nano* **2021**, *15*, 18822–18847. [CrossRef]
152. Wang, Y.; Li, B.; Tian, T.; Liu, Y.; Zhang, J.; Qian, K. Advanced on-site and in vitro signal amplification biosensors for biomolecule analysis. *TrAC Trends Anal. Chem.* **2022**, *149*, 116565. [CrossRef]

153. Meng, X.; Yang, F.; Dong, H.; Dou, L.; Zhang, X. Recent advances in optical imaging of biomarkers in vivo. *Nano Today* **2021**, *38*, 101156. [CrossRef]
154. Lee, H.K.; Lee, Y.H.; Koh, C.S.L.; Phan-Quang, G.C.; Han, X.; Lay, C.L.; Sim, H.Y.F.; Kao, Y.C.; An, Q.; Ling, X.Y. Designing surface-enhanced Raman scattering (SERS) platforms beyond hotspot engineering: Emerging opportunities in analyte manipulations and hybrid materials. *Chem. Soc. Rev.* **2019**, *48*, 731–756. [CrossRef] [PubMed]
155. Ding, Q.; Wang, J.; Chen, X.; Liu, H.; Li, Q.; Wang, Y.; Yang, S. Quantitative and Sensitive SERS Platform with Analyte Enrichment and Filtration Function. *Nano Lett.* **2020**, *20*, 7304–7312. [CrossRef]
156. Fang, Y.; Seong, N.H.; Dlott, D.D. Measurement of the Distribution of Site Enhancements in Surface-Enhanced Raman Scattering. *Science* **2008**, *321*, 388–392. [CrossRef]
157. Liu, B.; Liu, X.; Shi, S.; Huang, R.; Su, R.; Qi, W.; He, Z. Design and mechanisms of antifouling materials for surface plasmon resonance sensors. *Acta Biomater.* **2016**, *40*, 100–118. [CrossRef]
158. Zhou, M.Z.; Wang, Z.Q.; Xia, D.Q.; Xie, X.Y.; Chen, Y.H.; Xing, Y.X.; Cai, K.Y.; Zhang, J.X. Hybrid nanoassembly with two-tier host-guest architecture and regioselective enrichment capacity for repetitive SERS detection. *Sens. Actuators B* **2022**, *369*, 132359. [CrossRef]
159. Zhai, J.; Li, X.; Zhang, J.; Pan, H.; Peng, Q.; Gan, H.; Su, S.; Yuwen, L.; Song, C. SERS/electrochemical dual-mode biosensor based on multi-functionalized molybdenum disulfide nanosheet probes and SERS-active Ag nanorods array electrodes for reliable detection of cancer-related miRNA. *Sens. Actuators B* **2022**, *368*, 132245. [CrossRef]
160. Yuwen, L.; Sun, Y.; Tan, G.; Xiu, W.; Zhang, Y.; Weng, L.; Teng, Z.; Wang, L. MoS_2@polydopamine-Ag nanosheets with enhanced antibacterial activity for effective treatment of Staphylococcus aureus biofilms and wound infection. *Nanoscale* **2018**, *10*, 16711–16720. [CrossRef]
161. Wu, Y.; He, Y.; Yang, X.; Yuan, R.; Chai, Y. A novel recyclable surface-enhanced Raman spectroscopy platform with duplex-specific nuclease signal amplification for ultrasensitive analysis of microRNA 155. *Sens. Actuators B* **2018**, *275*, 260–266. [CrossRef]
162. Lu, D.; Xia, J.; Deng, Z.; Cao, X.W. Detection of squamous cell carcinoma antigen in cervical cancer by surface-enhanced Raman scattering-based immunoassay. *Anal. Methods* **2019**, *11*, 2809–2818. [CrossRef]
163. Wang, J.J.; Xu, C.X.; Lei, M.L.; Ma, Y.; Wang, X.X.; Wang, R.; Sun, J.L.; Wang, R. Microcavity-based SERS chip for ultrasensitive immune detection of cardiac biomarkers. *Microchem. J.* **2021**, *171*, 106875. [CrossRef]
164. Huang, Z.; Zhang, R.; Chen, H.; Weng, W.; Lin, Q.; Deng, D.; Li, Z.; Kong, J. Sensitive polydopamine bi-functionalized SERS immunoassay for microalbuminuria detection. *Biosens. Bioelectron.* **2019**, *142*, 111542. [CrossRef] [PubMed]
165. Li, X.; Li, B.; Hong, J.; Zhou, X. Highly selective determination of acid phosphatase in biological samples using a biomimetic recognition-based SERS sensor. *Sens. Actuators B* **2018**, *276*, 421–428. [CrossRef]
166. Liu, X.; Wang, K.; Cao, B.; Shen, L.; Ke, X.; Cui, D.; Zhong, C.; Li, W. Multifunctional Nano-Sunflowers with Color-Magnetic-Raman Properties for Multimodal Lateral Flow Immunoassay. *Anal. Chem.* **2021**, *93*, 3626–3634. [CrossRef]
167. Xia, J.; Liu, Y.F.; Ran, M.L.; Lu, W.B.; Bi, L.Y.; Wang, Q.; Lu, D.; Cao, X.W. The simultaneous detection of the squamous cell carcinoma antigen and cancer antigen 125 in the cervical cancer serum using nano-Ag polydopamine nanospheres in an SERS-based lateral flow immunoassay. *RSC Adv.* **2020**, *10*, 29156–29170. [CrossRef]
168. Cheng, N.; Du, D.; Wang, X.; Liu, D.; Xu, W.; Luo, Y.; Lin, Y. Recent Advances in Biosensors for Detecting Cancer-Derived Exosomes. *Trends Biotechnol.* **2019**, *37*, 1236–1254. [CrossRef]
169. Hong, W.E.; Hsu, I.L.; Huang, S.Y.; Lee, C.W.; Ko, H.; Tsai, P.J.; Shieh, D.B.; Huang, C.C. Assembled growth of 3D Fe_3O_4@Au nanoparticles for efficient photothermal ablation and SERS detection of microorganisms. *J. Mater. Chem. B* **2018**, *6*, 5689–5697. [CrossRef]
170. D'Autréaux, B.; Toledano, M.B. ROS as signalling molecules: Mechanisms that generate specificity in ROS homeostasis. *Nat. Rev. Mol. Cell Biol.* **2007**, *8*, 813–824. [CrossRef]
171. Sun, Y.; Sun, X.; Li, X.; Li, W.; Li, C.; Zhou, Y.; Wang, L.; Dong, B. A versatile nanocomposite based on nanoceria for antibacterial enhancement and protection from aPDT-aggravated inflammation via modulation of macrophage polarization. *Biomaterials* **2021**, *268*, 120614. [CrossRef]
172. Winterbourn, C.C. Reconciling the chemistry and biology of reactive oxygen species. *Nat. Chem. Biol.* **2008**, *4*, 278–286. [CrossRef] [PubMed]
173. Murphy, M.P.; Holmgren, A.; Larsson, N.-G.; Halliwell, B.; Chang, C.J.; Kalyanaraman, B.; Rhee, S.G.; Thornalley, P.J.; Partridge, L.; Gems, D.; et al. Unraveling the Biological Roles of Reactive Oxygen Species. *Cell Metab.* **2011**, *13*, 361–366. [CrossRef] [PubMed]
174. Kumar, S.; Kumar, A.; Kim, G.H.; Rhim, W.K.; Hartman, K.L.; Nam, J.M. Myoglobin and Polydopamine-Engineered Raman Nanoprobes for Detecting, Imaging, and Monitoring Reactive Oxygen Species in Biological Samples and Living Cells. *Small* **2017**, *13*, 1701584. [CrossRef] [PubMed]
175. Cheng, D.S.; Bai, X.; He, M.T.; Wu, J.H.; Yang, H.J.; Ran, J.H.; Cai, G.M.; Wang, X. Polydopamine-assisted immobilization of Ag@AuNPs on cotton fabrics for sensitive and responsive SERS detection. *Cellulose* **2019**, *26*, 4191–4204. [CrossRef]
176. Sieste, S.; Mack, T.; Synatschke, C.V.; Schilling, C.; Meyer zu Reckendorf, C.; Pendi, L.; Harvey, S.; Ruggeri, F.S.; Knowles, T.P.J.; Meier, C.; et al. Water-Dispersible Polydopamine-Coated Nanofibers for Stimulation of Neuronal Growth and Adhesion. *Adv. Healthc. Mater.* **2018**, *7*, 1701485. [CrossRef] [PubMed]

MDPI
St. Alban-Anlage 66
4052 Basel
Switzerland
www.mdpi.com

Sensors Editorial Office
E-mail: sensors@mdpi.com
www.mdpi.com/journal/sensors

www.ingramcontent.com/pod-product-compliance
Lightning Source LLC
LaVergne TN
LVHW070420100526
838202LV00014B/1495
* 9 7 8 3 7 2 5 8 0 1 6 9 5 *